Energy and the Social Sciences:
A Bibliographic Guide
to the Literature

Also of Interest

* *Energy Futures, Human Values, and Lifestyles: A New Look at the Energy Crisis,* Richard C. Carlson, Willis W. Harman, Peter Schwartz, and Associates

Energy Transitions: Long-Term Perspectives, edited by Lewis J. Perelman, August W. Giebelhaus, and Michael D. Yokell

* *Renewable Natural Resources: A Management Handbook for the 1980s,* edited by Dennis L. Little, Robert E. Dils, John Gray

Energy from Biological Processes, Office of Technology Assessment, U.S. Congress

* *Living with Energy Shortfall: A Future for American Towns and Cities,* Jon Van Til

Energy, Economics, and the Environment: Conflicting Views of an Essential Interrelationship, edited by Herman E. Daly and Alvaro F. Umaña

* *Accident at Three Mile Island: The Human Dimensions,* edited by David L. Sills, C. P. Wolf, and Vivien B. Shelanski

The Forever Fuel: The Story of Hydrogen, Peter Hoffmann

Alcohol Fuels: Policies, Production, and Potential, Doann Houghton-Alico

Solar Energy in the U.S. Economy, Christopher J. Pleatsikas, Edward A. Hudson, and Richard J. Goettle IV

Critical Energy Issues in Asia and the Pacific: The Next Twenty Years, Fereidun Fesharaki, Harrison Brown, Corazon M. Siddayao, Toufiq A. Siddiqi, Kirk R. Smith, and Kim Woodard

OPEC: Twenty Years and Beyond, edited by Ragaei El Mallakh

Water and Western Energy: Impacts, Issues, and Choices, Steven C. Ballard, Michael D. Devine, and Associates

Energy Analysis and Agriculture: An Application to U.S. Corn Production, Vaclav Smil, Paul Nachman, and Thomas V. Long II

Eating Oil: Energy Use in Food Production, Maurice B. Green

Energy Analysis: A New Public Policy Tool, edited by Martha W. Gilliland

Energy Conservation and Economic Growth, edited by Charles Hitch

* *Climate Change and Society: Consequences of Increasing Atmospheric Carbon Dioxide,* William W. Kellogg and Robert Schware

* *The Economics of Environmental and Natural Resources Policy,* edited by J. A. Butlin

* Available in hardcover and paperback.

Westview Special Studies in
Natural Resources and Energy Management

Energy and the Social Sciences:
A Bibliographic Guide to the Literature
Ernest J. Yanarella and Ann-Marie Yanarella

This comprehensive guide to the burgeoning energy literature fills a void in the bibliographic resources heretofore available by offering a broad range of relevant references to both U.S. and cross-national sources on energy, energy technologies and alternatives, and energy policy. The book presents more than 4,000 entries organized into major categories. It provides a framework of interconnected topics — arranged by theme and geographic area — that allows social scientists, students, and librarians to become quickly acquainted with a wide spectrum of works from the technical and social science literature relevant to their concerns.

In addition to a wealth of references to books, monographs, articles, and bibliographies, the book offers bibliographic essays to introduce each section, a special section describing energy journals and other information sources, and an annotated list of books essential to a new or expanding library collection on energy.

Ernest J. Yanarella is an associate professor in the Department of Political Science at the University of Kentucky. **Ann-Marie Yanarella** is a staff assistant in the Pulmonary Division of the Department of Medicine at the University of Kentucky's Medical Center. They have collaborated to produce a number of books on energy-related issues.

Energy and the Social Sciences: A Bibliographic Guide to the Literature

Ernest J. Yanarella
and Ann-Marie Yanarella

Westview Press / Boulder, Colorado

Westview Special Studies in Natural Resources and Energy Management

Copyright © 1982 by Westview Press, Inc.

Published in 1982 in the United States of America by
  Westview Press, Inc.
  5500 Central Avenue
  Boulder, Colorado 80301
  Frederick A. Praeger, President and Publisher

Library of Congress Catalog Card Number: 82-050070
ISBN 0-86531-304-0

Composition for this book was provided by the authors.
Printed and bound in the United States of America.

Contents

Preface and Acknowledgments

Preparing a bibliography based on a broad topic is in several respects akin to following the plot of Thomas Pynchon's book, V. In both cases, the exercise threatens to break down into incoherence due to the sheer volume of names and materials one tries to assimilate; simultaneously, one is driven by the perverse temptation to incorporate more and more into one's range of awareness. What makes the bibliographer's task even more difficult than that of the fiction-writer is that the former must come to grips with the fact that the literature being surveyed and organized almost daily undergoes expansion and, consequently, the groundwork for the obsolescence of the compiler's completed product is continuously being laid by fellow researchers and writers.

While accepting the limitations inherent in such a work, we were prompted to undertake this bibliographical guide out of the belief that such a research guide would be a useful contribution to the social science literature and the conviction that we possessed some of the requisite qualities for bringing such a project to fruition. We freely admit that we are not professional librarians with formal training in the compilation of bibliographies. Thus, some of the norms and guidelines for preparing such works have undoubtedly been violated. On the other hand, given our involvement in the study of energy policy for the last few years and our concern with generating broader interest among social scientists in the social science/energy nexus, we hope our background has been an asset in putting together this guide. We would hasten to add that, as a result of producing this work, we have developed a new-found appreciation for the efforts of other bibliographers.

As in any sizeable project, many cooperative individuals helped in translating this work from a roughly-hewn idea to a finished product. The assistance of the reference division of the Margaret I. King Library at the University of Kentucky must be acknowledged with gratitude. Among these members of the library staff, Alexander "Sandy" Gilchrist must be singled out for his cheerful manner and dogged persistence in tracking down information on especially difficult references. We would be negligent if we did not thank Dr. John Walker for his recognition of the value of this work and for his generosity in making available to us funds necessary for the typing and composition work on this final version of the book. In expressing our gratitude to him, we wish to thank the Office of Surface Mining, U.S. Department of the Interior, the funding agency from which this financial support came. The University of Kentucky Research Foundation must be acknowledged for granting us supplemental financial assistance to pay for the book's final composition.

A special acknowledgment is due Patricia Owens, who worked against adversity and competing responsibilities to prepare the camera-ready copy of the final manuscript. We are indebted to Dr. Ira Ross, Chairman of the Department of Agricultural Engineering, for the use of his Department's IBM composer to type the final copy of this guide. We are grateful to Lynne Rienner, our editor at Westview Press, for her enthusiastic support and timely editorial suggestions. To Dean Birkenkamp, her associate, goes the credit for recommending the idea of an annotated list of "must" books for any energy library.

Finally, we express our debt to our two children, Lisa and John Michael, who in their early adulthood will inhabit the energy future of the twenty-first century which is being fashioned today. To them, this is dedicated with love and hope.

As is customary, we assume responsibility for any errors of fact and interpretation contained in this book.

E.J.Y.
A.M.Y.

Overview

In recent years, a special need has emerged in the social sciences for a detailed and comprehensive guide to energy research and policies in the United States and around the world. The rise and consolidation of the Organization of Petroleum Exporting Countries and the accompanying dramatic consequences of this cartel for economic development and continuing economic growth have made energy a salient public policy issue for the nations of the world community, rich and poor alike. In addition, and as a direct consequence of its occurrence, energy policy – at least since the 1973-74 Arab oil embargo – has crystallized into a significant realm of government policymaking and administrative reorganization. Finally, events and trends internal and external to the social science community have prompted researchers in more and more disciplines to redirect social science research and teaching into areas of public policy, such as the energy domain.

If the confluence of these and other events has made the need for a bibliographical guide to this subject a compelling one, the customary sources for such a work have been slow in responding. Indeed, with few exceptions, those bibliographies which have been compiled by government agencies, environmental groups, library associations, and scholars have suffered from a number of inadequacies. Some have focused upon only a single facet of energy development or policymaking; others have been compiled in a largely haphazard, indiscriminate fashion from diverse and uneven sources; still others, which have been put together with an eye toward social science interests, have lacked the structured, systematic character so necessary to assist graduate students and field researchers in surveying specific topics and themes.

The following bibliography attempts to speak to the distinctive needs of social scientists entering the miasma of energy policy studies, while transcending at least some of the shortcomings of previous efforts. Designed specifically for the developing interests in energy, energy research, and energy policy within the social science community, it is intended to offer political scientists, policy analysts, economists, and sociologists a coherent framework of interconnected topics which allows both the budding researcher and graduate student and the acknowledged expert in the energy policy realm to become quickly acquainted with a broad spectrum of relevant works from the technical and social literature.

More concretely, the preparation of this work was animated by two basic concerns. In the first place, this bibliography reflects the felt need for a guide which draws particularly from the burgeoning literature on energy policy which has appeared

1

in the wake of the Arab embargo, but which also honors the research and writings pre-dating that cataclysmic event in order to put into proper historical context the post-embargo occurrences and to explain, among other things, the failure of the nations of the world to anticipate the energy crisis and prepare for the dawn of the post-petroleum era. While the bibliographical guide will not eliminate the need for literature searches for more specialized materials and those most recently published, it is hoped that it will be an initial source to which students and scholars may turn for references to the major books, monographs, and articles dealing with social scientific and public policy perspectives on energy research, policy, and sources.

No less important has been our desire to assist in the transcendence of policy parochialism in the study of energy by social scientists in particular. In part because of their over-fascination with the technical/methodological aspects of social science re-search, in part because of an insularity of scope and vision growing out of certain traits distinctive to U.S. political culture, social scientists in the United States have been prone to project unacknowledged American assumptions and prejudices on to other countries or to reduplicate the traditionalism and conventionalism of earlier phases in the development of their respective disciplines in modern scientific garb. Current energy policy research risks repeating past errors of analysis by not treating the energy crisis within a comparative framework and in terms of its international dimensions. For this reason, this work incorporates a lengthy section on features of energy policies around the globe.

Every bibliography, however apparently arbitrary or haphazard, rests upon an underlying outlook on the subject matter it surveys and the world it inhabits. This bibliographical guide is informed by the belief that every significant facet of energy policy involves fundamental questions of political economy, social values, and cultural transformation. The soft path/hard path debate within the energy field, whose terms were largely cast by Amory Lovins in his celebrated Foreign Affairs article, has en-gendered widespread recognition of the ethical and lifestyle dimensions to the energy debate; and recent theoretical studies by Leon Lindberg and others have begun to alert energy policy analysts to the implications of alternative energy futures for political economy. Yet, a majority of political scientists and sociologists in the United States remain attached to systems and other mechanistic approaches to public policy. In order to compensate for this bias, an effort has been made to highlight in one section those writings which treat issues of the state and economy and political economy and planning as integral elements of a theoretical approach to policy analysis.

The political economy approach to social and economic analysis has a time-honored heritage which goes back at least to the eighteenth and nineteenth centuries and became the center of controversy in the 1840's in the debate between Karl Marx and the bourgeois political economists over the laws and dimensions of political economy. An overview of the history of this approach need not detain us here. Suffice it to say that, in trying to rehabilitate this tradition of discourse and analysis as a critical approach to analyzing social problems. contemporary scholars have sought to demonstrate the ways in which it makes problematic and opens up for renewed investi-gation dimensions of social and political problems which other approaches or frame-works leave bracketed or unexplored. By viewing public policy as residing inevitably within a complex matrix of dynamic and changing relationships between state and economy and between nation-state and international system, the standpoint of politi-cal economy places old questions in a new light and generates new questions about a previously taken-for-granted outlook. It also rejoins the relationship between economics and ethics previously severed by the rise of the school of positive economics by affirming that every political economy is a _moral_ economy.

The ramifications of critical political economy for public policy analysis generally, and for energy policy analysis specifically, are manifold. For public policy analysis, it rekindles the political imagination, refounds fundamental questions of public policy squarely within the ambiguous and open-ended terrain of politics, and revives the theoretical mode as an ineluctable element of informed political action. For energy policy inquiry, it raises anew the question of the nature of power in advanced industrial societies, explores the complex relationships among trends in energy development at the regional, national, and international levels as parts of a single problem, and shows how national strategies of energy autonomy or self-sufficiency in an era of international energy interdependence involve fundamental issues of social justice, political power, ecological concern, international political economy, and global equity.

Such an approach asks questions like: What are the causes and consequences of increasing economic concentration within and across energy industries? Has the coalescence of OPEC into a world cartel diminished the power of the multinational oil corporations, or has it forged a duopoly? To what extent was the advancement of civilian atomic energy a result of the crystallization of an atomic-industrial complex combining public and private agencies and organizations into an influential power cluster? What institutional and ideological barriers flowing from political economy and culture obstruct the emergence of an energy future based upon conservation and renewable energy sources? Why have three successive American presidents appealed to an Apollo/Manhattan Project organizational model for mobilizing the American public behind a crash program for achieving a technological solution to the energy problem?

The primary organizational implication for this bibliography of its political-economic framework is that the major divisions ought not to be viewed in isolation from one another, but should instead be seen in terms of their theoretical and existential interconnections. The analytical categories and substantive themes which define the section and subsection divisions offer a convenient and useful means of organizing the mass of materials in the technical and social science literature on energy and energy-related matters. As we have argued above, however, the substantively significant and theoretically interesting consequences of the energy crisis are best captured by looking at the interstices of various themes and their complex and often obscure connections with the political economy at national and international levels. All this said, it must be acknowledged that both theoretically- and empirically-oriented social scientists will find much to mine in this bibliography. Moreover, as the organization of the major divisions discloses, some fairly standard and traditional categories and themes have been selected to subsume key references.

A noteworthy feature of this bibliographical guide is the inclusion of bibliographical essays which introduce the key themes, issues, and concepts addressed in the books and articles referenced. While these essays eschew any disembodied notion of value neutrality, they do attempt to illuminate competing perspectives on controversial energy issues. Students of energy policy should find the essays valuable resources for highlighting the many complex issues inhabiting the energy domain and for reviewing the major books, articles, and monographs analyzing those matters. Researchers interested in the nexus of two or more key themes or topics are urged to examine the

3

relevant bibliographical essays and parts or chapters bearing on those topics. For example, the relevant literature on the role of electric utilities in supporting decentralized solar technologies will be found both in the chapter on the electric utilities industry and the one on solar energy, while guides to the solar/utility interface will be offered in the bibliographical essays prefacing the parts where these chapters are located. Energy analysts in search of research and writings not covered in this bibliography are directed to the chapter of additional bibliographical sources in part VIII.

After beginning with an introduction (Part I) presenting a general overview on energy and the themes of political economy and planning, the bibliography turns its attention in Part II to works centering on a number of political and administrative areas. Because politics, the state, and state administration are neither mere epiphenomena of larger economic forces nor yet controlling factors in the shaping of energy policy and our collective energy future, the primacy accorded to <u>politics</u> in this bibliographical guide is awarded for its potential as a mediating agency and as the inevitable plane upon which the battle for an energy policy for future generations must be fought. Energy policy at the national and subnational levels is treated comprehensively because of the developing interests of social scientists and policy analysts in these domains and because of the felt conviction that realities indigenous to the American federal system in combination with values convergent with soft energy paths and the desire for regional autonomy make compelling the movement toward a decentralized energy system which would be nationally coordinated but regionally and locally administered.

The inclusion of sections on the historical embodiments of state administration in the United States (AEC/ERDA/DOE) and the key technoscientific vehicle in this country (the network of national scientific laboratories arising from the Manhattan Project) marks a recognition of the influence of state administration and technoscience from World War II to the present in directing the United States along the hard energy path through its devotion to high technology/capital intensive energy options. Moreover, were the role of public opinion neglected, an important obstacle to, and a necessary prerequisite for, a national energy policy would be slighted. Most students of public opinion in our view have yet to read the public temperament on energy issues with the subtlety and sophistication which the subject deserves. Finally, passing consideration is given to the writings and studies on the international and foreign policy dimensions to energy politics in America.

The third part of this guide attends to the <u>economic</u>, <u>social</u>, and <u>environmental</u> dimensions of energy policy and technology. The economic side of the energy picture unfolds in the sections dealing broadly with economic costs of various energy technologies and futures and with the structure and dynamics of the energy industry. Economics may still be regarded as a dismal science by many undergraduate students and even a few of its practitioners, but in the energy field questions of energy production and use, the role of and changes in different elements of the energy industry, and the comparative economic costs and consequences of diverse energy alternatives are crucial to the resolution of the political controversies over energy, nationally and internationally. The lively, often unpredictable, and sometimes unsettling interplay between energy and social structure and between energy production and use and the environment are highlighted in the next two sections. The notion of social structure is used in a broad sense to encompass both the impact of energy upon community and society and the effects of social and class structure upon energy. Likewise, the relationship between energy and the environment is understood in terms of the same kind of dialectical interaction. Individually and collectively, the exploration of economic, social, and environmental facets of energy perforce opens up major issues involving energy and ethics. It is a tribute to social scientists and theologians that these ethical questions are being asked; yet, the brevity of this section is a commentary on how much more thought and effort must still be given to these matters.

In the final analysis, the debate over alternative energy options turns on a basic dipute over our <u>energy future</u> – the outline of the future society, and, indeed, the future global community, that we wish to bequeath to subsequent generations. In other words, it opens up a question pondered by philosophers from Plato and Aristotle to Marx and Tawney and beyond – i.e., what is the good society and how can it be achieved? Part IV is organized around a number of topics dovetailing with this overriding theme. In illuminating the subject of energy and the future, this part collects those writings by social scientists, political theorists, and philosophers which seek to extrapolate or speculate from the past or present on the state of our energy situation at some definable point in the future – 1980, 2000, 2025, or beyond. While some of these works are content to offer a snapshot of a plausible energy future based upon assumptions and projections, others show a keen interest in sketching out the best or most viable means of getting from here to there. The emergence of energy futuristics is an event of some moment for political decision-making and political action, and social scientists would be remiss in failing to contribute to its advancement or in neglecting to reflect seriously on its authoritarian and democratic possibilities.

Two methods or techniques increasingly employed by energy analysts in expanding the realm of human choice in fashioning our energy future are energy modeling and risk analysis. Energy modeling has become an increasingly popular tool for energy policy advisers in charting out hypothetical models of alternative energy futures on the basis of the identification of strategic variables and their use in the construction of dynamic interaction processes. Risk analysis is a relatively new field springing from the efforts of decision-makers interested in long-range planning to deal with risk, uncertainty, and decision in a variety of fields. In the energy policy domain, it has served as a method – and an extremely controversial one at that -- for assessing the comparative risks of different sources of energy production.

Fundamental to the forging of our energy future are the role, assumptions, and institutional context of energy research and development (R&D). Consequently, the final section of this part of the bibliography focuses on works clarifying these topics. Social scientists who have eschewed the temptations of the Technocratic Dream and who remain aware of the open-endedness of the future and the role of political judgment, vision, and action in its shaping will appreciate the scope and limits of energy modeling, risk analysis, and energy R&D for energy analysis and energy policy-making.

The next two parts of the bibliography, dealing with various energy sources and technologies, are organized along fairly traditional lines. Part V surveys works specifically dealing with the <u>conventional energy sources</u> of our petroleum era and with the so-called "<u>swing</u>" <u>fuels</u> (coal and/or nuclear power) which most energy policy experts have argued should and could serve as the bridge to the year 2000, when alternative energy technologies would mature and take over the task of meeting our energy needs. In the case of coal and nuclear power, a number of issues associated with their use are delineated. For example, the massive literature on coal is organized into several subsections which take into account coal's potential both as a bridge fuel and as an alternative source of oil and gas (synthetic fuels), and which highlight the nagging problems associated with coal, including its impact upon the environment and miner health and safety. Similarly, in the section on nuclear power, the entire nuclear fuel cycle and its associated problems is considered.

Perhaps the most noteworthy feature of Part VI (alternative energy sources) is the placement and priority given to conservation in this category. This decision reflects our firm conviction, supported by soft energy paths advocates (e.g., Amory Lovins and Robert H. Williams) and by reform-minded corporate liberals (e.g., Daniel Yergin), that conservation strategies and technologies may be one of the most important energy sources for alleviating the energy crisis and perhaps for building a conserver society of the future. Of course, from our viewpoint, the necessity for cultural transformation and political change in the United States in order to realize the promise of conservation cannot be glossed over or dismissed. In this part, too, we adhere to the developing convention of interpreting solar energy not only in its narrowest sense, as those sources and technologies generating energy directly from the sun's rays, but also in its broader sense, which includes biomass, geothermal, wind, tidal, ocean thermal energy conversion, and so forth.

To avoid misinterpretation, we wish to underscore the point that, while solar energy and conservation topics have been assigned organizationally to the division on alternative energy sources, it would be incorrect to assume that these technologies have significant potential only in the long-term. Many energy specialists and solar advocates have made a strong case for the near-term feasibility and utility of some conservation measures and solar technologies; and the potential contribution of each alone or in combination to satisfying our energy needs by the year 2000 is a highly contentious issue within the national energy debate. As for the last three more esoteric alternative energy technologies covered in this part, no debate over their short-term roles exists. Fusion power and hydrogen especially are alternative energy sources whose technological feasibility remains in doubt and whose development as mature energy technologies will come, if at all, some time around the middle or the latter half of the twenty-first century.

The impetus and rationale for incorporating into the bibliography a major division on energy policies in international perspective (Part VII) have already been articulated. As evidenced by the table of contents, the architecture of this part conforms to major continents, regions, and nations of the world. Ideally, perhaps, this comparative part of the guide should have been organized topically like the earlier sections, which dealt almost exclusively with the American context; or, better yet, it might have been integrated into the previous subdivisions for a truly comparative structure. Alas, the occasion for either format is not yet at hand, given the unevenness and paucity of available energy studies of other nations and regions, the relative underdevelopment of comparative public policy in the social sciences, and the newness of the field of energy policy analysis here and abroad. It is our hope that the publication of this bibliography, with its fairly lengthy enumeration of references to non-American energy research and policy, will assist the present generation of energy policy specialists in overcoming these limitations and deficiencies.

No bibliographical guide can hope to exhaust all the relevant writings of an increasingly specialized field. This work is no exception. For this reason and others, another part (VIII) has been added which offers a list of other bibliographies covering energy topics, a section of basic sources compiled by researchers, agencies and institutes, and governmental and non-governmental organizations, as well as a section enumerating key journals and other periodicals which either publish energy research or report on important energy developments. The first should supplement the treatment given to specialized topics in the present volume; the second should at least temporarily slake the unquenchable thirst of empirically-minded social analysts for sources of energy statistics; and the last should be helpful to specialists and generalists in the energy field who wish to keep abreast of frontier research and breaking events.

The bibliography closes with a division (Part IX) presenting an annotated listing of 75 essential books which we recommend for any serious energy library. This part is directed particularly to acquisitions and general librarians who face the often monumental task of deciding which works in the evergrowing literature on energy to purchase for a new or expanding energy collection. It is our hope that, in providing a fairly lengthy, balanced, and wide-ranging selection of important and influential energy books, their responsibilities for acquiring informative works on energy topics will be made less imposing and their choices among the many such books currently available will be better informed.

Part I
Introduction

The first part of this bibliographical guide is divided into two chapters. The initial chapter offers a potpourri of books and articles presenting a general perspective for social scientists on energy and the energy crisis. Essential to any understanding of the social and political dimensions of the energy crisis is an awareness of the cardinal role of energy in human affairs. Many of the writings collected here (Abelson, 1974, 1975; Abelson and Hammond, 1978; Deju, 1974; Grenon, 1975; Goldsmith, 1976; Hottel and Howard, 1972; Hubbert, 1971; Priest, 1975; Ruedisili and Firebaugh, 1978; Shepard, et al., 1976) seek to explore scientific and technical facets of energy – including its diverse forms, estimated availability, extraction costs, and projected requirements. Still others offer an interdisciplinary view of energy and human affairs which shows the interconnectedness of energy, humanity, society, and environment (Commoner, 1976; Cook, 1976; Garvey, 1972; Kranzberg and Hall, 1980; Reed, 1975; Ridgeway and Conner, 1975; Soneblum, 1978). For the social scientist, such an approach discloses the mazelike character of public policy-making in the energy realm, since a synoptic and integrated orientation to energy policy demonstrates the complicated ways in which policy initiatives in the energy sphere inevitably have important impacts upon economy, society, and ecology, just as policy actions in these other spheres inevitably affect energy.

If the technical literature shows that energy is an ineluctable element of human survival and social betterment, other works of a more political and sociological view ask the more basic question of the nature of the present energy crisis. Indeed, the character of our prevailing energy woes manifested domestically and internationally is itself a key issue in the energy debate. There are certainly many signs of energy crisis in our midst: spiraling energy prices, double-digit inflation, governmental policies instituting new regulations in some areas (like fuel efficiency standards) and removing old laws in other areas (like de-regulation of natural gas and oil), as well as increasingly massive public subsidies for the development of new energy technologies for our energy future. Yet, various studies included here ask: Is the situation we face real or contrived (Coyne and Coyne, 1977; Doolittle, 1977; Foley, 1976; Metzger, 1977; Mondale, et al., 1974; Rocks and Runyan, 1972)? Is it a genuine crisis, or a government-induced and/or corporate-inspired conspiracy (Commoner, 1979; Ridgeway and Conner, 1975; Storms, 1974)? Is it both (Barnet, 1980)? Or, more modestly, is it a technical problem requiring a technological solution (Doolittle, 1977; Hottel and Howard, 1972; Priest, 1975)? Will it require a basic transformation of our institutions and our lifestyles

(Commoner, 1976, 1979; Mazur and Rosa, 1974), or will a combination of competent public administration and good old Yankee ingenuity allow us to produce it away (Ezra, 1978; Friedman, 1979; Lilienthal, 1980; Teller, 1979)?

In this domestic debate over the character and duration of America's energy problems, one fact that stands out is that the United States, by virtue of its advanced industrial and technological status, its wide-ranging and still sizable natural resources, and its remaining reserves of oil, coal, and natural gas, possesses a much broader array of energy options than most of the rest of the world. In addition, the United States can play a valuable role in helping to shape the broad design of a global response to the international energy crisis. By its choices, the United States will inevitably mold the contours of that response. For this reason, the quest for new and alternative energy sources in the emerging post-petroleum world of the twenty-first century is a focus of a number of these books and essays (Goldsmith, et al., 1976; Hayes, 1977; Israel, 1974). So, too, is the concern for analyzing the resolution of earlier energy crises (Fisher, 1974; Nef, 1977).

What is less well-recognized is that the energy crisis is an economic crisis, and that its resolution will inevitably have fundamental repercussions for the shape of our political economy, as well as the political economies of other nations and, indeed, the international order (Commoner, 1976; Gorz, 1980; Hammarlund and Lindberg, 1976; Lindberg, 1977; Tietenberg, 1975). For energy policies for future society and the global order involve traditional questions of distribution, pricing, and commercialization, as well as questions of a more ethical sort, such as equity, justice, and power. To repair this deficiency, a special chapter on the state, political economy, and planning is included. Insofar as a vision of the future political economy can be drawn from the competing positions voiced in the contemporary energy controversy, some interesting paradoxes and apparent mismatches result.

Many political analysts and cultural critics have observed how politically regressive America's ideological systems of thought have been in comparison to the character of its economic institutions and accompanying social relations. Terms like free enterprise and free market system are still bandied about as representations of the political-economic order in political discussion in the United States, despite the massive internal transformations which have taken place in Western capitalist systems since the mid-nineteenth century. Many of these changes have involved modifications in state-economic relations, while others have been centered on instituting a limited planning role for state agencies (Alford, 1975; Bell, 1973; Best and Connolly, 1976; Cohen, 1969; Goldstein, 1978; Lindblom, 1977; Martin, 1973; Shonfield, 1965). Although these changes have affected all state capitalist systems in the West, their impact and meaning have been reflected ideologically only in Western Europe.

If we examine the landscape of America's political economy today, we find that it is best characterized as an economic system divided into three levels or plateaus: (1) a quasi-free market level populated by small businesses, generally adhering to more traditional economic laws and norms of pricing and competition, but increasingly being crowded out by the other two levels; (2) an oligopolistic system peopled by megacorporations, which typically allows for marginal competition and modest price differentiation in a highly concentrated market; and (3) a techno-corporate system constituted by a handful of corporate giants heavily dependent upon the government for contracts and subsidies (O'Connor, 1973; Habermas, 1973; and Yanarella and Ihara, 1978). Concomitantly, the role of the state at each level is variable, serving as provider of a stable monetary system and promoter of a stable and positive business

9

environment in one, acting as a bureaucratic regulator, welfare state manager, and active economic stimulator in the second, and playing the role of consumer, subsidizer, and weak administrator in the third. Historically, these three levels and their accompanying business-government relationships have been sedimented into our socio-economic order by virtue of the internal transformation which our capitalist system has undergone over its history (Wolfe, 1977).

Increasingly, our political economy is approximating the image of a corporate state (Fusfield, 1972; Miller, 1976) which, though structurally differentiated and multi-layered, is increasingly being dominated by the proximate needs and economic capabilities of a relatively few "megacorporations," administered in special areas by state mechanisms, and skewed in its general "social" priorities by a militarized environment distinctive to American state capitalism. This central tendency presents some interesting problems and paradoxes for different schools of thought and competing parties in the energy policy debate. Among those politicians and policy-makers regarding themselves as pragmatists, who attempt to straddle the hard path/soft path debate, there is a propensity to accept the processes of the interest group struggle as the most legitimate means for contending with energy matters while embracing a corporate-liberal image of the economy roughly equivalent to the basic features of the oligopolistic system (e.g., Stobaugh and Yergin, 1979). Increasingly, this school and its interest group strategy carries with it its own seeds of transformation. For example, in seeking to mediate the energy/environment dispute in its various forms, its favored policy option of promulgating governmental regulations, while alleviating the environmental costs of energy production, tends to price small businesses out of the market and to wed the federal government to large energy corporations who can afford the capital costs of regulation. In the process, corporate liberalism digs its own grave and paves the way for techno-corporate institutions.

Meanwhile, many soft energy path advocates are embracing a laissez-faire capitalist philosophy (e.g., Amory Lovins' avowal of a "neo-capitalist energy manifesto" [1978] and John Gofman's conversion to libertarianism [1980]), due, in part, to their awareness of the fact that the really novel innovations in solar, wind, and other soft energy technologies are emerging from individual garage inventors and small laboratories rather than from the laboratories of the federal government or the energy corporations (see Barnet and Muller, 1974; and Blumberg, 1975). The problems with this identification of the philosophy of the soft energy path with laissez-faire economics are many. In the first place, it ignores or at least fails to make a political problem for theory and practice the general techno-corporate direction of our political economy. Secondly, it idealizes the role of the small entrepreneur and the laissez-faire phase of America's capitalist economy through a mythical rewriting of nineteenth century economic and political history. And, thirdly, its embrace of an economic philosophy grounded in an image of man as possessive individual contradicts the social vision and communitarian ethos which appeals to so many who have made the soft energy path school into an impressive social movement. The operative question for people like Amory Lovins and John Gofman is: Are there ideological and cultural resources in the American economic heritage which justify this reversion to an apparently outmoded economic system? Or is this attachment to an antiquarian system merely a symptom of a profound failure of political imagination?

Of course, the hard technology school promises to leave our economic system as it is (or as it was), and merely to discover and build those necessary technological means for reviving the American spirit and assuring our economic well-being. But, as

10

many critics of the hard technology path have noted and as the Carter National Energy Plan II (synfuels plus Energy Mobilization Board plus Energy Security Administration) testifies to, the hard technology path would entail vast amounts of mobile capital, a continuing spiral of energy prices, uncertain costs to public health and the environment, and increasingly stringent governmental administration and economic centralization in the energy realm. Whether the American people would come to regard a society with such a high degree of elite governance, social discipline, income inequality, and health and environmental risks as the good society or the friendly fascist (Gross, 1980) one is uncertain. What is evident is that the hard energy path, too, would reinforce tendencies toward the crystallization of the techno-corporate phase of our political economy and the realization of the corporate state.

Of the major public advocates of a preferred energy future for American society, only Barry Commoner (1976, 1979) has clearly recognized the economic dimensions of the energy crisis. In his books, The Poverty of Power (1976) and The Politics of Energy (1979), he has criticized the inflationary, anti-employment consequences of all hard path scenarios of our energy future. Moreover, he has responded to indictments of the soft energy alternative as unrealistic and utopian by charting out a strategy for carrying out the transition to a solar future. Simultaneously, Commoner has broken with the neo-laissez faire economic vision of Amory Lovins by advocating in a modest and muted way democratic socialism. What he has failed to do thus far is to integrate his economic socialist loyalties with his soft path critique of high technology options in energy policy. For instance, save for his recommendations for regulating oil corporations on the model of the electric utilities industry, he has not really shown why his solar transition blueprint is incompatible with our capitalist economic framework. In his latest book, especially, his call for social governance is too weak, and its underlying political philosphy too closely akin to the Madisonian principles and assumptions of our present constitutional system to serve as a compelling alternative to liberal capitalism in its many guises and modifications. What is needed of Commoner is a critique of political economy incorporating both demystifying and reconstructive moments combined with a more detailed strategy of the political tasks and social resources for activating the call for the solar transition as a democratic social movement beyond liberalism and capitalism. (For beginning steps, see Reid and Ihara, 1978; Jezer, 1977; and Worthington, 1978 and 1980.)

1. General Perspectives on the Energy Crisis

Abelson, Philip H. Energy for Tomorrow. Seattle, Washington: University of Washington Press, 1975.

_____, ed. Energy: Use, Conservation, and Supply. Washington, D.C.: American Association for the Advancement of Science, 1974.

_____ and Hammond, Allen L., eds. Energy II: Needs, Conservation, and Supply. Washington, D.C.: American Association for the Advancement of Science, 1978.

Barnet, Richard J. The Lean Years: Politics in the Age of Scarcity. New York: Simon and Schuster, 1980.

Commoner, Barry. The Poverty of Power: Energy and the Economic Crisis. New York: Alfred A. Knopf, 1976.

Conant, Melvin A. and Gold, Fern B. The Geopolitics of Energy. Boulder, Colorado: Westview Press, 1978.

Congressional Quarterly. Continuing Energy Crisis in America. Washington, D.C.: Congressional Quarterly, Inc., 1975.

Cook, Earl. "The Flow of Energy in an Industrial Society." Scientific American, 225 (September 1971), 135-144.

_____. Man, Energy, and Society. San Francisco, California: W. H. Freeman Company, 1976.

Copulos, Milton. Energy Perspectives. Washington, D.C.: Heritage Foundation, 1978.

Coyne, John R. and Coyne, Patricia. The Big Breakup: Energy in Crisis. Mission, Kansas: Sheed, Andrews & McMeel, 1977.

Doolittle, Jesse S. Energy: A Crisis, A Dilemma, or Just Another Problem? Champaign, Illinois. Matrix Publishers, 1977.

"The Energy Crisis: Reality or Myth." [Special Issue] The Annals of the American Academy of Political and Social Science, 410 (November 1973).

"Energy. Facing Up to the Problem, Getting Down to Solutions." [Special Report] National Geographic, 159 (February 1981).

Evans, Douglas. The Politics of Energy: The Emergence of the Superstate. Toronto, Ontario: Macmillan of Canada, 1976.

Fisher, John C. Energy Crises in Perspective. New York: Wiley Interscience, 1974.

Foley, Gerald. The Energy Question. Baltimore, Maryland: Penguin Books, 1976.

Garvey, Gerald. Energy, Ecology, Economy. New York: W. W. Norton & Company, 1972.

Goldsmith, Marc W., et al. New Energy Sources: Dreams and Promises. Framingham, Massachusetts: Energy Research Group, 1976.

Gordon, Lincoln. "Energy Development: Crisis and Transition." Bulletin of the Atomic Scientists, 37 (April 1981), 24-29.

Grenon, M., ed. Energy Resources. Laxenburg, Austria: International Institute for Applied Systems Analysis, 1975.

Hayes, Denis. Rays of Hope: The Transition to a Post-Petroleum World. New York: W. W. Norton & Company, 1977.

Holdren, John and Herrera, Philip. Energy. San Francisco, California: Sierra Club Books, 1971.

Hollander, Jack M., et al., eds. Annual Review of Energy. Volumes 1-5, Palo Alto, California: Annual Reviews, Inc., 1976-1980.

Hottel, H. C. and Howard, J. B. New Energy Technology - Some Facts and Assessments. Cambridge, Massachusetts: M.I.T. Press, 1972.

Hubbert, M. King. "The Energy Resources of the Earth." Scientific American, 225 (September 1971), 61-70.

Israel, Elaine. The Great Energy Search. New York: Julian Messner, 1974.

Kiefer, Irene. Energy for America. New York: Atheneum Publishers, 1979.

Knowles, Ruth S. America's Energy Famine: Its Cause and Cure. Norman, Oklahoma: University of Oklahoma Press, 1980.

13

Kranzberg, Melvin and Hall, Timothy, eds. Energy and the Way We Live: A Courses by Newspaper Reader. San Francisco, California: Boyd & Fraser Publishing Company, 1980.

Krenz, Jerrold H. Energy: From Opulence to Sufficiency. New York: Praeger Publishers, 1980.

Landsberg, Hans H. "Low-Cost, Abundant Energy: Paradise Lost?" Science, 184 (April 19, 1974), 247-253.

Loftness, R. L. Energy Handbook. New York: Van Nostrand Reinhold, 1979.

McMullan, J. T.; Morgan, R.; and Murray, R. B. Energy Resources and Supply. New York: Wiley Interscience, 1976.

Marion, Jerry B. Energy in Perspective. New York: Academic Press, 1974.

Messel, Harry, ed. Energy for Survival. Elmsford, New York: Pergamon Press, 1979.

Metzger, Norman. Energy - The Continuing Crisis. New York: Thomas Y. Crowell, 1977.

Mondale, Walter F., et al. Is the Energy Crisis Contrived? Washington, D.C.: American Enterprise Institute for Public Policy Research, 1974.

Morgan, M. Granger, ed. Energy and Man: Technical and Social Aspects of Energy. New York: Institute of Electrical and Electronics Engineers, 1975.

Nef, John U. "An Early Energy Crisis and Its Consequences." Scientific American, 237 (November 1977), 140-151.

Parker, Sybil P., ed. McGraw-Hill Encyclopedia of Energy. 2nd ed. New York: McGraw-Hill, 1981.

Penner, S. S. and Icerman, L. Energy. Reading, Massachusetts: Addison-Wesley Publishing Company, 1974.

_____. Energy. Volume 2: Non-Nuclear Technologies. Reading, Massachusetts: Addison-Welsley Publishing Company, 1975.

Priest, Joseph. Energy for a Technological Society: Principles, Problems, Alternatives. Reading, Massachusetts: Addison-Wesley Publishing Company, 1975.

Reed, C. B. Fuels, Minerals, and Human Survival. Ann Arbor, Michigan: Ann Arbor Science Publishers, 1975.

Regens, J. L., ed. Energy Issues and Options. Athens, Georgia: University of Georgia Press, 1979.

Ridgeway, James and Conner, Bettina. New Energy: Understanding the Crisis and a Guide to Alternative Energy System. Scranton, Pennsylvania: Beacon Press, 1975.

Rocks, Lawrence and Runyon, Richard P. The Energy Crisis. New York: Crown Publishers, 1972.

Ruedisili, L. C. and Firebaugh, M. W. Perspectives on Energy. New York: Oxford University Press, 1978.

Shepard, Marion; Chaddock, J. B.; Cocks, Franklin H.;and Harman, C. M. Introduction to Energy Technology. Ann Arbor, Michigan: Ann Arbor Science Publishers, 1976.

Simon, Andrew L. Energy Resources. Elmsford, New York: Pergamon Press, 1975.

Sobel, Lester A., ed. Energy Crisis. Volume 1: 1969-1973; Volume 2: 1974-1975; Volume 3: 1975-1977. New York: Facts on File, 1974, 1975, and 1977, respectively.

Starr, Chauncey. Current Issues in Energy. Elmsford, New York: Pergamon Press, 1979.

_____. "Energy and Power." Scientific American, 225 (September 1971), 37-49.

_____. "Realities of the Energy Crisis." Bulletin of the Atomic Scientists, 29 (September 1973), 15-20.

Steinhart, John and Steinhart, Carol. Energy: Sources, Uses, and Role in Human Affairs. Belmont, California: Duxbury Press, 1974.

Stoker, Howard Stephen, et al. Energy from Source to Use. Glenview, Illinois: Scott, Foresman Company, 1975.

Storms, Ray E. Myths and Realities of the Energy Shortage: Contrivance by the Companies or Bungling by the Government. Hicksville, New York: Exposition Press, 1974.

Teller, Edward. Energy from Heaven to Earth. San Francisco, California: W. H. Freeman Company, 1979.

Tussing, Arlon R. "Three Classes of Energy Resorces." Energy Policy, 2 (September 1974), 179-188.

Wells, Malcolm. Energy Essays. Barrington, New Jersey: Edmund Scientific Company, 1976.

2. The State, Political Economy, and Planning

Alford, Robert. "Planning versus the Market: Introduction." In Leon N. Lindberg, et al. Stress and Contradiction in Modern Capitalism. Lexington, Massachusetts: D. C. Heath and Company, 1975, pp. 3-11.

Baran, Paul and Sweezy, Paul. Monopoly Capital. New York: Monthly Review Press, 1966.

Barnet, Richard J. and Muller, Ronald. Global Reach: The Power of the Multinational Corporations. New York: Simon and Schuster, 1974.

Bell, Daniel. The Coming of Post-Industrial Society: A Venture in Social Forecasting. New York: Basic Books, 1973.

Best, Michael H. and Connolly, William F. The Politicized Economy. Lexington, Massachusetts: D. C. Heath and Company, 1976, 1981 [rev. ed.].

Blissett, Marlan. Politics in Science. Boston, Massachusetts: Little, Brown and Company, 1972.

Blumberg, Philip. The Megacorporation in American Society: The Scope of Corporate Power. Englewood Cliffs, New Jersey: Prentice-Hall, 1975.

Brzezinski, Zbigniew. Between Two Ages: America's Role in the Technetronic Era. New York: Viking Press, 1970.

Bupp, Irvin C. "Energy Policy Planning in the United States: Ideological BTU's." In Leon N. Lindberg, ed. The Energy Syndrome: Comparing National Responses to the Energy Crisis. Lexington, Massachusetts: D. C. Heath and Company, 1977, 285-324.

Caldwell, Martha and Wolley, John T. "Energy Policy and the Capitalist State." In Jeffrey Hammarlund and Leon Lindberg, eds. The Political Economy of Energy Policy: A Perspective from Capitalist Society. Madison, Wisconsin: Institute for Environmental Studies, University of Wisconsin, Madison, December 1976, pp. 110-153.

Calleo, David P. and Rowland, Benjamin M. America and the World Political Economy. Bloomington, Indiana: Indiana University Press, 1977.

Cohen, Stephen S. Modern Capitalist Planning: The French Model. Cambridge, Massachusetts: Harvard University Press, 1969.

Friedman, Milton. Capitalism and Freedom. Chicago, Illinois: University of Chicago Press, 1952.

Funigiello, Philip J. Toward a National Power Policy: The New Deal and the Electric Utility Industry, 1933-1941. Pittsburgh, Pennsylvania: University of Pittsburgh Press, 1973.

Galbreath, John Kenneth. The New Industrial State. Rev. ed. New York: Signet Books, 1972.

Goldstein, Walter, ed. Planning, Politics, and the Public Interest. New York: Columbia University Press, 1978.

Gorz, Andre. Ecology as Politics. Boston, Massachusetts: South End Press, 1980.

Graham, Otis L., Jr. Toward a Planned Society: From Roosevelt to Nixon. New York: Oxford University Press, 1976.

Greenberg, Edward S. Serving the Few: Corporate Capitalism and the Bias of Government Policy. New York: John Wiley & Sons, 1974.

_____. Understanding Modern Government: The Rise and Decline of the American Political Economy. New York: John Wiley & Sons, 1979.

Gross, Bertram. Friendly Fascism: The New Face of Power in America. New York: Maurice Evans, 1980.

_____. "Planning in an Era of Social Revolution." Public Administration Review, 31 (May/June 1971), 259-297.

Habermas, Jurgen. Legitimation Crisis. Boston, Massachusetts: Beacon Press, 1973.

Hammarlund, Jeffrey R. and Lindberg, Leon N., eds. The Political Economy of Energy Policy: A Projection for Capitalist Society. [IES Report 70] Madison, Wisconsin, Institute for Environmental Studies, University of Wisconsin, Madison, December 1976.

Herman, Edward S. Corporate Control, Corporate Power. New York: Cambridge University Press, 1981.

Hoos, Ida. Systems Analysis in Public Policy: A Critique. Los Angeles, California: University of California Press, 1972.

Levinson, Charles. Capital, Inflation, and the Multinationals. New York: Macmillan, 1971.

Lindberg, Leon N., ed. The Energy Syndrome: Comparing National Responses to the Energy Crisis. Lexington, Massachusetts: D. C. Heath and Company, 1977.

_____. Politics and the Future of Industrial Society. New York: David McKay Company, 1976.

_____, et al. Stress and Contradiction in Modern Capitalism. Lexington, Massachusetts: D. C. Heath and Company, 1975.

Lindblom, Charles E. Politics and Markets: The World's Political-Economic Systems. New York: Basic Books, 1977.

Lowi, Theodore J. The End of Liberalism. New York: W. W. Norton and Company, 1969.

Martin, Andres. The Politics of Economic Policy in the United States: A Tentative View from a Comparative Perspective. Beverly Hills, California: Sage Publications, 1973.

Melman, Seymour. Pentagon Capitalism: The Political Economy of War. New York: McGraw-Hill, 1970.

_____. The Permanent War Economy: American Capitalism in Decline. New York: Simon and Schuster, 1974.

Miliband, Ralph. "The Capitalist State -- Reply to Nicos Poulantzas." New Left Review, No. 59 (January - February 1970), 53-60.

_____. "Poulantzas and the Capitalist State." New Left Review, No. 82 (November - December 1973), 83-92.

_____. The State in Capitalist Society. New York: Basic Books, 1969.

O'Connor, James. The Corporations and the State. New York: Harper & Row, Publishers, 1974.

_____. The Fiscal Crisis of the State. New York: St. Martin's Press, 1973.

Offe, Claus. "The Theory of the Advanced Capitalist State and the Problem of Policy Formation." In Leon Lindberg, et al. Stress and Contradiction in Modern Capitalism. Lexington, Massachusetts: D. C. Heath and Company, 1975, pp. 125-144.

Polanyi, Karl. The Great Transformation: The Political and Economic Origins of Our Time. Boston, Massachusetts: Beacon Press, 1957.

Poulantzas, Nicos. Political Power and Social Classes. London, England: New Left Books and Sheed & Ward, 1973.

_____ . "The Problem of the Capitalist State." New Left Review, No. 58 (November - December 1969), 67-78.

_____ . State, Power, Socialism. [Trans. by Patrick Camiller] London, England: New Left Books, 1978.

Sharkansky, Ira. Whither the State? Politics and Public Enterprise in Three Countries. Chatham, New Jersey: Chatham House Publishers, 1979.

Sherman, Howard. Radical Political Economy: Capitalism and Socialism from a Marxist-Humanist Perspective. New York: Basic Books, 1972.

Shonfield, Andrew. Modern Capitalism: The Changing Balance of Public and Private Power. New York: Oxford University Press, 1965.

Tietenberg, Thomas H. Energy Planning and Policy: The Political Economy of Project Independence. Lexington, Massachusett: D. C. Heath and Company, 1976.

Vanek, Jaroslav. The Participatory Economy: An Evolutionary Hypothesis and a Strategy for Development. Ithaca, New York: Cornell University Press, 1971.

Wilson, David E. The National Planning Idea in U.S. Public Policy: Five Alternative Approaches. Boulder, Colorado: Westview Press, 1980.

Wolfe, Alan. The Limits of Legitimacy. New York: Basic Books, 1977.

Yanarella, Ernest J. and Ihara, Randal H. "The Military/Energy Connection: The Institutionalization of the 'Technological Breakthrough' Approach to Energy R & D." Northeast Peace Science Review, 1 (1978), 187-207.

19

Part II
Politics

    Tension marks the relationship between politics and administration in the energy policy realm, as even a cursory survey of the social scientific literature on energy subjects reveals. Although many issues of the twentieth century have been institutionalized into huge governmental agencies and bureaucracies without thereby being resolved (e.g., the environmental crisis), the technocratic vision of an administered state, a planned economy, and a post-industrial society stands out as more dream than reality. Politics in a very basic sense remains irrepressible. In part, the technocratic dream has been frustrated by the continuing corporate bias of public policy; in part, it remains an illusion because public policy issues such as energy policy are ultimately political and ethical problems subject to resolution neither by bureaucratic nor technical means. Consequently, politics will continue to be that medium through which the shape of energy policy now and in the future is forged, either through open, democratic means or through veiled, mystified means or through processes somewhere in between these two poles.

    This part of the bibliography gives expression to the primacy of the political realm by collecting a sizable number of references around a variety of political topics. A central issue in the energy debate revolves around the necessity and character of a national energy policy for the United States. The standard view is that the United States has for decades operated without a coherent energy plan or policy; rather, energy policy-making has been allowed to take place in a fragmented and decentralized manner in any of a host of agencies and bureaus scattered throughout the federal bureaucracy (Anthrop, 1974; Davis, 1974; Henderson, 1978; Mancke, 1974, 1976; Rycroft, et al., 1978). In this view, the stimuli of the Arab oil embargo and the appearance of OPEC as an international cartel have spurred the United States and other industrial nations to elevate energy policy planning to a visible and high-level focus of public policy-making and have prompted recent presidents from Nixon to Carter to frame such federally-directed and unified energy plans in order to overcome the often contradictory and "sub-system-dominated" nature of past energy policies (Anthrop, 1978; Bupp, 1977; Commoner, 1977; Congressional Quarterly, Inc., 1979; Executive Office, 1977; Meade, 1977; Tietenberg, 1976; ERDA, 1976; U.S. FEA, 1974).

    Challenging this interpretation are two other, less orthodox views. One school of thought, represented in the writings of scholars like Henry Nash (1976), Leon Lindberg (1977), and Robert Engler (1977), has emphasized the deeper truth that the United States has long had a discernible and consistent national energy policy organized around governmental subservience to oil industry (and other energy corporation) needs. These scholars have clarified how important aspects of energy policy in the domestic and foreign policy arenas have been all but ceded to these major corporate actors (see also Wright, 1978). On the other hand, a small, but increasingly influential, school stressing the importance of free-market economics and supply-side solutions to public policy problems (Adelman, et al., 1975; Friedman, 1979; Institute for Contemporary Studies, 1977; Meade, 1979; Stockman, 1977) has advanced the position that the effort to fashion national energy policies thwarts the built-in regulatory and adjustment mechanisms of the laissez-faire capitalist system which it seeks to recreate in American society. In truth, however, the role of public policy and the state will remain significant whatever direction or course future energy policy takes, since the interest-group nature of American politics will remain strong at the middle levels of power, and public policies and procedures will have to be instituted to structure relative priorities among various policy goals (energy, environment, agriculture, foreign policy, and economic policy) and to establish or strengthen market incentives for various parties in the policy game.

Concerning the question of how the United States has fared in resolving this issue, the answer of some writers and policy analysts seems to be: "not too well" (Commoner, 1979; Daneke and Lagassa, 1980; Goldstein, 1978; Henderson, 1978; Sachs, 1980). For, nationally, energy policy-making and planning, particularly in the United States and several other advanced capitalist societies, bears all of the earmarks of a policy quagmire. A number of students of energy policy (Bupp, 1977; Bupp and Derian, 1978, 1980; Lindberg, 1977) believe strongly that the energy programs of the industrialized West are at an impasse -- a policy stalemate which is only apparently belied by the billions of dollars, hundreds of studies, and countless ideas contributed to the solution of the energy crisis. This energy policy immobilism is evidenced most dramatically by the tendency of public debate over alternative energy strategies to divide into two highly principled and seemingly irreconcilable positions. (For a compilation of articles and views fueling this debate, see U.S. Senate Select Committee on Small Business, 1977.) On the one hand, there are proponents of the "hard technology" school of thought (Lilienthal, 1980; Srouji, 1977; Starr, 1979; Teller, 1975, 1979), composed primarily of key Congressmen, nuclear engineers, energy corporation executives, and government bureaucrats. Basically, this school promotes a hard technology energy path emphasizing high technology/capital-intensive centralized power systems, like atomic reactors, breeder reactors, fusion power, and synfuels plants. And, as a corollary, because the energy crisis is seen in traditional technological terms, it is viewed as essentially a supply-side problem to be solved by the following industrial imperative: "produce, produce." At least since World War II and the beginnings of the Manhattan Project, the members of this school have populated key positions of authority and influence within the energy complex (i.e., the network of public and private institutions which has shaped America's energy policies) and, until recently, their decisions and actions have been unchallenged (Yanarella, 1981).

Emerging out of the environmental movement of the late sixties and taking its grounding image, philosophy, and economics from the writings of Amory Lovins (1973, 1976, 1977a and b, 1980), E.F. Schumacher (1973, 1979), and Herman Daly (1977), a second group of energy visionaries – the soft energy path school of thought -- has gained a wide and increasingly influential following at the elite and mass levels. In contrast to their hard path competitors, soft energy path advocates look to appropriate or low technology/labor intensive energy options as well as broad conservation measures

21

as the preferred means to solving the energy crisis. Moreover, the energy problem is interpreted as more of a demand-side problem and is understood more overtly in political and cultural terms rather than merely in technological terms.

The ideological/institutional context of energy policy debate and decision-making, having now shifted from the province of elite prerogative to the sphere of domestic politics, has produced a number of consequences for energy politics. What Antonio Gramsci said of the crisis of capitalism in the early twentieth century is equally applicable to the crisis of energy policy today: "the old is dying and the new cannot be born; in this interregnum a great variety of morbid symptoms appears." In other words, given the ideological and institutional lags in our political system in coming to grips with new realities presaged by the energy crisis, neither the hard nor the soft energy path proponents have been able to muster sufficient power to win the day in the energy debate. Under these circumstances, everything is up for grabs, since no part of the debate is mutually agreed upon. At issue in the energy controversy are: the definition of the energy crisis itself, the appropriateness of new energy technologies, the efficacy and ethics of various public policy measures, the propriety and relevance of different ethical and political standards, and the very legitimacy of the views of the various participants in the on-going energy debate. What we are confronting on the domestic political scene, then, is a bewildering, contentious controversy of uncertain scope and duration, in which the credibility of the arguments of any one side is automatically put in doubt by other parties in the dispute.

One response to the energy policy stalemate at the national level has been the work of grassroots organizations, regional committees, and state offices to design subnational energy policies more reflective of local or regional conditions, needs, and opportunities. The results of these experiments have been mixed. As a rule, state energy programs (Blissett, et al., 1975; Carter, 1974, 1978; Council of State Governments, 1974; Freeman, 1979; National Conference of State Legislatures, 1975; Worthington, 1980) have shown the least creativity because they have often been constrained by the limits and biases of entrenched political interests and established industrial power; consequently, such offices have been underfunded and undermanned, and their policies have been tailored to mesh with state power constellations. Frequently, these offices have become little more than publicly financed lobbying agencies for major state industries tied into the federal pork-barrel. Moreover, as some analysts have observed, the energy crisis and the competition for scarce resources (energy, water, land) have placed heavy strains on intergovernmental relations between and among states and between states and the federal government (Carter, 1977; Council on Economic Priorities, 1975; Gilbreath, 1974; Harris, 1974; Laird, 1976; Lamm, 1976; Light, 1976; Mitzman, 1978; Shapiro, 1976).

There are exceptions to the general rule noted above, the most notable of which is energy planning and programming in California (Gardner, 1975; Jarret and Howard, 1976; Solarcal Council, 1979). Most significant in the area of state energy policy innovation is Solarcal Council's energy program. While still bearing upper middle-class biases which beset most solar programs to date, it has shown a degree of visionary planning and programmatic novelty which suggests that state energy policies may yet come into their own (Yergin, 1979).

Even more impressive are those model programs generated by citizen initiatives at the local and county levels (Allen, 1975; Bronfman, et al., 1979; Ridgeway, 1979; Sharp and Bruner, 1977). In a policy area crowded with overcentralized administration and high-technology proposals, local energy programs in places like Davis, California, Portland, Oregon, and Franklin County, Massachusetts, have come to symbolize the extent to which decentralized energy systems can respond actively and with palpable results to national policy crises. Combining a broad range of conservation and solar options in a context of extensive community involvement in all phases of planning and implementation, these community energy projects provide important lessons and models for other cities and communities embued with Jeffersonian values and intent on achieving a measure of energy self-reliance. Concomitantly, it must be acknowledged that the ideology and strategy of energy policy localism can obscure and mystify the degree to which the energy crisis is linked to larger national and even international political and economic problems which simultaneously must be addressed in politically novel and democratic ways if these local efforts are not to be overwhelmed and if a long-range solution to the global energy crisis is to be found (Worthington, 1980).

If community energy programs are struggling to generate a more decentralized energy future based upon conservation and renewable energy resources, administration over energy policy at the federal level has remained for the most part fixated on variations of the hard energy path. Until the end of World War II, state administration over the energy arena tended to be extremely fragmented and disjointed, being scattered over a patchwork of uncoordinated executive offices, sub-agencies, and commissions charged with the responsibility of supervising or supporting industries which were highly specific to individual energy sources. Out of the Manhattan Project and its organizational apparatus designed to develop and produce an atomic weapon before Nazi Germany and out of other post-war efforts of civilian leaders in the Executive and the Congress to put atomic energy development on a firm civilian foundation, however, there emerged a state administrative center over nuclear matters (the Atomic Energy Commission) which, through successive administrative reorganizations, was to be transformed into a state administrative agency (the Energy Research and Development Administration and now the Department of Energy), spanning the entire energy spectrum.

The story of the key stages in the institutional development of this administrative apparatus from the early forties to the present constitutes a long and fascinating chapter in the complex and often mysterious history of the evolving relationships among the federal government, some of our major corporate businesses, and modern science and technology (Allardice and Trapnell, 1974; Bradley and Althoff, 1973; Golay, 1980; Yanarella, 1981). Given the short-lived nature of ERDA and the relative infancy of DOE, few writings have shed light on the dynamics of these energy agencies. A few sources do exist, however, and they are worth mining, especially as they relate to the hard path bias of these organizations and to their continuing subservience to corporate priorities (Ogden, 1978; Smith, 1975). The stormy history and politics of the Atomic Energy Commission from its birth in 1947 to its demise in 1975 are extensively recorded and analyzed (Hewlett and Duncan, 1969; Metzger, 1972). If a single generalization about the state administrative apparatus over energy can be hazarded, it is that since the decade of the forties the atomic energy establishment has never transcended the principal characteristics and key contradictions of its origins in the war-time program to develop an operating nuclear weapon and its immediate aftermath (Teich, 1977; Yanarella and Ihara, 1978).

In important respects, the Department of Energy represents the end point of the telos latent in the origins of the AEC and more evidently expressed by the Institutionalization of energy R&D in ERDA. Its overriding function remains continuous

23

with the early role of the AEC as a small administrative agency charged with managing and funding energy R&D through the prototype stage for eventual production and sale of viable reactor designs by private enterprise. Moreover, it has not shed the high technology/predominantly nuclear bias of the Atomic Energy Commission, which was originally structured into the organizational framework of the AEC by a combination of the decision to house both civilian and military R&D in the nuclear field under the institutional umbrella of the AEC and the force of the organizational imperatives of the ring of national scientific laboratories serving as its scientific and technological arm. Aside from its broader authority to explore alternative energy sources and to promote a coherent energy policy, the Department of Energy has broken with the central tendencies and characteristics of the AEC most significantly by asserting a tighter, more centralized management style and a more hierarchical and functional organization. Consequently, with its wider legislative mandate and its technocratic style and ethos, the Department of Energy has the potential for taking on the role of state administrative head of the incipient energy complex, although the "Reagan revolution" may return executive direction of energy matters unambiguously to the corporate boards of the energy industry.

Centralization of administration over the energy arena also brought with it the development of an organized and influential set of national scientific laboratories to further nuclear energy R&D in both the civilian and military realms and later to explore more complex civilian reactor designs and military weapons programs (Greenbaum, 1971; Hammond, 1972; Mitchell, 1970; York, 1975). Historically, this network of national laboratories (Los Alamos, Lawrence-Berkeley, Lawrence-Livermore, Brookhaven, Pacific Northwest, Argonne, Oak Ridge, and Sandia) grew out of the Manhattan Project's scientific-technical need for large laboratories to divide up the labors of investigating issues of theory and application in regard to atomic weapons technology (Teich, 1977). Given the loose, decentralized organizational structure of the Atomic Energy Commission and the policy of its civilian administrators to immunize the labs from the vicissitudes of changing trends and shifting priorities in governmental funding, the stability of the techno-scientific component of the atomic-industrial complex was firmly established. Two other consequences were the success of the scientific labs in carving out broad spheres of authority in areas of management and program development in relation to the civilian agency and the ability of the labs to have their highest normative activity (the development of capital-intensive nuclear technology) identified as the highest status activity of the larger organization. In practical terms, this meant that the nuclear power programs within AEC/ERDA/DOE – typified by the big science/ high technology strategy – have maintained a position of dominance vis-a-vis the other programs (coal research, solar, etc.), while the latter programs have remained strictly ancillary to nuclear technology and have not, at least until recently, been characterized by a capital intensive/technological breakthrough approach (Comptroller General, 1978; MacDonald, 1972; Sachs, 1972; U.S. J.C.A.E., 1960). With the massive infusion of public capital into energy R&D by ERDA and the Department of Energy, the role of the national laboratories in shaping our energy future is expected to remain significant (Comptroller General, 1978; U.S., Congress, House, 1978).

Since the Arab oil embargo of 1973, the role of public opinion in the shaping of national energy policy has been on the ascendency. While the civics book view of the function of an enlightened public in a democratic society has stressed its centrality, prior to 1973, energy policy-making tended to be the special province of elite decision-making so long as the cost of energy made it almost a "free good." Now, with energy prices skyrocketing in an inflationary economy, public perceptions have become a key element in fueling the energy debate and in widening its boundaries.

The energy crisis and the public's role in resolving it have presented social scientists with a challenge, and public opinion experts have grappled with that challenge by training their array of surveys, polls, and computers on the public's psyche (see, e.g., Bultena, 1976; Farhar, et al., 1979b, 1980; Harris, 1979; Melber, et al., 1977). Despite the massive number of attitudinal studies on energy topics generated and the continuing profusion of such surveys, analyses and interpretations of public opinion and energy have tended to fall short of the promise of this speciality and the supposed sophistication of its techniques. In part, the shortcomings of this research can be attributed to the failure of social scientists to link their empirical studies to a broader theoretical or normative framework which explicitly embraces a set of democratic values supportive of a more active role for public opinion in energy policy-making and a more significant place for mutual political education between experts and the larger citizenry. In part, the woefully inadequate nature of the public opinion literature stems from the willingness of social scientists, operating under the cloak of scientific neutrality, to become servants of corporate and political power by generating knowledge whose express purpose is to remold and manipulate the public's views and attitudes on energy questions.

At the very least, a more critical orientation of social scientists toward the role and content of public opinion in the energy debate seems warranted. Insofar as a collective portrait of the public's views on energy issues can be derived from these studies, it is generally convergent with many of the interpretations offered by political analysts sensitive to the political economy of energy in corporate America. According to this profile, the public has tended to regard the energy crisis as largely a contrivance of Big Oil and the federal government (Gallup, December 1977; Harris, 1979; Rosa, 1978). It is willing to buy the rhetoric of national sacrifice and reduced living standards only to the extent that these effects of policy actions are equitably shared (e.g., Harris, December 3, 1973; Harris, 1979). Moreover, it is far more sympathetic toward solar energy and other renewable energy sources (Farhar, et al., 1979a; Harris, February 17, 1977), and it is skeptical of optimistic forecasts of the energy picture in the near term, feeling that things will get worse before they get better (Harris, December 29, 1977). Finally, there is a pervasive public suspicion of economic concentration (understood simply as antagonism toward bigness) in the energy field, and widespread, but diffuse, political anger toward the course of these developments (Hummel, 1978). At the same time, of course, much of the public remains abysmally ignorant about specific details of America's energy situation (Gallup, June 1977), and many elements in the public maintain a nostalgic longing for the way things were in the halcyon days of cheap energy and its wasteful consumption (Harris, August 4, 1975; and Richman, 1979).

The point to be stressed is that social scientists investigating public attitudes toward a variety of energy topics, by characterizing the mass public as the main enemy to a rational energy policy and by allowing their talents, skills, and knowledge to be put in the service of technocratic and corporate elites wielding irresponsible power, are doing their fellow citizens a great disservice and neglecting to advance in their professional work the largely unfulfilled democratic promise of American politics. A far more sophisticated understanding of the role of public opinion in resolving the energy crisis would see it as both part of the problem and part of the solution.

Energy is not merely a problem of domestic politics; it also has many serious implications for <u>international affairs</u>. In recent years, scholars have begun to explore a host of topics pertaining to the international consequences of the energy crisis (Hardesty, 1974; White, 1976). Energy policy has long been put into a national security context in the United States, and so it should occasion no surprise that social scientists of late have become concerned with energy policy and national defense (Bucknell, 1976; CED, 1976; Sakharov, 1978). The increasing awareness of the geopolitical dimensions of global energy production and distribution resulting from the transformation of OPEC into a more-or-less tightly-knit cartel, has sparked a review of the foreign policy and diplomatic implications of energy (Coan, <u>et al.</u>, 1974; Hunter, 1973; Nau, 1974-75; Szyliowicz and O'Neill, 1975; Willrich, 1975; Yager and Steinberg, 1975), particularly in relation to American interests in the Middle East (Crabb, 1973; Griffith, 1974; Klebanoff, 1974; Nordlinger, 1974). Finally, the emergence of an international market for civilian nuclear reactors and the problems of nuclear proliferation attendant from this global development have prompted inquiry into the perilous progress of the "peaceful atom" (Baker, 1975; Ebinger, 1978; Feiveson and Taylor, 1976; Johnson, 1977; Kramish, 1963). With the emergence of global energy resources as a rapidly increasing focus of international competition and possible armed conflict, the literature on this topic is likely to become more extensive in succeeding years.

1. Energy Policy

Adelman, Morris, et al. No Time to Confuse. San Francisco, California: Institute for Contemporary Studies, 1975.

Ahmed, S. Basheer. Nuclear Fuel and Energy Policy. Lexington, Massachusetts: Lexington Books, 1979.

Aman, Alfred C., Jr. "Institutionalizing the Energy Crisis: Some Structural and Procedural Lessons." Cornell Law Review, 65 (April 1980), 491-598.

American Association for the Advancement of Science, et al. The Proceedings of the Conference on National Energy Policy. Washington, D. C.: American Association for the Advancement of Science, 1977.

Anthrop, Donald F. "The Carter Energy Plan and the American West." Bulletin of the Atomic Scientists, 34 (January 1978), 27-33.

_____. "The Need for a Long-Term Policy." Bulletin of the Atomic Scientists, 30 (May 1974), 33-38.

Aronofsky, J. S., et al., eds. Energy Policy. New York: Elsevier-North Holland Publishing Company, 1979.

Auer, Peter L. "Energy Self-Sufficiency." In Jack M. Hollander and Melvin K. Simmons, eds. Annual Review of Energy. Volume 1. Palo Alto, California: Annual Reviews, Inc., 1976, pp. 685-714.

Bailey, James. Energy Systems: An Analysis for Engineers and Policy Makers. New York: Marcel Dekker, 1978.

Ball, Ben C., Jr. "Energy: Policymaking in a New Reality." Technology Review, 80 (October/November 1977), 48-51.

Battelle Memorial Institute. An Analysis of Federal Incentives Used to Stimulate Energy Production. Springfield, Virginia: National Technical Information Service, March 1978.

Bayraktar, B. A.; Chernavsky, E. A.; Laughton, M. A.; and Ruff, L. E., eds. Energy Policy Planning. New York: Plenum Press, 1981.

Benedict, Manson. "U.S. Energy: The Plan That Can Work." Technology Review, 78 (May 1976), 52-59.

Bethe, Hans. "The Case for Coal and Nuclear Energy." The Center Magazine, 13 (May - June 1980), 14-22.

Bezdek, Roger H. and Cone, Bruce W. "Federal Incentives for Energy Development." Energy - The International Journal, 5 (May 1980), 389-406.

Boshier, John F. "Can We Save Energy by Taxing It?" Technology Review, 80 (August/ September 1978), 62-71.

Breyer, Stephen G. and MacAvoy, Paul W. Energy Regulation by the Federal Power Commission. Washington, D.C.: The Brookings Institution, 1974.

Buggey, J. The Energy Crisis: What Are Our Choices? Englewood Cliffs, New Jersey: Prentice-Hall, 1976.

Bupp, Irvin C. "Energy Policy Planning in the United States: Ideological BTU's." In Leonard N. Lindberg, ed. The Energy Syndrome: Comparing National Responses to the Energy Crisis. Lexington, Massachusetts: Lexington Books, 1977, pp. 285-324.

Burton, Dudley J. The Governance of Energy: Problems, Prospects and Underlying Issues. New York: Praeger Publishers, 1980.

Carron, Andrew S. "Congress and Energy: A Need for Policy Analysis and More." Policy Analysis, 2 (Spring 1976), 283-298.

Chancellor, W. J. and Goss, J. R. "Balancing Energy and Food Production, 1975-2000." Science, 192 (April 16, 1976), 213-218.

Chatterji, M. and Rompuy, P. Van, eds. Energy, Regional Science, and Public Policy. New York: Springer Verlag New York, 1976.

Christian, W. T. and Kovall, G. E. "Simplifying the Federal Permitting Process - The Proposed Energy Mobilization Board Legislation." Natural Resources Lawyer, 12 (1979), 721-726.

Committee for Economic Development. Achieving Energy Independence. New York: Committee for Economic Development, 1974.

Commoner, Barry. The National Energy Plan: A Critique. Washington, D.C.: The National League of Cities, June 1977.

_____ . The Politics of Energy. New York: Alfred A. Knopf, 1979.

_____ , and Boksenbaum, Howard, eds. Energy and Human Welfare. Volumes 1-3. New York: Macmillan Information, 1975.

Congressional Quarterly, Inc. Energy Policy. 2nd ed. Washington, D.C.: Congressional Quarterly, Inc., March 1981.

Cook, Lesley P. and Surrey, John. Energy Policy: Strategies for Uncertainty. Totowa, New Jersey: Biblio Distribution Centre, 1978.

Daneke, Gregory A. and Lagassa, George K., eds. Energy Policy and Public Administration. Lexington, Massachusetts: Lexington Books, 1980.

David, Edward E., Jr. "Energy: A Strategy of Diversity." Technology Review, 75 (June 1973), 26-31.

Davis, David H. Energy Politics. New York: St. Martin's Press, 1974.

De Volpi, A. "Energy Policy Decision-Making: The Need for Balanced Input." Bulletin of the Atomic Scientists, 30 (December 1974), 29-33.

Doub, William O. "Energy Regulation: A Quagmire for Energy Policy." In Jack M. Hollander and Melvin K. Simmons, eds. Annual Review of Energy. Volume 1. Palo Alto, California: Annual Reviews, Inc., 1976, pp. 715-726.

Dowall, David F. "U.S. Land Use and Energy Policy - Assessing Potential Conflicts." Energy Policy, 8 (March 1980), 50-60.

Ducsik, Dennis W. "Citizen Participation in Power Plant Siting: Alladin's Lamp or Pandora's Box?" Journal of the American Planning Association, 47 (April 1981), 154-166.

Edison Electric Institute, ed. The Transitional Storm. New York: Edison Electric Institute, 1977.

Engler, Robert. The Brotherhood of Oil: Energy Policy and the Public Interest. Chigaco, Illinois: University of Chicago Press; New York: New American Library, 1977.

Eppen, Gary D., ed. Energy: The Policy Issues. Chicago, Illinois: University of Chicago Press, 1975.

Erickson, Edward W. and Waverman, Leonard, eds. The Energy Question: An International Failure of Policy. Volume 2. North America. Buffalo, New York: University of Toronto Press, 1974.

Executive Office of the President of the United States. The National Energy Plan 1977. Cambridge, Massachusetts: Ballinger Publishing Company, 1977.

Ferrar, Terry A. and Clemente, Frank. Public Policy and Energy Development. Ann Arbor, Michigan: Ann Arbor Science Publishers, 1978.

Foley, Gerald, et al. The Energy Question. Baltimore, Maryland: Penguin Books, 1976.

Freeman, S. David. Energy: The New Era. New York: The Twentieth Century Fund, 1974.

_____, et al. A Time to Choose. [Final Report of the Ford Foundation Energy Policy Project] Cambridge, Massachusetts: Ballinger Publishing Company, 1974.

Friedland, R. J. "Energy Policy, the Dollar, and the U.S. Political System." Energy Policy, 7 (December 1979), 295-306.

Friedman, Milton. The Economics of Freedom. Columbus, Ohio: SOHIO, 1979.

Goldstein, Walter. "The Political Failure of U.S. Energy Policy." Bulletin of the Atomic Scientists, 34 (November 1978), 17-19.

_____. "U.S. Energy Policy - The Continuing Failure." Energy Policy, 7 (December 1979), 275-294.

Goodwin, Craufurd D., ed. Energy Policy in Perspective: Today's Problems, Yesterday's Solutions. Washington, D.C.: The Brookings Institution, 1981.

Gordon, Howard and Meador, Roy, eds. Perspectives on the Energy Crisis. Volumes I and II. Ann Arbor, Michigan: Ann Arbor Science Publishers, 1977.

Gordon, Richard R. "Mythology and Reality in Energy Policy." Energy Policy, 2 (September 1974), 189-203.

Gray, John E. Energy Policy: Industry Perspectives. Cambridge, Massachusetts: Ballinger Publishing Company, 1975.

Hafele, Wolf and Manne, Alan S. "Strategies for a Transition from Fossil to Nuclear Fuels." Energy Policy, 3 (March 1975), 3-23.

Hagel, John, III. Alternative Energy Strategies: Constraints and Opportunities. New York: Praeger Publishers, 1976.

Hammond, Allen L. "Individual Self-Sufficiency in Energy." Science, 184 (April 19, 1974), 278-282.

Henderson, Carter. "The Tragic Failure of Energy Planning." Bulletin of the Atomic Scientists, 34 (December 1978), 15-19.

Hottell, Hoyt C. and Howard Jack B. "An Agenda for Energy." Technology Review, 74 (January 1972), 38-48.

Institute for Contemporary Studies. Options for U.S. Energy Policy. San Francisco, California: Institute for Contemporary Studies, 1977.

Jones, Charles O. "American Politics and the Organization of Energy Decision-Making." In Jack M. Hollander, et al., eds. Annual Review of Energy. Volume 4. Palo Alto, California: Annual Reviews, Inc., 1979, pp. 89-121.

Kalt, Joseph P. and Stillman, Robert S. "The Role of Governmental Incentives in Energy Production: An Historical Overview." In Jack M. Hollander, et al., eds., Annual Review of Energy. Volume 5. Palo Alto, California: Annual Reviews, Inc., 1980, pp. 1-32.

Kalter, Robert J. and Vogely, William A., eds. Energy Supply and Government Policy. Ithaca, New York: Cornell University Press, 1976.

Katz, Milton. "Decision-Making in the Production of Power." Scientific American, 225 (September 1971), 191-200.

Knorr, Klaus. Toward a U.S. Energy Policy. [Agenda Paper No. 2] New York: National Security Information Center, 1975.

Krutilla, John V. and Page, R. Talbot. "Towards a Responsible Energy Policy." Policy Analysis, 1 (Winter 1975), 77-100.

Kursunoglu, B. and Perlmutter, A. Directions in Energy Policy: A Comprehensive Approach to Energy Resource Decision-Making. Cambridge, Massachusetts: Ballinger Publishing Company, 1980.

Laird, Melvin R. Energy - A Crisis in Public Policy. Washington, D.C.: American Enterprise Institute for Public Policy Research, 1977.

Landsberg, Hans H. "Let's All Play Energy Policy!" Daedalus, 109 (Summer 1980), 71-84.

Lawrence, Robert, ed. Energy Policy Issues. Urbana, Illinois: Policy Studies Organization, University of Illinois at Urbana-Champaign, 1978.

_____. New Dimensions to Energy Policy. Lexington, Massachusetts: Lexington Books, 1979.

League of Women Voters Education Fund. Energy Dilemmas: An Overview of U.S. Energy Problems and Issues. Washington, D.C.: League of Women Voters of the U.S., 1977.

League of Women Voters Education Fund. Energy Options: Examining Sources and Defining Government's Role. Washington, D.C.: League of Women Voters of the U.S., 1977.

Lindberg, Leon N., ed. The Energy Syndrome: Comparing National Responses to the Energy Crisis. Lexington, Massachusetts: Lexington Books, 1977.

Littrell, W. B. and Sjoberg, G. Current Issues in Social Policy. Beverly Hills, California: Sage Publications, 1976.

Lovins, Amory B. "The Case for Long-Term Planning." Bulletin of the Atomic Scientists, 30 (June 1974), 38-50.

_____. "Energy: Some Constraints and Opportunities." Ambio, 3 (1974), 123-125.

_____. Soft Energy Paths: Towards a Durable Peace. Cambridge, Massachusetts: Published for Friends of the Earth International by Ballinger Publishing Company; New York: Harper & Row, Publishers, 1979.

McDonald, Stephen L. The Leasing of Federal Lands for Fossil Fuels Production. Baltimore, Maryland: The Johns Hopkins University Press, 1979.

McFarland, Andrew S. Public Interest Lobbies: Decision-Making on Energy. Washington, D.C.: American Enterprise Institute for Public Policy Research, 1976.

MacAvoy, Paul W.; Stangle, Bruce E.; and Tepper, Jonathan B. "The Federal Energy Office as Regulator of the Energy Crisis." Technology Review, 77 (May 1975), 38-45.

Maddox, John Royden. Beyond the Energy Crisis: A Global Perspective. New York: McGraw-Hill, 1975.

Mancke, Richard B. The Failure of U.S. Energy Policy. New York: Columbia University Press, 1974.

_____. Squeaking By: U.S. Energy Policy Since the Embargo. New York: Columbia University Press, 1976.

Meade, Walter J. "An Economic Appraisal of President Carter's Energy Program." Science, 197 (July 22, 1977), 340-345.

_____. U.S. Energy Policy: Errors of the Past, Proposals of the Future. Cambridge, Massachusetts: Ballinger Publishing Company, 1979.

Mitchell, Edward J., ed. Perspectives on U.S. Energy Policy: A Critique of Regulation. New York: Praeger Publishers, 1976.

Mitchell, Edward J. U.S. Energy Policy: A Primer. Washington, D.C.: American Enterprise Institute for Public Policy Research, 1974.

Moore, John R. and Schlottman, Alan. Some Comments on the ERDA Plan and Program. Knoxville, Tennessee: Appalachian Resources Project, University of Tennessee, 1976.

Murray, Francis X. Energy: A National Issue. Washington, D.C.: Center for Strategic and International Studies, Georgetown University, 1976.

_____. Where We Agree: Report of the National Coal Policy Project. 2 Volumes. Boulder, Colorado: Westview Press, 1978.

Naill, Roger F. Managing the Energy Transition: A System Dynamic Search for Alternatives to Oil and Gas. Cambridge, Massachusetts: Ballinger Publishing Company, 1977.

_____ and Backus, George A. "Evaluating the National Energy Plan." Technology Review, 79 (July/August 1977), 51-55.

Orr, David W. "U.S. Energy Policy and the Political Economy of Participation." Journal of Politics, 41 (November 1979), 1027-1056.

Pirages, Dennis Clarke, ed. The Sustainable Society: Implications for Limited Growth. New York: Praeger Publishers, 1977.

"Proceedings of a Conference on Regional Energy Problems and National Energy Policies." Growth and Change, 10 (January 1979), 1-142.

Risser, Herbert E. The U.S. Energy Dilemma: The Gap Between Today's Requirements and Tomorrow's Potential. [Environmental Geology Notes, No. 64] Urbana, Illinois: Illinois State Geological Survey, July 1973.

Rose, David J. "Energy Policy in the U.S." Scientific American, 230 (January 1974), 20-29.

Rosenbaum, Walter A. Energy, Politics, and Public Policy. Washington, D.C.: Congressional Quarterly Press, 1981.

Ross, Marc H. and Williams, Robert H. Our Energy: Regaining Control. New York: McGraw-Hill, 1981.

Rycroft, Robert W., et al. Energy Policy-Making. Norman, Oklahoma: University of Oklahoma Press, 1978.

Sachs, Robert G., ed. National Energy Issues - How Do We Decide? Plutonium as a Test Case. Cambridge, Massachusetts: Ballinger Publishing Company, 1980.

Scarborough, Alexander A. Undermining the Energy Crisis. 4th ed. Milledgeville, Georgia: Ander Publications, 1978.

Scheffer, Walter F., ed. Energy Impacts on Public Policy and Administration. Norman, Oklahoma: University of Oklahoma Press, 1976.

Schmalz, Anton B., ed. Energy: Today's Choices, Tomorrow's Opportunities: Essential Dimensions in Thinking for Energy Policy. Washington, D.C.: World Future Society, 1974.

Smernoff, Barry J. "Energy Policy Interactions in the United States." Energy Policy, 1 (September 1973), 136-153.

Sporn, Philip. Energy in an Age of Limited Availability and Delimited Applicability. Elmsford, New York: Pergamon Press, 1977.

Stern, P.C. and Gardner, G.T. "Psychological Research and Energy Policy." American Psychologist, 36 (April 1981), 329-342.

Stiefel, Michael. "Soft and Hard Energy Paths: The Road Not Taken?" Technology Review, 82 (October 1979), 56-66.

Stobaugh, Robert and Yergin, Daniel. "After the Second Shock: Pragmatic Energy Strategies." Foreign Affairs, 57 (Spring 1979), 836-871.

_____. "Energy: The Telescoped Emergency." Foreign Affairs, 58 (1979), 563-595.

Stockman, David. "The Wrong War? The Case Against National Energy Policy." Public Interest, No. 53 (Fall 1978), 3-44.

Tavoulareas, William. A Debate on A Time to Choose. Cambridge, Massachusetts: Ballinger Publishing Company, 1977.

Teller, Edward. Energy - A Plan for Action. New York: Commission on Critical Choices for Americans, 1975.

_____, et al. Power and Security. Lexington, Massachusetts: Lexington Books, 1976.

Tietenberg, Thomas H. Energy Planning and Policy: The Political Economy of Project Independence. Lexington, Massachusetts: Lexington Books, 1976.

Twentieth Century Fund, Inc. Providing for Energy: Report of the Twentieth Century Fund Task Force on United States Energy Policy. New York: McGraw-Hill, 1977.

Udall, Stewart, et al. The Energy Balloon. New York: McGraw-Hill, 1974; Baltimore, Maryland: Penguin Books, 1975.

34

United States. Congress. House of Representatives. Committee on Interstate and Foreign Commerce. Middle and Long-Term Energy Policies and Alternatives. Washington, D.C.: U.S. Government Printing Office, 1976.

United States. Congress. Office of Technology Assessment. Comparative Analysis of the 1976 ERDA Plan and Program. Washington, D.C.: U.S. Government Printing Office, May 1976.

United States. Congress. Senate. Committee on Interior and Insular Affairs. Presidential Energy Statements. Washington, D.C.: U.S. Government Printing Office, 1973.

United States. Department of Energy. Securing America's Energy Future: The National Energy Policy Plan. [DOE/S-008] Washington, D.C.: U.S. Department of Energy, July 1981.

United States. Energy Research and Development Administration. A National Plan for Energy Research, Development, and Demonstration: Creating Energy Choices for the Future. Volumes I and II. Washington, D.C.: U.S. Government Printing Office, 1976.

United States. Federal Energy Administration. Project Independence: Final Task Force Report on Coal. Washington, D.C.: U.S. Government Printing Office, November 1974.

Warkov, Seymour, ed. Energy Policy in the United States: Social and Behavioral Dimensions. New York: Praeger Publishers, 1978.

Weidenbaum, Murray L., et al. Government Credit Subsidies for Energy Development. Washington, D.C.: American Enterprise Institute for Public Policy Research, 1976.

Wellhofer, E. Spencer. "The Politics of Energy Policy Choices: Germany, Sweden, and the United States." In Robert M. Lawrence and Morris O. Heisler, eds. International Energy Policy. Lexington, Massachusetts: Lexington Books, 1980, pp. 145-161.

White, Irvin L. "Energy Policy-Making: Limitations of a Conceptual Model." Bulletin of the Atomic Scientists, 27 (October 1971), 20-26.

Wright, Arthur W. "The Case of the United States: Energy as a Political Good." Journal of Comparative Economics, 2 (1978), 144-176.

Yanarella, Ernest and Reid, Herbert G. "The Politics of Energy [A Critique of Reagan's Energy Program]." Fellowship, 47 (July/August 1981), 12-13, 29.

2. Subnational Energy Policies

Allen, Edward H. Handbook of Energy Policy for Local Governments. Lexington, Massachusetts: Lexington Books, 1975.

Aron, Joan. "Decision Making in Energy Supply at the Metropolitan Level: A Study of the New York Area." Public Administration Review, 35 (July-August 1975), 340-345.

Ballard, Steven C. and Hall, Timothy A. Managing Growth in Communities Impacted by Energy Development: Problems of Tax Distribution and Timing. Norman, Oklahoma: Science and Public Policy Program, University of Oklahoma, 1976.

Blissett, Marlan; Davis, Bob; and Hahn, Harriet. "Energy Policy in Texas: State Problems and Responses." Public Affairs Commentary, 21 (May 1975), 1-6.

Boffey, Philip M. "Energy: Plan to Use Peat as Fuel Stirs Concern in Minnesota." Science, 190 (December 12, 1975), 1066-1070.

_____. "Solar Research Sweepstakes: States Vie for a Place in the Sun." Science, 190 (October 10, 1975), 128-130.

Bronfman, Benson, et al. Decentralized Energy Technology Assessment Research Program. Oak Ridge, Tennessee: Oak Ridge National Laboratory, May 1979.

Brooks, Gary H. "The Utility of Downs' Analysis of Bureau Territoriality for Policy Evaluation: The Case of Kansas Energy Policy." Midwest Review of Public Administration, 8 (April 1974), 178-190.

Brunner, Ronald D. "Decentralized Energy Policies." Public Policy, 28 (Winter 1980), 71-92.

Calderon, Cinda M. and McKenna, David. Energy and Local Government. Arlington, Texas: Institute of Urban Studies, University of Texas at Arlington, 1974.

California Energy Commission. Energy Choices for California: Looking Ahead. San Mateo, California: Solar Energy Information Services, 1979.

Calzonetti, Frank J.; Eckert, Mark S.; and Malecki, Edward J. "Siting Energy Facilities in the USA: Policies for the Western States." Energy Policy, 8 (June 1980), 138-152.

Carter, Luther J. "Con Edison: Endless Storm King Dispute Adds to Its Troubles." Science, 184 (June 28, 1974), 1353-1358.

_____. "Florida: An Energy Policy Emerges in a Growth State." Science, 184 (April 19, 1974), 302-305.

_____. "New York Puts Together Its Own State Energy Policy and Plan." Science, 199 (February 24, 1978), 864-868.

Center for Governmental Responsibility. Energy: The Power of the States. Gainesville, Florida: Center for Governmental Responsibility, University of Florida, 1975.

Christiansen, Bill and Clack, Theodore H., Jr. "A Western Perspective on Energy: A Plea for Rational Energy Planning." Science, 194 (November 5, 1976), 578-584.

Cigler, Beverly A. "Directions in Local Energy Policy and Management." The Urban Interest, 2 (Autumn 1980), 32-42.

Coates, Gary J., ed. Resettling America: Energy, Ecology, and Community. Andover, Massachusetts: Brick House Publishing Company, 1981.

Community Energy Self-Reliance: Proceedings of the First Conference on Community Renewable Energy Systems. [SERI/CP-354-421] Golden, Colorado: Solar Energy Research Institute, July 1980.

Comptroller General. Region at the Crossroads: The Pacific Northwest Searches for New Sources of Electric Energy. [EMD-78-76] Washington, D.C.: General Accounting Office, 1978.

Cortner, Hanna J. Energy Policy Planning, Administration, and Coordination in the Four Corner States: A Report. Farmington, New Mexico: Four Corners Regional Commission, March 1977.

Council of State Governments. Energy Conservation: Policy Considerations for the States. Lexington, Kentucky: Council of State Governments, 1976.

_____. State Responses to the Energy Crisis. Lexington, Kentucky: Council of State Governments, 1974.

Council on Economic Priorities. Leased and Lost: A Study of Public and Indian Coal Leasing in the West. New York: Council on Economic Priorities, 1975.

37

Cunningham, William H. and Lopreato, Sally C. Energy Use and Conservation Incentives: A Study of the Southwestern United States. New York: Praeger Publishers, 1977.

Darmstadter, Joel. Conserving Energy: Prospects and Opportunities in the New York Region. Baltimore, Maryland: The Johns Hopkins University Press, 1975.

Devine, Michael D.; Ballard, Steven C.; and White, Irvine L. "Energy from the Western States of the USA: Conflicts and Constraints." Energy Policy, 8 (September 1980), 229-244.

Energy and Local Government: The Price of Crisis. Arlington, Texas: Institute of Urban Studies, University of Texas at Arlington, 1975.

Feehan, John G. "The Energy Crisis and the Consumer States." Natural Resources Lawyer, 6 (Fall 1973), 495-502.

Finder, Alan. "State Responses to Energy Problems." State Government, 49 (Summer 1976), 161-165.

Franklin County Energy Study: A Renewable Energy Future. Amherst, Massachusetts: University of Massachusetts, Future Studies Program, May 1979.

Freeman, Patricia K. "The States' Response to the Energy Crisis: An Evaluation of Innovation." In Robert M. Lawrence, ed. New Dimensions to Energy Policy. Lexington, Massachusetts: Lexington Books, 1979, pp. 201-207.

Gardner, Neely. "California Jousts with the Energy Crisis." Public Administration Review, 35 (July - August 1975), 336-340.

Governor's Commission on Cogeneration. Cogeneration: Its Benefits to New England. Boston, Massachusetts: Commonwealth of Massachusetts, 1978.

Gunn, Anita. A Survey of Model Programs: State and Local Solar/Conservation Projects. Washington, D.C.: Center for Renewable Resources, 1977.

Harris, W. R. Intergovernmental Relations in Energy Policy, or How to Get Along with the In-Laws. Santa Monica, California: RAND Corporation, 1974.

Hartnett, J. P., ed. Energy and Conservation Policy Options for Illinois. Washington, D.C.: Hemisphere Publishing Corporation, 1974.

Hirst, E. and Armstrong, R. "Managing State Energy Conservation Programs: The Minnesota Experience." Science, 210 (November 14, 1980), 740-744.

Hollon, Jennifer K. Solar Energy for California's Residential Sector: Progress, Problems, and Prospects. Berkeley, California: Institute for Governmental Studies, 1980.

Jarrett, James and Howard, Dick. State Energy Management: The California Energy Resources Conservation and Development Commission. Lexington, Kentucky: Council of State Governments, 1976.

Krass, Allan S.; Donovan, Christine T.; Ford, Lucia M.; and Small, Sandra C. "Energy Self-Sufficiency in a Small City." Bulletin of the Atomic Scientists, 37 (March 1981), 34-38.

Laird, Melvin, et al. Energy Policy: A New War Between the States? Washington, D.C.: American Enterprise Institute for Public Policy Research, 1976.

Lamm, Richard D. "State Rights vs. National Energy Needs." Natural Resources Lawyer, 9 (1976), 41-48.

Langdon, Jim C. "The Energy Crisis and the Producer States." Natural Resources Lawyer, 6 (Fall 1973), 485-494.

Lawrence, Andy and Daneke, Gregory. "Issues Affecting Decentralization of Energy Supply." In Gregory Daneke and George Lagassa, eds. Energy Policy and Public Administration. Lexington, Massachusetts: Lexington Books, 1980, pp. 63-79.

Lieberman, Marvin S. "The Energy Quandary: State Attitudes, Activities, and Concerns." Public Utilities Fortnightly, 94 (August 15, 1974), 25-29.

Light, Alfred R. "Federalism and the Energy Crisis: A View from the States." Publius, 6 (Winter 1976), 81-96.

Lippitt, Henry F., II. "State and Federal Regulatory Agencies – Conflict or Cooperation?" Public Utilities Fortnightly, 85 (March 26, 1970), 33-38.

Living Systems, Inc. Davis Energy Conservation Report; Practical Use of the Sun. [Report No. DAC PL 79-101] Woodland, California: Living Systems, Inc., March 1977.

Marts, M. E. Electric Power and the Future of the Pacific Northwest. Seattle, Washington: University of Washington Press, 1980.

Meshenberg, M. J. Community Energy Plans and Planning Methodologies: A Preliminary Bibliography. [ANL/CNSV-TM-10] Argonne, Illinois: Argonne National Laboratory, June 1979.

Mitchell, Edward J. Energy: Regional Goals and the National Interest. Washington, D.C.: American Enterprise Institute for Public Policy Research, 1976.

Mitzman, Barry. "War Between the States." Environmental Action, 10 (May 6, 1978), 4-7, 15.

Murray, T. J. "California's Solar Push." Dun's Review, 116 (November 1980), 98-106.

Muschett, F. Douglas. Coal Development in Montana: Economic and Environmental Impacts. Ann Arbor, Michigan: Department of Geography, University of Michigan, 1977.

National Conference of State Legislatures. Energy: The States' Response. 3 Volumes. Washington, D.C.: National Conference of State Legislatures, 1975.

Nehring, Richard and Zycher, Benjamin. Coal Development and Government Regulation in the Northern Great Plains: A Preliminary Report. Santa Monica, California: RAND Corporation, 1976.

New England Energy Congress. Final Report of the New England Energy Congress: A Blueprint for Energy Action. Somerville, Massachusetts: The New England Energy Congress, 1979.

"Nuclear Power and the Straitjacketed State." Technology Review, 78 (March/April 1976), 18-19.

Okagaki, Alan and Benson, Jim. County Energy Plan Guidebook: Creating a Renewable Energy Future. Fairfax, Virginia: Institute for Ecological Policies, 1979.

Oklahoma. University. Science and Public Policy Program. Energy from the West. Irwin White, Project Director. [Prepared for the Office of Research and Development, U.S. Environmental Protection Agency] Washington, D.C.: Environmental Protection Agency, March 1979.

Patton, Janet W. "Administering State Energy Programs." Public Affairs Analyst, 2 (1975), 2-4.

Pushkarev, Boris S. "Energy in the New York Region." Proceedings of the Academy of Political Science, 31 (December 1973), 13-23.

Ridgeway, James. Energy Efficient Community Planning. Erasmus, Pennsylvania: J. G. Press, 1979.

Shapiro, Michael E. "Energy Development on the Public Domain: Federal/State Cooperation and Conflict Regarding Environmental Land Use Control." Natural Resources Lawyer, 9 (1976), 397-440.

Sharp, Phillip and Bruner, Ronald. Local Energy Policies. [Discussion Paper No. 127] Ann Arbor, Michigan: Institute for Public Policy Studies, University of Michigan, July 1978.

Talley, William. "The Role of the State in Energy." In Walter F. Scheffer, ed. Energy Impacts on Public Policy and Administration. Norman, Oklahoma: University of Oklahoma Press, 1976, pp. 185-199.

Thorup, A. Robert. "Electric Range War in Texas: A Case Study in Federal/State Energy Regulation." The George Washington Law Review, 48 (March 1980), 392-432.

United States. Congress. House of Representatives. Committee on Interstate and Foreign Commerce. Subcommittee on Energy and Power. Hearings on Local Energy Policies. Washington, D.C.: U.S. Government Printing Office, May 22, June 5 and 9, 1978.

White, M. D. and Barry, H. J., III. "Energy Development in the West: Conflict and Cooperation of Governmental Decision-Making." North Dakota Law Review, 52 (Spring 1976), 451-528.

Williams, John; Kruvant, William J.; and Newman, Dorothy K. Metropolitan Impacts of Alternative Energy Futures. Washington, D.C.: Metrostudy Corporation, 1976.

Worthington, Richard. "The Politics of Energy Self-Sufficiency in American States and Regions." In Gregory Daneke and George Lagassa, eds. Energy Policy and Public Administration. Lexington, Massachusetts: Lexington Books, 1980, pp. 37-61.

Yanarella, Ernest J. and Yanarella, Ann-Marie, eds. Energy Development in Kentucky: Its Impact Upon Community Life and Higher Education. [Proceedings of a Symposium, University of Kentucky, March 28, 1979] Lexington, Kentucky: Social Science/Technology Development Group, University of Kentucky, April 1979.

3. AEC/ERDA/DOE

Allardice, Corbin and Trapnell, Edward R. The Atomic Energy Commission. New York: Praeger Publishers, 1974.

Brady, David and Althoff, Philip. "The Politics of Regulation: The Case of the Atomic Energy Commission and the Nuclear Industry." American Politics Quarterly, 1 (July 1973), 361-384.

Golay, Michael W. "How Prometheus Came to be Bound: Nuclear Regulation in America." Technology Review, 83 (June/July 1980), 28-41.

Green, Alex E. S. "The Fundamental Nuclear Interaction." Science, 169 (September 4, 1970), 933-941.

Groueff, Stephane. The Manhattan Project. Boston, Massachusetts: Little, Brown and Company, 1967.

Groves, Leslie R. Now It Can Be Told: The Story of the Manhattan Project. New York: Harper & Brothers, 1962.

Hewlett, Richard G. and Anderson, Oscar E., Jr. A History of the United States Atomic Energy Commission: The New World, 1939-1946. University Park, Pennsylvania: Pennsylvania State University Press, 1962.

_____ and Duncan, Francis. Atomic Shield, 1947-1952: A History of the United States Atomic Energy Commission. University Park, Pennsylvania: Pennsylvania State University Press, 1969.

Lewis, Richard S. The Nuclear Power Rebellion: Citizen vs. the Atomic Industrial Establishment. New York: Viking Press, 1972.

MacAvoy, Paul W.; Stangle, Bruce E.; and Tepper, Jonathan B. "The Federal Energy Office as Regulator of the Energy Crisis." Technology Review, 77 (May 1975), 38-45.

Mancke, Richard B. Performance of the Federal Energy Office. Washington, D.C.: American Enterprise Institute for Public Policy Research, 1975.

Marinelli, Janet. "Energy Shortage at DOE." Environmental Action, 10 (November 4, 1978), 12-14.

Marshall, Eliot. "The Solar Institute: Hobbled by DOE?" Science, 203 (March 23, 1979), 1226-1228.

Metz, William D. "Basic Research Funding: ERDA De-Energizes Nuclear Science." Science, 191 (March 5, 1976), 931-933.

_____ . "Department of Energy: Opposition Rises as the Plan Leaks Out." Science, 197 (September 16, 1977), 1166-1167.

_____ . "Energy: ERDA Stresses Multiple Sources and Conservation." Science, 189 (August 1, 1975), 369-370.

Metzger, H. Peter. The Atomic Establishment. New York: Simon and Schuster, 1972.

Mitchell, Edward J., ed. Perspectives on U.S. Energy Policy: A Critique of Regulation. New York: Praeger Publishers, 1976.

Natchez, Peter B. and Bupp, Irvin C. "Policy and Priority in the Budgetary Process." American Political Science Review, 67 (September 1973), 951-963.

Ogden, Daniel M., Jr. "Protecting Energy Turf: The Department of Energy Organization Act." Natural Resources Journal, 18 (October 1978), 845-857.

Rycroft, Robert W. "Energy Policy Feedback: Bureaucratic Responsiveness in the Federal Energy Administration." Policy Analysis, 5 (Winter 1979), 1-20.

Smith, A. Robert. "ERDA: The New Glamor Agency." Bulletin of the Atomic Scientists, 31 (January 1975), 29-31.

Stephenson, Lee. "Energy Czars: Cheaper by the Dozen?" Environmental Action, 5 (January 19, 1974), 9-13.

United States. Congress. Office of Technology Assessment. Conservation and Solar Energy Programs in the Department of Energy: A Critique. [E-120] Washington: D.C.: U.S. Government Printing Office, June 1980.

Yanarella, Ernest J. and Ihara, Randal H. "The Military/Energy Connection: The Institutionalization of the 'Technological Breakthrough' Approach to Energy R&D." Northeast Peace Science Review, 1 (1978), 187-207.

4. National Scientific Laboratories

Carter, Luther J. "Swords into Plowshares: Hanford Makes the Switch." Science, 167 (March 6, 1970), 1357-1361.

Comptroller General. The Multiprogram Laboratories: A National Resource for Non-nuclear Energy Research, Development, and Demonstration. Washington, D.C.: General Accounting Office, May 22, 1978.

Day, Samuel H., Jr. "The Nuclear Weapons Labs." Bulletin of the Atomic Scientists, 33 (April 1977), 21-32.

Greenbaum, Leonard. A Special Interest: The Atomic Energy Commission, Argonne National Laboratory, and the Midwestern Universities. Ann Arbor, Michigan: The University of Michigan Press, 1971.

Hammond, Allen L. "Los Alamos Scientific Laboratory: Weapons Are Still the Focus." Science, 178 (December 15, 1972), 1180-1183.

_____. "Los Alamos: 30 Years After, Life Begins in Earnest." Science, 178 (December 8, 1972), 1075-1078.

Macdonald, J. Ross. "Federal Laboratories and National Policy." Technology Review, 74 (July/August 1972), 10-11.

Metz, William D. "National Laboratories: Focused Goals and Field Work Hinted Under DOE." Science, 198 (December 2, 1977), 901-904.

Mitchell, Charles I. "Los Alamos: From Weapon Shop to Scientific Laboratory." Bulletin of the Atomic Scientists, 26 (November 1970), 24-27.

Moravczik, Michael J. "Reflections on National Laboratories." Bulletin of the Atomic Scientists, 26 (February 1970), 11-15.

Orlans, Harold. Contracting for Atoms. Washington, D.C.: The Brookings Institution, 1967.

Sachs, Robert G. "The National Laboratories." Bulletin of the Atomic Scientists, 28 (June 1972), 51-55.

Schwartz, Charles. "The Berkeley Controversy Over Nuclear Weapons." Bulletin of the Atomic Scientists, 34 (September 1978), 20-24.

Smith, Alice Kimball. "Los Alamos: Focus on an Age." Bulletin of the Atomic Scientists, 26 (June 1970), 15-20.

Teich, Albert H. "Bureaucracy and Politics in Big Science: Relations Between Headquarters and the National Laboratories in AEC and ERDA." [A paper prepared for delivery at the 1977 annual meeting of the American Political Science Association, Washington, D.C., September 1-4, 1977]

United States. Congress. House. The Role of the National Energy Laboratories in ERDA and Department of Energy Operations: Retrospect and Prospect. Washington, D.C.: U.S. Government Printing Office, 1978.

United States. Congress. Joint Committee on Atomic Energy. The Future Role of the Atomic Energy Commission Laboratories. Washington, D.C.: U.S. Government Printing Office, October 1960.

Walsh, John. "Debate on the Future of Weapons Labs Widens." Science, 204 (May 4, 1979), 481-484.

_____ . "ERDA Laboratories: Los Alamos Attracts Some Special Attention." Science, 196 (May 13, 1977), 743-745.

_____ . "Livermore and Los Alamos: Another Look at the UC Link." Science, 199 (March 31, 1978), 1418-1422.

York, Herbert F. "The Origins of the Lawrence Livermore Laboratory." Bulletin of the Atomic Scientists, 31 (September 1975), 8-14.

5. Energy and Public Opinion

Angell and Associates, Inc. Qualitative Study of Consumer Attitudes Toward Energy Conservation. Washington, D.C.: Published for the Federal Energy Administration, Office of Energy Conservation and Environment, by Angell and Associates, Inc., November 1975.

Becker Research Corporation. The Electric Utility Industry: A National Survey of Public Knowledge and Attitudes. Boston, Massachusetts: Becker Research Corporation, July 1974.

Benedict, Robert; Bone, Hugh; Levell, Willard; and Ross, Rice. "The Voters' Attitudes Toward Nuclear Power: A Comparative Study of 'Nuclear Moratorium' Initiatives." Western Political Quarterly, 33 (March 1980), 7-23.

Bultena, Gordon. Public Response to the Energy Crisis: A Study of Citizens' Attitudes and Beliefs in the Southwest. Austin, Texas: Center for Energy Studies, The University of Texas at Austin, 1976.

Darmstadter, Joel and Rattien, Stephen. "Social and Institutional Factors in Energy Conservation." In Jack M. Hollander and Melvin K. Simmons, eds. Annual Review of Energy. Volume 1. Palo Alto, California: Annual Reviews, Inc., 1976, pp. 535-551.

Farhar, Barbara C.; Unseld, Charles T.; Caputo, Richard; and Easterling, James. Citizen Participation in the Domestic Policy Review of Solar Energy. [SERI/TR-53-126] Golden, Colorado: Solar Energy Research Institute, January 1979a.

_____; Vories, Rebecca; and Crews, Robin. "Public Opinion About Energy." In Jack M. Hollander, et al., eds. Annual Review of Energy. Volume 5. Palo Alto, California: Annual Reviews, Inc., 1980, pp. 141-172.

_____; Weiss, Patricia; Unseld, Charles; and Burns, Barbara. Public Opinion About Energy: A Literature Search. [SERI/TR-53-155] Golden, Colorado: Solar Energy Research Institute, June 1979b.

Foreman, Harry, ed. Nuclear Power and the Public. Minneapolis, Minnesota: University of Minnesota Press, 1970.

Gallup, George. "Index: Energy Controls, Energy Crisis." The Gallup Poll: Public Opinion, 1972-1977. 2 Volumes, Wilmington, Delaware: Scholarly Resources, 1978.

_____ . "Index: Energy Situation." The Gallup Poll: Public Opinion, 1978. Wilmington, Delaware: Scholarly Resources, 1979.

_____ . "Index: Energy Situation." The Gallup Poll: Public Opinion, 1979. Wilmington, Delaware: Scholarly Resources, 1980.

Gallup International, Inc. The Gallup Opinion Index. Princeton, New Jersey: American Institute of Public Opinion. [Monthly]

Gottlieb, David and Matre, Mare. Sociological Dimensions of the Energy Crisis: A Follow-Up Study. Houston, Texas: The Energy Institute, University of Houston, 1976.

Hannon, Bruce. "Energy Conservation and the Consumer." Science, 189 (July 11, 1975), 95-102.

Harris, Louis. "Index: Energy and Environment." ABC News-Harris Survey. New York: Chicago Tribune - N.Y. News Syndicate, Inc. [periodic].

_____ . "Index: Energy and Environment." The Harris Survey. New York: Chicago Tribune - N.Y. News Syndicate, Inc. [periodic].

Harris, Louis and Associates, Inc. A Survey of Public and Leadership Attitudes Toward Nuclear Power Development in the United States. New York: Louis Harris and Associates, Inc., August 1975.

_____ . A Second Survey of Public and Leadership Attitudes Toward Nuclear Power Development in the United States. New York: Louis Harris and Associates, Inc. November 1976.

_____ . A Survey of the Public, State Regulators, Educators, and the Media on Energy. New York: Louis Harris and Associates, Inc., 1979.

Hastings, M. and Cawley, M. E. "Community Leaders' Perspectives on Socio-Economic Impacts of Power Plant Development. Energy - The International Journal, 6 (May 1981), 447-456.

Hummel, Carl F., et al. "Perceptions of the Energy Crisis: Who is Blamed and How Do Citizens React to Environment-Lifestyle Trade-Offs?" Environment and Behavior, March 10, 1978, 37-87.

47

International Atomic Energy Agency. Nuclear Power and Its Fuel Cycle: Nuclear Power and Public Opinion and Safeguards. Volume 7. New York: Unipub, 1978.

Kefalas, A. G. and Mehra, S. "U.S. Business and the Energy Crisis - A Survey of Managers." Energy Policy, 7 (September 1979), 260-264.

King, Jill. Impact of Energy Price Increases on Low Income Families. Washington, D.C.: Published for the Federal Energy Administration by Mathematica, Inc., December 1976.

Lopreato, Sally Cook and Meriweather, Marian Wossum. Energy Attitudinal Surveys: Summary, Annotations, Research Recommendations. Austin, Texas: Published for the Energy Research and Development Administration by the Center for Energy Studies, University of Texas at Austin, November 1976.

Melber, Barbara; Nealey, Stanley; Hammersla, Joy; and Rankin, William. Nuclear Power and the Public: Analysis of Collected Survey Research. Seattle, Washington: Battelle Memorial Institute, November 1977.

Milstein, Jeffrey. Attitudes, Knowledge, and Behavior of American Consumers Regarding Energy Conservation with Some Implications for Governmental Action. Washington, D.C.: Office of Energy Conservation, Federal Energy Administration, October 1976.

_____. "Soft and Hard Energy Paths: What People on the Streets Think." Washington, D.C.: Office of Energy Conservation and Solar Applications, U.S. Department of Energy, March 1978.

Mitchell, Robert C. "Public Opinion and Nuclear Power Before and After Three Mile Island." Resources, 60 (January - April 1980), 5-7.

_____. The Public Response to Three Mile Island: A Compilation of Public Opinion Data About Nuclear Energy. [D-58] Washington, D.C.: Resources for the Future, 1979.

Murray, James R., et al. "Evolution of Public Response to the Energy Crisis." Science, 184 (April 19, 1974), 257-263.

National Opinion Research Center. The Impact of the 1973-1974 Oil Embargo on the American Household. Chicago, Illinois: National Opinion Research Center, 1974.

Newman, Dorothy and Day, Dawn. The American Energy Consumer. Cambridge, Massachusetts: Ballinger Publishing Company, 1975.

O'Brien, Robert M. and Kamieniecki, Sheldon. "An Exploratory Study of Social Class and Energy Issues." Political Behavior, 2 (1980), 371-384.

Opinion Research Corporation, Princeton, New Jersey. Highlight Reports. Vols. I-XXIV [Public Opinion Polls on Various Energy Topics for the Federal Energy Administration] Springfield, Virginia: National Technical Infomation Service, 1975-1977.

Otway, Harry J. "Public Attitudes Toward Nuclear Power." International Atomic Energy Agency, 18 (1976), 53-59.

Perlman, Robert and Warren, Roland L. Families in the Energy Crisis: Impacts and Implications for Theory and Policy. Cambridge, Massachusetts: Ballinger Publishing Company, 1977.

Phillips, Kevin. "The Energy Battle: Why the White House Misfired." Public Opinion, 1 (1978), 9-140.

Richman, Al. "The Polls: Public Attitudes Toward the Energy Crisis [1977-79]." Public Opinion Quarterly, 43 (Winter 1979), 576-585.

Roper Organization, Inc. Roper Reports. New York: Roper Organization, Inc. [Monthly]

Rosa, Eugene. "The Public and the Energy Problem." Bulletin of the Atomic Scientists, 34 (April 1978), 5-7.

Schneider, William. "Public Opinion and the Energy Crisis." In Yergin, Daniel, ed. The Dependence Dilemma: Gasoline Consumption and America's Security. Cambridge, Massachusetts: Harvard Center for International Affairs, 1980, pp. 153-156.

Schutz, Howard; Groth, Alexander and Vine, Edward. "A Selected Review of Energy Attitudinal Surveys." In U.S. Department of Energy, Assistant Secretary for Environment, Office of Technology Impacts. Distributed Energy Systems in California's Future. Vol. II. [HCP/P7405-03] Washington, D.C.: U.S. Government Printing Office, May 1978, pp. 239-264.

Stearns, Mary D. The Social Impacts of the Energy Shortage: Behavioral and Attitude Shifts. Washington, D.C.: U.S. Department of Transportation, 1975.

Thomas, Kerry, et al. A Comparative Study of Public Beliefs about Five Energy Systems. Laxenburg, Austria: International Institute for Applied Systems Analysis, 1980.

_____ ; Swaton, Elizabeth; Fishbein, Martin; and Otway, Harry. "Nuclear Energy: The Accuracy of Policy Makers' Perceptions of Public Beliefs." Behavioral Science, 25 (September 1980), 332-344.

United States. Energy Research and Development Administration. Proceedings of the Public Meeting to Review the Status of the Inexhaustible Energy Resources Study. Springfield, Virginia: National Technical Information Service, July 1977.

Yankelovich, Daniel, et al. The Impact of Scarcities: The United States, Sweden, and the Federal Republic of Germany. New York: Published for the Ford Foundation by the Graduate Faculty of Political and Social Science, New School for Social Research, February 1977.

Zinberg, Dorothy. "The Public and Nuclear Waste Management." Bulletin of the Atomic Scientists, 35 (January 1979), 34-39.

6. Energy and International Affairs

Adelman, Morris A. "American Import Policy and the World Oil Market." Energy Policy, 1 (September 1973), 91-99.

_____. "Foreign Oil: A Political-Economic Problem." Technology Review, 76 (March-April 1974), 43-47.

Akins, James E. "The Oil Crisis: This Time the Wolf is Here." Foreign Affairs, 51 (1973), 462-490.

Alm, A. L. "Energy Supply Interruptions and National Security." Science, 211 (March 27, 1981), 1379-1384.

Bucknell, Howard, III. Energy and the National Defense. Lexington, Kentucky: The University Press of Kentucky, 1981.

_____. Energy Policy and Naval Strategy. Beverly Hills, California: Sage Publications, 1976.

Cleveland, Harlan. "World Energy and U.S. Leadership." Atlantic Community Quarterly, 13 (Spring 1975), 23-45.

Committee for Economic Development. Nuclear Energy and National Security. New York: Committee for Economic Development, 1976.

Crabb, Cecil V., Jr. "The Energy Crisis, the Middle East, and American Foreign Policy." World Affairs, 136 (Summer 1973), 48-73.

Deese, David A. and Nye, Joseph S. Energy and Security: A Report of Harvard University's Energy and Security Research Project. Cambridge, Massachusetts: Ballinger Publishing Company, 1980.

Ebinger, Charles K. The Critical Link: Energy and National Security. Cambridge, Massachusetts: Ballinger Publishing Company, 1981.

Ebinger, Charles K. International Politics of Nuclear Energy. Beverly Hills, California: Sage Publications, 1978.

Energy and Defense Project. Energy, Vulnerability, and War: Dispersed, Decentralized, and Renewable Energy Sources: Alternatives to National Vulnerability and War. Wilson Clark, Project Director. Washington, D.C.: Federal Emergency Management Agency, December 1980. [Published in paperback as: Clark, Wilson and Page, John. Energy, Vulnerability, and War: An Alternative for America. New York: W. W. Norton & Company, 1981.]

Fulda, Michael. Oil and International Relations: Energy, Trade, Technology, and Politics. Ed. by Stuart Bruchey. New York: Arno Press, 1979.

Gordon, Richard L. Coal and Canada - U.S. Energy Relations. Washington, D.C.: Canadian - American Committee, 1976.

Griffith, William E. "The Fourth Middle East War, the Energy Crisis, and U.S. Policy." Orbis, 17 (Winter 1974), 1161-1188.

Hardesty, C. H., Jr. "U.S. Energy Situation and Its International Implications." Natural Resources Lawyer, 7 (Fall 1974), 69-80.

Hunter, Robert E. The Energy 'Crisis' and U.S. Foreign Policy. New York: Foreign Policy Association, 1973.

Irwin, J. N., II. "International Implications of the Energy Situation." U.S. Department of State Bulletin, 66 (May 1, 1972), 626-631.

Klebanoff, Shoshana. Middle East Oil and U.S. Foreign Policy. New York: Praeger Publishers, 1974.

Kramish, Arnold. The Peaceful Atom in Foreign Policy. New York: Harper & Row, Publishers, 1963.

Kuenne, Robert E., et al. "A Policy to Protect the U.S. Against Oil Embargoes." Policy Analysis, 1 (Fall 1975), 571-598.

Moore, Arnold B. "U.S. Energy Policy in the World Context." In Morris Adelman, et al. No Time to Confuse. San Francisco, California: Institute for Contemporary Studies, 1975, pp. 89-103.

Moorsteen, Richard. "Action Proposal: OPEC Can Wait -- We Can't." Foreign Policy, 18 (Spring 1975), 3-11.

Nau, Henry R. "U.S. Foreign Policy in the Energy Crisis." Atlantic Community Quarterly, 12 (Winter 1974-75), 426-439.

Nordlinger, Stephen. "Our Blundering Oil Diplomacy: The 'National Security' Cartel." Nation, 218 (April 27, 1974), 523-527.

Olsen, Leif H. "The Energy Crisis and the Balance of Payments." Bulletin of the Atomic Scientists, 30 (March 1974), 26-30.

Oppenheim, V. H. "Why Oil Prices Go Up: The Past: We Pushed Them." Foreign Policy, 25 (Winter 1976-77), 24-57.

Ramberg, Bennett. The Destruction of Nuclear Energy Facilities in War. Lexington, Massachusetts: Lexington Books, 1980.

Rustow, Dankwart. "U.S. - Saudi Relations and the Oil Crises of the 1980's." Foreign Affairs, 55 (April 1977), 494-516.

Starr, Chauncey. "International Realities of Energy Crisis." U.S. Department of State Newsletter, 141 (January 1973), 20-25.

Stewart, Charles F. "Energy and the Balance of Payments." Proceedings of the Academy of Political Science, 31 (December 1973), 137-147.

Stoff, Michael B. Oil, War, and American Security: The Search for a National Policy on Foreign Oil, 1941-1947. New Haven, Connecticut: Yale University Press, 1980.

Szyliowicz, J. S. and O'Neill, B. E. The Energy Crisis and U.S. Foreign Policy. New York: Praeger Publishers, 1975.

Teller, Edward, et al. Power and Security. Lexington, Massachusetts: Lexington Books, 1976.

Treverton, Gregory. Energy and Security. New York: Allanheld, Osmun & Company, 1981.

United States. Congress. House. Committee on Foreign Affairs. Foreign Policy Implications of the Energy Crisis. Washington, D.C.: U.S. Government Printing Office, 1972.

United States. Congress. Senate. Committee on Foreign Relations. Energy and Foreign Policy. Washington, D.C.: U.S. Government Printing Office, 1974.

White, Irvin L. "Energy Impacts on Domestic and International Priorities and Policies." In Walter F. Scheffer, ed. Energy Impacts on Public Policy and Administration. Norman, Oklahoma: University of Oklahoma Press, 1976, pp. 93-114.

Willrich, Mason. Energy and World Politics. New York: The Free Press, 1975.

Yager, Joseph A. and Steinberg, Eleanor B. Energy and U.S. Foreign Policy. Cambridge, Massachusetts: Ballinger Publishing Company, 1975.

Part III
Economics, Society,
and Environment

The pivotal nature of energy for human affairs and social well-being is reflected in the many ways in which energy concerns influence, and in tandem are influenced by, the realms of economics, society, and the environment – the organizing theme of the third part of this bibliography. Prior to the emergence of widespread environmental consciousness and the development of the international duopoly uniting the OPEC cartel with the multinational oil corporations, the apparently limitless and inexpensive character of energy tended to obscure problems attendant upon its production and use. Today, energy can no longer be viewed as a virtual "free good," and its social and environmental costs are increasingly being recognized and internalized into its price. Furthermore, within major segments of the populace, the ethical dimensions of energy choices have surfaced and have been expressed as themes and issues in the continuing debate over energy policy and our collective energy future.

While the energy controversy cannot be reduced to questions of economic calculations, the economics of energy remains a highly significant dimension of energy policy-making in an era where continuing economic concentration in the energy sector and the real prospect of the eventual depletion of the world's stock of fossil fuels confront humanity with a condition of resource scarcity. The uncertain duration of this situation has prompted renewed interest in general issues of energy and economics (Askin, 1978; Daly and Umano, 1981; Dix, 1977; Grayson, 1975; Schurr and Netschert, 1960; Slesser, 1978; and Webb and Ricketts, 1980) and has spawned within academic economics a specialized field devoted to energy economics (Gardel, 1980; Georgescu-Roegen, 1977; Griffin and Steele, 1980; Merklein and Hardy, 1977). A whole body of research has developed, and the scope of this literature now spans both traditional economic issues characteristic of this "dismal science" and more contemporary ones distinctive to the new age of ecological scarcity. Thus, traditional concerns for economic growth (Abrahamson, 1978; Banks, 1977; and Kannan, 1979), energy use (Basile, 1976; Center for Advanced Computation, 1976; Darmstadter, 1978; Darmstadter, et al., 1977; Maddala, et al., 1978; National Research Council, 1979), and energy markets (Pindyck, 1979) have been explored by economic analysts, while simultaneously new concerns over the employment impacts of alternative energy sources (Cogan, 1976; Grossman and Daneker, 1979; Nordlung and Thayne, 1980; Starr and Field, 1979), the capital costs of individual and alternative power-producing facilities (I.A.E.A., 1970; Komanoff, 1976, 1979; Miller, 1976; Moore, et al., 1973; Shaw, 1979; Pelley, et al., 1976), the real or illusory link between energy growth and economic growth (Bettling, et al., 1976; Hitch, 1978; and Soneblum, 1978), and the nagging problem of anticipating or coping with energy crisis in an era of stagflation (Appleby, 1976; Chaplian, 1974a and b, 1975; Foster Associates, 1974; Hill, 1980; Mork, 1980; and Sander, 1976) are commanding the time, talents, and energies of energy economists. Public policy aspects of the nexus between energy and economics (Brannon, 1974; Griffin and Steele, 1980; Henderson, 1978; Thompson, et al., 1978; and Weidenbaum, 1976), as well as the continuing attraction of energy self-sufficiency (M.I.T. Energy Laboratory, 1974a and b), have also drawn the attention of economists and other social scientists.

The energy industry in general and the individual corporate elements within it have provided another focus for scholarly analysis and investigation. The introductory section of this chapter offers a general perspective on levels of concentration and trends in the energy sector of the United States economy. Generally speaking, while each industrial element possesses its own unique history grounded in past and prevailing social relationships with government, trade unions, and possibly foreign countries, many of these works point to the fact that the energy industry as a whole is highly concentrated economically (Mulholland and Webbink, 1974; Netschert, 1971; and Pratt and Ward, 1978). Moreover, as other studies reveal (e.g., Duchesneau, 1975; Sanger and Mason, 1977, 1979; and U.S., Congress, Senate, 1977), a variety of factors - including corporate strategies and international events – have accelerated the process of concentration and control across resource-specific industries and into alternative energy sources. What is highly contested is the impact of these trends upon pricing and interfuel competition within the industry (Norman, 1977; U.S., Congress, House, 1978; U.S., Congress, Senate, 1975). Beyond these concerns, the study by Gray (1975) illuminates corporate views on energy policy, while Herman and Cannon's (1977) work analyzes the character of new energy technologies being developed in laboratories and testing facilities of major energy and other corporations.

When nuclear critics and proponents speak of the nuclear power industry, they usually have in mind either the electric utilities industry or the atomic reactor builders or both. In actuality, the nuclear industry encompasses a wide variety of other elements spanning the entire nuclear fuel cycle, including uranium milling and mining, fuel enrichment and fabrication, waste storage and disposal management, and perhaps even unions (like the powerful International Oil, Chemical, and Atomic Workers Union). In the following section, the history and structure of the nuclear industry are examined by close students of this multi-faceted enterprise, both generally (Bupp and Derian, 1975; Burn, 1978; Cruetz, 1970; Elliot, 1978; Metzger, 1972; and Robbins, 1970) and in terms of its more particular industrial components (Ezell, 1979; Kostuik, 1976; Mandelbaum, 1977; and Taylor and Yokell, 1979). Other historical studies reflect on the high tide of industry expectations (Burness, et al., 1980) and on the uncertain future of the industry (Lonnroth and Walker, 1980; and Joskow and Baughman, 1976). Still others direct attention to issues of industry regulation (Brady and Althoff, 1973; Metzger, 1972; and Montgomery and Rose, 1979). In combination with the works collected in a later chapter (Part 5, Chapter 3), these writings lend insight into the background of the present nuclear stalemate and the causes of the stagnation of the nuclear industry.

At least since the 1973 Arab oil embargo and the subsequent quadrupling of domestic gas prices punctuated by periodic supply shortages, the structure of the oil industry and the extent of its economic and political power have become subjects of growing scrutiny. General surveys of the rise and influence of the multinational oil

55

corporations (Engler, 1977; Giddens, 1938, 1975; Ickes, 1975; Kaufman, 1978; Sampson, 1976; and Williamson, 1963), as well as histories of individual oil companies oftentimes building upon reprinted classics from the early twentieth century (Continental Oil Company, 1975; Gibb and Knowlton, 1975; Hidy and Hidy, 1955; Larson, 1959; Lloyd, 1973; Montague, 1973; Tarbell, 1965; and White, 1962), have contributed to our knowledge of the growth of the industry to multinational status in the world and to pre-eminent rank within the American political economy. Many economists and social analysts have explored a host of issues relating to the industry's structure (Blair, 1976; Ikard, 1975; Johnson, 1975; Mancke, 1974; Medvin, 1974; Mitchell, 1976; Pratt, 1980; Riddick, 1973; Stocking, 1976; Uhl, 1976; and Wilson, 1976) and a number of Congressional inquiries and government studies have been stimulated to examine the public policy implications of its complex organization (U.S., Congress, 1973, 1974; U.S., F.E.A., 1975a and b). Along similar lines, the relationship between state power and government policy, on the one hand, and the oil multinationals, on the other, and the implications of this long-term marriage have prompted important scholarly analysis (Blair, 1976; Engler, 1977; Mead, 1970; Nash, 1968; Pikl, 1970; and Pratt, 1980). Although the shape and power of the oil industry have remained hotly disputed issues within the literature, the call for general public control over the multinational oil corporations has led to considerations of the merits of the policies of horizontal and vertical divestiture within the industry (Hobbie, 1977; Johnson and Messnick, 1976; and Moore, 1977).

The efforts of oil multinationals to extend their "global reach" for new markets in the industrialized and, especially, the underdeveloped regions of the world and to stake out and control the remainder of the world's energy sources (Ridgeway, n.d.; Tanzer, 1969; Vernon, 1977) have stimulated past and continuing exploration of the political economy of international oil (Al-Otaiba, 1975; Fischer, 1976; Hartshorn, 1962; Jacoby, 1974; Khan, 1979; Krueger, 1975; McAfee, 1974; O'Connor, 1955, 1962; and Solberg, 1976). Meanwhile, indications of a trend within the industry toward renewed interest in refocusing exploration efforts domestically and in exploiting wider segments of the United States – including the virgin territory of Alaska, the ranges and hilly landscapes of the Midwest, and the offshore areas of the West Coast – have been a stimulus to investigations into the oil pricing practices and profits of the "Seven Sisters" (Ben-Shaar, 1976; Loomis, 1974; Sonder, 1973). These writings suggest that the term, "oil corporation," is rapidly becoming a misnomer and is perhaps better replaced by the term, "energy corporation," as the Seven Sisters extend their investments into aspects of the coal, nuclear, and natural gas production process in an effort to maintain their power and position in energy and economy in the post-petroleum world of the not-too-distant future.

Some light is shed on the turbulent history (Newcomb, 1978; Thomas, 1980; and Thompson, 1980) and the uncertain future of the coal industry (Cohn, et al., 1975; U.N., 1978; and U.S., Library of Congress, 1978) by studies offered in the next section. As with the other industrial components of the energy sector of the American economy, considerable analysis has been undertaken to clarify the structure of the industry and to determine the degree to which economic concentration within the industry itself and horizontal integration across industries constituting the corporate segment of our energy economy have affected the economic behavior (including competition and prices) of the coal industry (Attorney General, 1978; Fisher and James, 1955; Giffen, 1972; Hinson, 1970, 1979; Kahn and Hand, 1976; Lawrence, 1975; Moyer, 1964; National Coal Association, 1977; U.S., F.T.C., 1978; and Walls, 1979).

Given the continued reliance of the United States on coal and the often promised acceleration of its use in the near term, the issues of coal industry and labor (Baratz, 1972) and the relationship between firm size and technological innovation in the industry (Mansfield, 1975) are certain to attract greater interest and inquiry among economists and social scientists than heretofore has been the case.

The electric utility industry – the focus of the succeeding section – has clearly come upon hard times. Whereas prior to 1973, the industry could comfortably expect steady growth, relatively short construction lead times, guaranteed fuel supplies, and easily obtainable capital financing, in the years since the Arab oil embargo, with the precipitate decline in consumer demand, the rise of widespread anti-nuclear opposition, and the materialization of stagflation, this is no longer the case. The industry has become embroiled in a series of public controversies involving the demand for reforms in prevailing rate structures (Levy, 1978; Miller, 1970; Mitchell and Manning, 1978; Sichel, 1978; and Stelzer, 1980), the escalating capital costs and the environmental and public safety liabilities of electric-generating atomic power plants (Council on Economic Priorities, 1977; Lowe, 1972; Quarles, 1974; Ramsay, 1979; and Young, 1977), and the role of utility financing and administration of conservation and decentralized solar applications (Bossong, 1979; Dixon, et al., 1977; Hamlen and Tschirhart, 1980; Institute of Gas Technology, 1978; Maidique and Woo, 1980; Meadows, 1980; and Smackey, 1980). Even more ominous, recent economic forecasts depict a cloudy future for the industry, and some studies question the very financial solvency of the enterprise in its present form in a no-growth or low-growth scenario in the near to middle term (Bupp, 1979; Hub, 1973; and Marsh, 1980). What few utility managers appear willing to concede is that most of these emerging public policy issues, which have placed the electric utilities at the center of the storm, go to the very heart of the structure of the industry (Messing, 1979; Novick, 1973; and Sporn, 1971).

Several historical studies do much to illuminate the crystallization of the industry's "natural monopoly" structure fashioned out of the ideas and energies of its first generation managerial elite (Funigiello, 1973; Hounshell, 1980; and, especially, Novick, 1975, 1979). Still other historical treatments help to shed light on the tension and conflict between public and private conceptions of electric power production which developed during the early New Deal period (Brown, 1980; Doyle, 1979; McCraw, 1971; and Waltrip, 1979) and their implications for the course of the debate over centralized vs. decentralized systems of electric power production in the future (Lagassa, 1980; and U.S., Library of Congress, 1979). Equally critical to the future shape of the industry will be the resolution of the controversy over governmental regulation of electrical utilities (Berlin, 1974; Cichetti and Jurewitz, 1975; and Comptroller General, 1981) and the determination of the capacity of the electric utilities industry to face up to public and environmental challenges confronting it (Casper and Wellstone, 1981; and Robert and Bluhm, 1981).

The chapter on energy and social structure is intended as a repository for a multitude of works focusing on the broadest social considerations bearing upon energy production and use. At the most general level, it includes books and articles introducing the reader to the whole spectrum of issues relating to energy and society (Boulding, 1974; Curran and Curran, 1980; Diesendorf, 1980; Healy, 1976; Morgan, 1975; National Research Council, 1980; Nebins, et al., 1960; and Thring, 1974). Other works open up inquiry into the social context of energy production (Althouse, 1974; Bowman, 1977; Blumer, 1975; Burpy and Bell, 1978; Caldwell, 1976; Ergood and Kuhre, 1976; and Lantz, 1958). In recent years, sociologists, economists, and political scientists have begun to probe the socioeconomic and sociopolitical impacts of energy development generally (Budnitz and Holdren, 1976; Centaur Management Consultants, 1978; Garrett, 1977; Rieff, 1976; and Unseld, et al., 1979) and in terms of the social impacts of specific energy sources, such as coal (Baillet, 1978; Bryant, 1976; Caudill, 1962, 1971; Erickson, 1976; Lewis, et al., 1978), nuclear power (Klineberg, 1964; Nelkin, 1981; Sagan, 1972, 1974; Sherfield, 1972; and Weinberg, 1972), and oil (Pfuhl, 1980). These studies have been supplemented by even more focused research on energy and labor (Arble, 1976; Coleman, 1969; and Densmore, 1977), energy and the consumer (Newman and Day, 1975; and Perlman and Warren, 1977), and energy and native Indians (Callaway, 1976; and Nelkin, 1981).

No less important are those books, monographs, and articles dealing with a panoply of issues and problems relating to energy and the environment, which are collected in the subsequent chapter. The development of environmental consciousness and ecological sensitivity permeating broad segments of the American populace has brought about a growing recognition of the degradation of the globe's ecosystem and of its various components (human beings, air, water, and land) attendant from social and industrial processes. As a consequence, the environment can no longer be looked upon as a convenient dumping ground for industrial and human wastes without objection or protest. A large and growing corpus of general studies has highlighted the interaction of energy and the environment (Ashley, et al., 1976; Axelrod, 1980; Budnitz and Holdren, 1976; El-Hinnawi, 1980a; Evans, 1980; Fowler, 1975; Garvey, 1972; Goodwin, 1974; Kwee and Mullender, 1972; Miller, 1980; Odum, 1971; Seale and Sierka, 1973; Theodore, et al., 1980; Van Tassel, 1975; and Wilson and Jones, 1974). Another part of this literature has been concerned with defining and analyzing environmental considerations related to specific energy sources, such as nuclear power (Eicholz, 1976; El-Hinnawi, 1980b; Glasstone and Jordan, 1980; I.A.E.A., 1975; Karam and Morgan, 1976; Karam, et al., 1977; Keating, 1979; Pentreath, 1980; and Weinberg, 1970), coal (Appalachian Regional Commission, 1976; Atwood, 1975; Austin, 1973; Austin and Borelli, 1971; Beray, 1966; Camplin, 1965; Caudill, 1971; Doyle, 1976; Greenburg, 1973; Hardt, 1978; Muschett, 1977; Rall, 1977; Randall, et al., 1978; Reitze, 1971; Stacks, 1972; Toole, 1976; U.S., Department of the Interior, 1967), and renewable energy resources (Holdren, et al., 1980).

Research devoted to particular environmental problems and threats is similarly large and includes at least the following matters: the "greenhouse effect" or the threat to the globe's climatic patterns (Base, et al., 1976; Bernard, 1980; Bishop, 1980; Damon and Kunen, 1976; Kellogg, 1978; National Research Council, 1977; Schneider and Bennett, 1975; Woodwell, 1978; Woodwell, et al., 1979), land use and reclamation (Box, et al., 1974; Environmental Studies Board, 1974; Rowe, 1979; Schlottman and Spore, 1976; Thames, 1977; U.S., Congress, 1973; Wali, 1975; Wiener, 1980; Wright, 1978; and Yanarella and Yanarella, 1979), air pollution (Jimeson and Spindt, 1973; and Navarro, 1981), water pollution (Harte and El-Gasseir, 1978), and acid rain (Coffin and Knelson, 1976), and, not least of all, human health (Finkel, 1973). The search for political means for repairing or alleviating these problems has precipitated studies concerned with laws, organizations, and processes for advancing environmental protection (Ackerman and Hassler, 1981; Council on Environmental Quality, 1979; Foell, 1979; Ford, 1979; Landy, 1976, and Navarro, 1981; Schlottman, 1977; and Steinman, 1979), as well as with environmentalism as a movement for change and citizen action as a vehicle for ecological renewal (Caldwell, 1976; Cook, 1980; Lenzer, et al., 1978; Talbot, 1972; and Vietor, 1980). Insofar as recent public opinion polls are accurate in demonstrating the durability of environmental concern within the American public, it would seem that the difficult problems and continuing dilemmas associated with the twin goals of promoting energy development and enhancing the environment will be salient political issues through the dawn of the twenty-first century and beyond.

Though brief, the chapter on energy and ethics acknowledges the fundamental ethical nature of so many apparently economic or technical questions involved in the energy debate. In an era where rising energy prices threaten to worsen the maldistribution of wealth nationally and globally and to endanger the very survival of segments of whole populations in less developed countries, the issues of energy and equity (Abbate, 1978; Cose, 1979; Fritsch, 1980; Illich, 1973; Kannan, 1979; Morrison, 1978; and Reader, 1977) seem an entirely appropriate, even compelling, matter for philosophical reflection and social research. So, too, is a consideration of the question of what kind of energy future we owe to succeeding generations (Partridge, 1981; Routley and Routley, 1978). It has been the environmental/ecological movement, sometimes refracted through the no-nukes movement, which has stimulated a growing concern for the notion of generativity in the context of the debate over energy and environmental policies. Implied here is an ethical concern that human beings must develop a sense of care for or stewardship over nature and the world for future generations. Finally, theologians and religious organizations, too, have injected themselves into the energy controversy, offering religious perspectives on energy issues, such as nuclear power development and plutonium production (Boffey, 1976; Cesaretti, 1980; Gremillion, 1978; Harnik, 1979; Hessel, 1979; and National Council of Churches, 1979).

1. The Economics of Energy

Abrahamson, Bernard J., ed. Conservation and the Changing Direction of Economic Growth. Boulder, Colorado: Westview Press, 1978.

Appleby, A. J. "Energy Costs and Society: The High Price of Future Energy." Energy Policy, 4 (June 1976), 87-97.

Askin, A. Bradley, ed. How Energy Effects the Economy. Lexington, Massachusetts: Lexington Books, 1978.

Ayres, Ronald F. Coal: New Markets/New Prices - Ramifications of the Federal Coal Conversion Program. New York: McGraw-Hill, 1977.

Balliet, Lee. 'A Pleasing Tho' Dreadful Sight': Social and Economic Impacts of Coal Production in the Eastern Coalfields. Washington, D.C.: Office of Technology Assessment, 1978.

Bankers Trust Company. U.S. Energy and Capital. New York: Bankers Trust Company, Energy Group, 1978.

Banks, Ferdinand E. Scarcity, Energy, and Economic Progress. Lexington, Massachusetts: Lexington Books, 1977.

Basile, Paul, ed. Energy Demand Studies: Major Consuming Countries. [First Technical Report of the Workshop on Alternative Energy Strategies (WAES)] Cambridge, Massachusetts: M.I.T. Press, 1976.

Bettling, David J., Jr.; Dullien, Robert; and Hudson, Edward. The Relationship of Energy Growth to Economic Growth Under Alternative Energy Policies. Brookhaven, Connecticut: Associated Universities, 1976.

Brannon, Gerard M. Energy Taxes and Subsidies. Cambridge, Massachusetts: Ballinger Publishing Company, 1974.

Burness, H. Stuart. The Effect of Uncertainty on the Supply of Coal. [Prepared for the Institute for Mining and Minerals Research] Lexington, Kentucky: Office of Research and Engineering Services, College of Engineering, University of Kentucky, 1976.

Carmen, D. G. "The Difficult Choice Between Coal and Nuclear Power." The Social Science Journal, 18 (January 1981), 19-40.

Centaur Management Consultants, Inc. Managing the Social and Economic Impacts of Energy Developments. Washington, D.C.: Energy Research and Development Administration, 1976.

Center for Advanced Computation. Energy Flow Through the United States Economy. Urbana, Illinois: University of Illinois Press, 1976.

Chapman, P. F. "Energy Budgets: 1. Energy Costs - A Review of Methods." Energy Policy, 2 (June 1974a), 91-103.

_____ . "Energy Budgets: 4. The Energy Costs of Materials." Energy Policy, 3 (March 1975), 47-57.

_____ ; Leach, G.; and Slesser, M. "Energy Budgets: 2. The Energy Cost of Fuels." Energy Policy, 2 (September 1974b), 231-243.

Cogan, John, et al. Energy and Jobs: A Long-Run Analysis. Edison, New Jersey: Green Hill Publishers, 1976.

Daly, Herman E. and Umana, Alvaro F., eds. Energy, Economics, and the Environment: Conflicting Views of an Essential Interrelationship. Boulder, Colorado: Westview Press, 1981.

Darmstadter, Joel. "Intercountry Comparisons of Energy Use: Any Lessons for the United States?" In Bernard J. Abrahamson, ed. Conservation and the Changing Direction of Economic Growth. Boulder, Colorado: Westview Press, 1978, pp. 69-78.

_____ ; Dunkerley, Joy; and Alterman, Jack. How Industrial Societies Use Energy: A Comparative Analysis. Baltimore, Maryland: The Johns Hopkins University Press, 1977.

Dix, Samuel M. Energy: A Critical Decision for the United States Economy. Grand Rapids, Michigan: Energy Education Publishers, 1977.

Foster Associates. Energy Prices, 1960-1973. Cambridge, Massachusetts: Ballinger Publishing Company, 1974.

Gardel, A. Energy: Economy and Prospective: A Handbook for Engineers and Economists. Elmsford, New York: Pergamon Press, 1980.

Georgescu-Roegen, Nicholas. Energy and Economic Myths. Elmsford, New York: Pergamon Press, 1977.

Grayson, Leslie E., comp. Economics of Energy: Readings on Environment, Resources, and Markets. Princeton, New Jersey: Darwin Press, 1975.

Griffin, James M. and Steele, Henry B. Energy Economics and Policy. New York: Academic Press, 1980.

Grossman, Richard and Daneker, Gail. Energy, Jobs and the Economy. Boston, Massachusetts: Carrier Pigeon Press, 1979.

Hammond, Ogden and Zimmerman, Martin B. "The Economics of Coal-Based Synthetic Gas." Technology Review, 77 (July/August 1975), 42-51.

Hannon, Bruce, et al. "Energy and Labor in the Construction Section." Science, 202 (November 24, 1978), 837-847.

Henderson, David Richard. The Economics of Safety Legislation in Underground Coal Mining. Ann Arbor, Michigan: University Microfilms International, 1978.

Hill, Lewis E. "Reflections on Stagflation and the Energy Crisis." Review of Social Economy, 38 (December 1980), 289-292.

Hitch, Charles, ed. Energy Conservation and Economic Growth. Boulder, Colorado: Westview Press for the American Association for the Advancement of Science, 1978.

Hyman, Mark, Jr. "Solar Economics Comes Home." Technology Review, 80 (February 1978), 28-35.

I.C.F. Incorporated. Energy and Economic Impacts of HR 13950 (Surface Mining Control and Reclamation Act of 1976, 94th Congress): Draft Final Report. [Report submitted to the Council on Environmental Quality and the Environmental Protection Agency] Washington, D.C.: Environmental Protection Agency, 1977.

International Atomic Energy Agency. Nuclear Energy Costs and Economic Development. New York: Unipub, 1970.

Jorgenson, D. W. Economic Studies of U.S. Energy Policy. New York: Elsevier - North Holland Publishing Company, 1976.

Kalt, Joseph P. The Economics and Politics of Oil Price Regulation. Cambridge, Massachusetts: M.I.T. Press, 1981.

Kannan, Narisimhan P. Energy, Economic Growth, and Equity in the U.S. New York: Praeger Publishers, 1979.

Komanoff, Charles. "Doing Without Nuclear Power." New York Review of Books, 26 (May 17, 1979), 14-17.

_____. Power Plant Cost Escalation. New York: Komanoff Energy Associates, 1981.

_____. Power Plant Performance: Nuclear and Coal Capacity Factors and Economics. New York: Council on Economic Priorities, 1976.

_____. Power Propaganda: A Critique of the Atomic Industrial Forum's Nuclear and Coal Power Cost Data for 1978. Washington, D.C.: Environmental Action Foundation, 1980.

Kreith, Frank and West R. E. Economics of Solar Energy and Conservation Systems. Volumes I - III. West Palm Beach, Florida: CRC Press, 1979.

Krutilla, John V., et al., eds. Economic and Fiscal Impacts of Coal Development: Northern Great Plains. Baltimore, Maryland: The Johns Hopkins University Press, 1978.

Luten, Daniel B. "The Economic Geography of Energy." Scientific American, 225 (September 1971), 165-175.

MacAvoy, Paul W. and Pindyck, Robert S. The Economics of the Natural Gas Shortage (1960-1980). New York: Elsevier - North Holland Publishing Company, 1975.

_____; Samuelson, Paul A.;and Thurow, Lester C. "The Economics of the Energy Crisis." Technology Review, 76 (March/April 1974), 49-59.

Maddala, G. S., et al. Economic Studies in Energy Demand and Supply. New York: Praeger Publishers, 1978.

Massachusetts Institute of Technology. Energy Laboratory. Policy Study Group. "Energy Self-Sufficiency: An Economic Evaluation." Technology Review, 76 (May 1974a), 22-58.

_____. Energy Self-Sufficiency: An Economic Evaluation. Washington, D.C.: American Enterprise Institute for Public Policy Research, 1974b.

Merklein, Helmut A. and Hardy, W. Carey. Energy Economics. Houston, Texas: Gulf Publishing Company, 1977.

Miernyk, William H., et al. Regional Impacts of Rising Energy Prices. Cambridge, Massachusetts: Ballinger Publishing Company, 1977.

Miller, Saunders. The Economics of Nuclear and Coal Power. New York: Praeger Publishers, 1976.

Moore, John R.., et al. Economics of the Private and Social Costs of Appalachian Coal Production: A Progress Report. Knoxville, Tennessee: Appalachian Resources Project, University of Tennessee, 1973.

Mork, Knut A., ed. Energy Prices, Inflation, and Economic Activity. Cambridge, Massachusetts: Ballinger Publishing Company, 1980.

Muschett, F. Douglas. Coal Development in Montana: Economic and Environmental Impacts. Ann Arbor, Michigan: Department of Geography, University of Michigan, 1977.

National Research Council. Report of the Supply and Delivery Panel to the Committee on Nuclear and Alternative Energy Systems. Washington, D.C.: National Academy of Sciences/National Research Council, 1979.

Newman, Monroe O. The Political Economy of Appalachia: A Case Study in Regional Integration. Lexington, Massachusetts: Lexington Books, 1972.

Nordlung, Willis J. and Thayne, R. Energy and Employment. New York: Praeger Publishers, 1980.

Pelley, William E.; Constable, Richard W.; and Krupp, Herbert W. "The Energy Industry and the Capital Market." In Jack M. Hollander and Melvin K. Simmons, eds. Annual Review of Energy. Volume 1. Palo Alto, California: Annual Reviews, Inc., 1976, pp. 369-390.

Pindyck, R. S. The Structure of Energy Markets. Greenwich, Connecticut: Jai Press, 1979.

Posner, Michael V. Fuel Policy: A Study in Applied Economics. New York: Macmillan, 1973.

Reische, Diana, ed. Energy Demand vs. Supply. Bronx, New York: H. W. Wilson, 1973.

Richardson, Harry W. Economic Aspects of the Energy Crisis. Lexington, Massachusetts: Lexington Books, 1975.

Risser, Hubert E. The Economics of the Coal Industry. Lawrence, Kansas: Bureau of Business Research, School of Business, University of Kansas, 1953; Westport, Connecticut: Greeenwood Press, 1976.

Rosen, Sumner M. Economic Power Failure: The Current American Crisis. New York: McGraw-Hill, 1975.

Rossin, A. D. and Rieck, T. A. "Economics of Nuclear Power." Science, 201 (August 18, 1978), 582-589.

Rybczynski, T. M., ed. The Economics of the Oil Crisis. London, England: Macmillan for the Trade Policy Research Centre, 1976.

Sander, Diana. "The Price of Energy." In Jack M. Hollander and Melvin K. Simmons, eds. Annual Review of Energy. Volume 1. Palo Alto, California: Annual Reviews, Inc., 1976, pp. 391-422.

Schlottman, Alan M. and Spore, Robert L. Economic Impacts of Surface Mine Reclamation. Knoxville, Tennessee: Appalachian Resources Project, University of Tennessee, 1976.

Schurr, Sam H. and Maraschak, J. Economic Aspects of Atomic Power. Princeton, New Jersey: Princeton University Press, 1950.

_____ and Netschert, Bruce C., et al. Energy in the American Economy, 1850-1975: An Economic Study of Its History and Prospects. Baltimore, Maryland: The Johns Hopkins University Press for Resources for the Future, 1960.

Shaw, K. R. "Capital Cost Escalation and the Choice of Power Stations." Energy Policy, 7 (December 1979), 321-328.

Sherfield, Lord, ed. Economic and Social Consequences of Nuclear Energy. London, England: Oxford University Press, 1972.

Slesser, Malcolm. Energy in the Economy. New York: St. Martin's Press, 1978.

Soneblum, Sidney. The Energy Connection: Between Energy and Economic Growth. Cambridge, Massachusetts: Ballinger Publishing Company, 1978.

Starr, Chauncey and Field, Stanford. "Economic Growth, Employment, and Energy." Energy Policy, 7 (March 1979), 2-22.

Thompson, Russell G., et al. The Cost of Energy and a Clean Environment. Houston, Texas: Gulf Publishing Company, 1978.

Tilton, John E. U.S. Energy R&D Policy: The Role of Economics. Washington, D.C.: Resources for the Future, 1974.

United States. Council on Environmental Quality. Coal Surface Mining and Reclamation: An Environmental and Economic Assessment of Alternatives. Washington, D.C.: U.S. Government Printing Office, 1973.

Vallenilla, Luis. Oil: The Making of a New Economic Order. New York: McGraw-Hill, 1975.

Webb, M. G. and Ricketts, M. J. The Economics of Energy. London, England: Macmillan, 1980.

Weidenbaum, Murray L., et al. Government Credit Subsidies for Energy Development. Washington, D.C.: American Enterprise Institute for Public Policy Research, 1976.

Weinberg, Alvin M., ed. Economic and Environmental Implications of a U.S. Nuclear Moratorium. Cambridge, Massachusetts: M.I.T. Press, 1979.

White, David C. "The Energy-Environment-Economic Triangle." Technology Review, 76 (December 1973), 10-19.

_____ . "Energy, the Economy, and the Environment." Technology Review, 74 (October/November 1971), 18-31.

Wright, David J. "Energy Budgets: 3. Goods and Services: An Input-Output Analysis." Energy Policy, 2 (December 1974), 307-315.

2. The Energy Sector in the American Economy

GENERAL

Duchesneau, Thomas. Competition in the U.S. Energy Industry. Cambridge, Massachusetts: Ballinger Publishing Company, 1975.

Gray, John E. Energy Policy: Industry Perspectives. Cambridge, Massachusetts: Ballinger Publishing Company, 1975.

Herman, Stewart W. and Cannon, James S. Energy Futures: Industry and the New Technologies. New York: INFORM, Inc., 1976; Cambridge, Massachusetts; Ballinger Publishing Company, 1977.

Jarman, Rufus. Energy Merchant. New York: Rosen, Richards Press, 1977.

Mulholland, Joseph P. and Webbink, Douglas W. Concentration Levels and Trends in the Energy Sector of the U.S. Economy. [Staff Report to the Federal Trade Commission] Washington, D.C.: U.S. Government Printing Office, March 1974.

Netschert, Bruce C. "The Energy Company: A Monopoly Trend in the Energy Markets." Bulletin of the Atomic Scientists, 27 (October 1971), 13-17.

Norman, Donald. The Performance of Oil Firm Affiliates in the Coal Industry. [Research Study No. 004] Washington, D.C.: American Petroleum Institute, March 1977.

Pelley, William E.; Constable, Richard W.; and Krupp, Herbert W. "The Energy Industry and the Capital Market." In Jack M. Hollander and Melvin K. Simmons, eds. Annual Review of Energy. Volume 1. Palo Alto, California: Annual Reviews, Inc., 1976, pp. 369-390.

Pratt, Raymond B. and Ward, Dwayne. "Corporations, the State, and Energy Development in the Northern Rockies." [A paper prepared for delivery at the annual meeting of the American Political Science Association, New York, August 30-September 3, 1978]

Sanger, Herbert S., Jr. and Mason, William E. <u>The Structure of the Energy Markets:</u>
<u>A Report of T.V.A.'s Antitrust Investigation of the Coal and Uranium Industries.</u>
Knoxville, Tennessee: Tennessee Valley Authority, June 14, 1977.

_____. <u>The Structure of the Energy Markets:</u>
<u>A Report of T.V.A.'s Antitrust Investigation of the Coal and Uranium Industries:</u>
<u>1979 Update</u>. Knoxville, Tennessee: Tennessee Valley Authority, Febraury 26,
1979.

United States. Congress. House of Representatives. Committee on the Judiciary. Subcommittee on Monopolies and Commercial Law. <u>Competitive Aspects of Oil</u>
<u>Company Expansion into Other Energy Sources</u>. Washington, D.C.: U.S. Govern-
ment Printing Office, November 1978.

United States. Congress. Senate. <u>Interfuel Competition</u>. [Hearings before the Sub-
committee on Antitrust and Monopoly of the Committee on the Judiciary,
June 17, 18, 19; July 14; October 21 and 22, 1975] Washington, D.C.: U.S.
Government Printing Office, 1975.

_____. <u>Petroleum Industry Involvement in Alternative</u>
<u>Sources of Energy</u>. [Prepared at the request of Frank Church for the Subcom-
mittee on Energy Research and Development of the Committee on Energy and
Natural Resources, Publication No. 95-54] Washington, D.C.: U.S. Government
Printing Office, September 1977.

NUCLEAR INDUSTRY

Barnaby, Wendy. "The Nuclear Industry and Its Public." <u>Bulletin of the Atomic Scien-</u>
<u>tists,</u> 37 (May 1981), 56-57.

Brady, David and Althoff, Philip. "The Politics of Regulation: The Case of the Atomic
Energy Commission and the Nuclear Industry." <u>American Politics Quarterly,</u>
1 (July 1973), 361-384.

Bupp, Irvin C. and Derian, Jean-Claude. "The Nuclear Power Industry." In <u>The Manage-</u>
<u>ment of Global Issues</u>. Appendix B of <u>The Commission on the Organization of</u>
<u>the Government for the Conduct of Foreign Policy</u>. Volume I. Washington,
D.C.: U.S. Government Printing Office, 1975.

Burn, Duncan. <u>Nuclear Power and the Energy Crisis: Politics and the Atomic Industry</u>.
New York: New York University Press, 1978.

Burness, H. Stuart; Montgomery, W. David;and Quirk, James P. "The Turnkey Era in
Nuclear Power." <u>Land Economics</u>, 56 (May 1980), 188-202.

Cruetz, Edward. "Nuclear Power: Rise of an Industry." <u>Bulletin of the Atomic Scien-</u>
<u>tists,</u> 25 (June 1970), 75-82.

Duffy, Gloria and Adams, Gordon. Power Politics: The Nuclear Industry and Nuclear Exports. New York: Council on Economic Priorities, 1978.

Elliot, David. The Politics of Nuclear Power. London, England: Pluto Press, 1978.

Ezell, J. S. Innovations in Energy: The Story of Kerr-McGee. Norman, Oklahoma: University of Oklahoma Press, 1979.

"An Industry Looks to Its Recovery." Nuclear News (Mid-February 1975), 33-64.

Joskow, P. L. and Baughman, M. L. "The Future of the U.S. Nuclear Industry." The Bell Journal of Economics, 7 (1976), 3-32.

Kostuik, John. "The Uranium Industry: Key Issues in Future Development." Energy Policy, 4 (September 1976), 212-224.

Lonnroth, Mans and Walker, William. The Viability of the Civil Nuclear Industry. New York: International Consultative Group on Nuclear Energy, 1980.

Mandel, Heinrich. "Construction Costs of Nuclear Power Sations."Energy Policy, 4 (March 1976), 12-24.

Mandelbaum, Michael. "A Nuclear Exporters Cartel." Bulletin of the Atomic Scientists, 33 (January 1977), 42-50.

Marcus, Gail H. "The Status of Women in the Nuclear Industry." Bulletin of the Atomic Scientists, 32 (April 1976), 34-39.

Metzger, H. Peter. The Atomic Establishment. New York: Simon and Schuster, 1972.

Montgomery, Timothy L. and Rose, David J. "Some Institutional Problems of the U.S. Nuclear Industry." Technology Review, 81 (March/April 1979), 53-62.

Robbins, Charles, et al. "U.S. Atomic Energy Industry." Nuclear Engineering International, 15 (November 1970), 905-929.

Taylor, June H. and Yokell, M. D. Yellowcake: The International Uranium Cartel. Elmsford, New York: Pergamon Press, 1979.

United States. Atomic Energy Commission. The Nuclear Industry - 1967. Washington, D.C.: U.S. Government Printing Office, November 6, 1967.

_____ . The Nuclear Industry - 1973. Washington, D.C.: U.S. Government Printing Office, 1973.

_____ . Nuclear Power Growth, 1974-2000. Washington, D.C.: U.S. Government Printing Office, 1974.

OIL INDUSTRY

Al-Otaiba, Mana Saeed. OPEC and the Petroleum Industry. New York: John Wiley & Sons, 1975.

Allvine, Fred and Patterson, James M. Competition, Ltd.: The Marketing of Gasoline. Bloomington, Indiana: Indiana University Press, 1972.

Ben-Shahar, Haim. Oil: Prices and Capital. Lexington, Massachusetts: D. C. Heath and Company, 1976.

Berry, Glyn R. "The Oil Lobby and the Energy Crisis." Canadian Public Administration, 17 (Winter 1974), 600-635.

Blair, John M. The Control of Oil. New York: Pantheon Books, 1976.

Clark, James A. and Halbouty, Michael T. The Last Boom. New York: Random House, 1972.

Continental Oil Company. CONOCO: The First One Hundred Years. New York: Dell, 1975.

Eichen, Marc. "The Reaction of U.S. Fuel Industries to a Range of Solar Energy Policies." Energy Policy, 7 (December 1979), 329-345.

Engler, Robert. The Brotherhood of Oil: Energy Policy and the Public Interest. Chicago, Illinois: University of Chicago Press, 1977.

Fischer, Louis. Oil Imperialism, the International Struggle for Petroleum. Westport, Connecticut: Hyperion Press, 1976.

Gibb, George Sweet and Knowlton, Evelyn H. History of Standard Oil Company (New Jersey): The Resurgent Years, 1911-1927. New York: Arno Press, 1975.

Giddens, Paul H. The Birth of the Oil Industry. New York: Macmillan, 1938.

_____ . Standard Oil Company (Indiana) - Oil Pioneer of the Middle West. New York: Appleton-Century-Crofts, 1975.

Glasscock, Carl Burgess. Then Came Oil. Westport, Connecticut: Hyperion Press, 1975.

Hanighen, Frank Clearly. The Secret War. Westport, Connecticut: Hyperion Press, 1975.

Harris, Fred R. "Oil: Capitalism Betrayed in Its Own Camp." Progressive, 37 (April 1973), 27-31.

Hartshorn, J. E. Politics and World Oil Economics. New York: Praeger Publishers, 1962.

Hidy, Ralph W. and Hidy, Muriel E. History of Standard Oil Company (New Jersey): Pioneering in Big Business, 1882-1911. New York: Arno Press, 1955.

Hobbie, Barbara. Oil Company Divestiture and the Press: Economic Vs. Journalistic Perceptions. New York: Praeger Publishers, 1977.

Ickes, Harold LeClaire. Fightin' Oil. Westport, Connecticut: Hyperion Press, 1975.

Ikard, Frank N. "Competition in the Petroleum Industry: Separating Fact from Myth." Oregon Law Review, 54 (1975), 583-605.

Independent Petroleum Association of America. The Oil Producing Industry in Your State. Washington, D.C.: IPAA Publications, 1977.

Jacoby, Neil H. Multinational Oil - A Study in Industrial Dynamics. New York: Macmillan, 1974.

Johnson, William A. Competition in the Oil Industry. Washington, D.C.: George Washington University Press, 1975.

_____ and Messnick, Richard E. "Vertical Divestiture of U.S. Oil Firms: The Impact on the World Oil Market." Law and Policy in International Business, 8 (1976), 963-989.

Kaufman, Burton I. Oil Cartel Case: A Documentary Study of Antitrust Activity in the Cold War Era. Westport, Connecticut: Greenwood Press, 1978.

Khan, Kabirur-Rahman. Law and Policy in Petroleum Development: Changing Relations Between Transnationals and Governments. London, England: Frances Pinter, 1979.

Krueger, Robert B., ed. The United States and International Oil: A Report for the Federal Energy Administration on U.S. Firms and Government Policy. New York: Praeger Publishers, 1975.

Larson, Henrietta M. History of Humble Oil and Refining Company. New York: Harper & Brothers, 1959.

Lloyd, Henry Demarest. Lords of Industry. New York: G. P. Putnam's Sons, 1910: Arno Press, 1973.

Loomis, Carol J. "How to Think About Oil Company Profits." Fortune, 89 (April 1974), 98-103.

McAfee, Jerry. "Government and the Oil Industry – Quo Vadis." Business Quarterly, 39 (Summer 1974), 56-61.

Mancke, Richard B. "Petroleum Conspiracy: A Costly Myth." Public Policy, 12 (Winter 1974), 1-13.

Markham, Jesse W.; Hovrihan, Anthony P.; and Sterling, Francis L. Horizontal Divestiture and the Petroleum Industry. Cambridge, Massachusetts: Ballinger Publishing Company, 1977.

Mead, Walter J. "The System of Government Subsidies to the Oil Industry." Natural Resources Journal, 10 (January 1970), 113-125.

Medvin, Norman. The Energy Cartel: Who Runs the American Oil Industry? New York: Vintage Books, 1974.

Mikesell, Raymond F. Foreign Investment in the Petroleum and Mineral Industries. Baltimore, Maryland: The Johns Hopkins University Press, 1971.

Mitchell, Edward J., ed. Vertical Integration in the Oil Industry. Washington, D.C.: American Enterprise Institute for Public Policy Research, 1976.

Mohr, Anton. Oil War. Westport, Connecticut: Hyperion Press, 1975.

Montague, Gilbert Holland. The Rise and Progress of Standard Oil Company. London, England: Harper, 1904; New York: Arno Press, 1973.

Moore, W. S., ed. Horizontal Divestiture: Highlights of a Conference on Whether Oil Companies Should be Prohibited from Owning Nonpetroleum Energy Resources. Washington, D.C.: American Enterprise Institute for Public Policy Research, 1977.

Nash, Gerald D. United States Oil Policy, 1890-1964. Pittsburgh, Pennsylvania: University of Pittsburgh Press, 1968.

Neurer, Edward J. The Natural Gas Industry: Monopoly and Competition in Field Markets. Norman, Oklahoma: University of Oklahoma Press, 1960.

"The New Shape of the U.S. Oil Industry." Business Week, No. 2316 (February 2, 1974), 50-58.

O'Connor, Harvey. The Empire of Oil. New York: Monthly Review Press, 1955.

_____. World Crisis in Oil. New York: Monthly Review Press, 1962.

Penrose, Edith T. The Large International Firms in Developing Countries: The International Petroleum Industry. Cambridge, Massachusetts: M.I.T. Press, 1968.

Pikl, I. J., ed. Public Policy and the Future of the Petroleum Industry. Laramie, Wyoming: University of Wyoming Press, 1970.

Pratt, Joseph A. "The Petroleum Industry in Transition: Antitrust and the Decline of Monopoly Control of Oil." The Journal of Economic History, 40 (December 1980), 815-836.

Riddick, Winston W. "The Nature of the Petroleum Industry." Proceedings of the Academy of Political Science, 31 (December 1973), 148-158.

Sampson, Anthony. The Seven Sisters: The Great Oil Companies and the World They Shaped. New York: Viking Penguin, 1976.

Solberg, Carl. Oil Power: The Rise and Imminent Fall of an American Empire. New York: New American Library, 1976.

Spence, Hartzell. Portrait in Oil. New York: McGraw-Hill, 1962.

Stern, Lawrence. "Oil: Our Private Government." Progressive, 37 (April 1973), 19-21.

Stobaugh, Robert. "The Oil Companies in the Crisis." Daedalus, 104 (Fall 1975), 179-202.

Stocking, George W. Oil Industry and the Competitive System. Westport, Connecticut: Hyperion Press, 1976.

Sunder, Shyam. Oil Industry Profits. Washington, D.C.: American Enterprise Institute for Public Policy Research, 1977.

Swanson, E. B. A Century of Oil and Gas in Books. New York: Appleton-Century-Crofts, 1960.

Tanzer, Michael. The Political Economy of International Oil and the Underdeveloped Countries. Boston, Massachusetts: Beacon Press, 1969.

Tarbell, Ida M. The History of Standard Oil Company. New York: Macmillan, 1925.

Teece, David J., ed. R&D in Energy: Implications of Petroleum Industry Reorganization. Stanford, California: Institute for Energy Studies, Stanford University, 1977.

Uhl, William, et al. Oil Industry U.S.A. Des Moines, Iowa: Wallace-Homestead Book Company, 1976.

United States. Congress. Senate. Governmental Intervention in the Market Mechanism: The Petroleum Industry. Parts I-III. [Hearings before the Subcommittee on Antitrust and Monopoly of the Committee on the Judiciary] Washington, D.C.: U.S. Government Printing Office, 1969.

United States. Congress. Senate. Committee on Interior and Insular Affairs. Preliminary Federal Trade Commission Staff Report on Its Investigation of the Petroleum Industry. Washington, D.C.: U.S. Government Printing Office, 1973.

_____ . Special Subcommittee on Integrated Oil Operations. The Energy Industry: Organization and Public Policy. Washington, D.C.: U.S. Government Printing Office, 1974.

United States. Federal Energy Administration. The Petroleum Industry: A Report on Corporate and Industry Structure and Ownership. 2 Volumes. Washington, D.C.: U.S. Government Printing Office, 1975.

_____ . The Relationship of Oil Companies and Foreign Governments. Washington, D.C.: U.S. Government Printing Office, 1956.

Vicker, Ray. The Kingdom of Oil. New York: Charles Scribner's Sons, 1974.

White, Gerald T. Formative Years in the Far West: A History of Standard Oil Company of California and Predecessors Through 1919. New York: Appleton-Century-Crofts, 1962.

Wilkins, Mire. "The Oil Crisis in Perspective: The Oil Companies." Daedalus, 104 (Fall 1975), 159-178.

Williamson, Harold F. The American Petroleum Industry - The Age of Energy, 1899-1959. Evanston, Illinois: Northwestern University Press, 1963.

Wilson, John W. "Competitive Market Structure and Performance in the Petroleum Industry." In Walter F. Scheffer, ed. Energy Impacts on Public Policy and Administration. Norman, Oklahoma: University of Oklahoma Press, 1976, pp. 35-75.

Yamani, A. Z. "Oil Industry in Transition." Natural Resources Lawyer, 8 (1975), 391-398.

COAL INDUSTRY

Attorney General of the United States. Competition in the Coal Industry. [Report of the U.S. Department of Justice pursuant to Section 8 of the Federal Coal Leasing Amendments Act of 1975] Washington, D.C.: U.S. Government Printing Office, May 1978.

Baratz, Morton S. The Union and the Coal Industry. New Haven, Connecticut: Yale University Press, 1955; Port Washington, New York: Kennikat Press, 1972.

Cohn, Elchanon, et al. The Bituminous Coal Industry, A Forecast: Manpower, Government Policy, Technology. University Park, Pennsylvania: Institute for Research on Human Resources, Pennsylvania State University, 1975.

Comptroller General. The State of Competition in the Coal Industry. [Report to the Congress, EMD-78-22] Washington, D.C.: U.S. General Accounting Office, December 30, 1977.

Fisher, Waldo E. and James, Charles M. Minimum Price Fixing in the Bituminous Coal Industry. Princeton, New Jersey: Princeton University Press, 1955.

Giffen, Philip E. Industrial Concentration and Firm Diversification in Bituminous Coal with Special Reference to the Southeastern United States, 1950-1970. [Appalachian Resources Project Report No. 7] Knoxville, Tennessee: Appalachian Resources Project, University of Tennessee, August 1972.

Hinson, William R. An Analysis of Anti-Trust Activity in the Coal Industry, 1964-1974. [Appalachian Resources Project Report No. 32] Knoxville, Tennessee: Appalachian Resources Project, University of Tennessee, 1970, 1979.

Kahn, Marvin H. and Hand, Robert. Implications of Ownership Patterns of Western Coal Reserves and Their Impact on Coal Development. McLean, Virginia: The MITRE Corporation, 1976.

Lawrence, Anthony G. The Causes and Consequences of the Changing Pattern of Mine Ownership: An A Priori Analysis. Lexington, Kentucky: Institute for Mining and Minerals Research, University of Kentucky, September 1975.

Mansfield, E. "Firm Size and Technological Change in the Petroleum and Bituminous Coal Industries." In Thomas D. Duchesneau, ed. Competition in the U.S. Energy Industry. Cambridge, Massachusetts: Ballinger Publishing Company, 1975, pp. 317-345.

Mattick, Paul and Adamic, Louis. Fighting for Survival: The Bootleg Coal Industry. Huntington, West Virginia: Appalachian Movement Press, 1973.

Moyer, Reed. Competition in the Midwestern Coal Industry. Cambridge, Massachusetts: Harvard University Press, 1964.

National Coal Association. Implications of Investments in the Coal Industry by Firms from Other Energy Industries. Washington, D.C.: National Coal Association, September 1977.

_____. Study of New Mine Additions and Major Expansion Plans of the Coal Industry and the Potential for Future Coal Production. Washington, D.C.: National Coal Association, November 1977.

Newcomb, Richard. "The American Coal Industry." Current History, 74 (May/June 1978), 206-209, 228.

Risser, Hubert E. The Economics of the Coal Industry. Lawrence, Kansas: Bureau of Business Research, School of Business, University of Kansas, 1958; Westport, Connecticut: Greenwood Press, 1976.

Thomas, Jerry Bruce. Coal Country: The Rise of the Southern Smokeless Coal Industry and Its Effect on Area Development, 1872-1910. [Thesis - University of North Carolina at Chapel Hill, 1971] Ann Arbor, Michigan: University Microfilms International, 1972.

Thompson, A. MacKenzie. The U.S. Coal Industry. New York: Garland Press, 1980.

Tomimatsu, T. T. and Johnson, Robert E. The State of the U.S. Coal Industry: A Financial Analysis of Selected Coal-Producing Companies with Observations on Industry Structure. [Bureau of Mines Information Circular 8707] Washington, D.C.: U.S. Government Printing Office, 1976.

United Nations. Coal: 1985 and Beyond: A Prospective Study of the Coal Industry in Europe and North America. Elmsford, New York: Pergamon Press, 1978.

United States. Federal Trade Commission. The Structure of the Nation's Coal Industry, 1964-1974. [Staff Report] Washington, D.C.: Federal Trade Commission, November 1978.

United States. Library of Congress. Congressional Research Service. The Coal Industry: Problems and Prospects - A Background Study. Washington, D.C.: U.S. Government Printing Office, 1978.

Walls, David S., et al. A Baseline Assessment of Coal Industry Structure in the Ohio River Basin Energy Study Region. Urbana, Illinois: Ohio River Basin Energy Study [ORBES], June 1979.

ELECTRIC UTILITIES

Allen, Howard P. "Electric Utilities: Can They Meet Future Power Needs?" Annals of the American Academy of Political and Social Science, 410 (November 1973), 86-96.

Altman, Manfred; Telkes, Maria;and Wolf, Martin. The Energy Resources and Electric Power Situation in the United States. Philadelphia, Pennsylvania: University of Pennsylvania Press, 1971.

Baughman, M. L. and Bottaro, D. J. Electric Power Transmission and Distribution Systems: Costs and Their Allocation. Austin, Texas: University of Texas at Austin, Center for Energy Studies, July 1975.

_____ ; Joskow, P. J. and Kamat, Dilip P. Electric Power in the United States: Models and Policy Analysis. Cambridge, Massachusetts: M.I.T. Press, 1979.

Berlin, Edward, et al. Perspective on Power: A Study of the Regulation and Pricing of Electric Power. Cambridge, Massachusetts: Ballinger Publishing Company, 1974.

Bodansky, David. "Electricity Generation Choices for the Near Term." Science, 207 (February 15, 1980), 721-727.

Bossong, Ken. The Case Against Utility Involvement in Solar Commercialization. Washington, D.C.: Citizens' Energy Project, 1979.

Brown, D. C. Electricity for Rural America: The Fight for REA. Westport, Connecticut: Greenwood Press, 1980.

Bupp, Irvin C., Jr. "A Dark Future for Utilities." Business Week, No. 2587 (May 28, 1979), 108-124.

Business Communications Company Staff. Future Utility Requirements. Rev. ed. Stamford, Connecticut: Business Communications Company, 1977.

Casper, Barry M. and Wellstone, Paul David. Powerline: The First Battle of America's Energy War. Amherst, Massachusetts: University of Massachusetts Press, 1981.

Chapman, Duane, et al. "Electricity Demand Growth and the Energy Crisis." Science, 178 (November 17, 1972), 703-708.

Cicchetti, Charles J. and Jurewitz, John, eds. Studies in Electric Utility Regulation. Cambridge, Massachusetts: Ballinger Publishing Company, 1975.

Comptroller General. The Effects of Regulation on the Electric Utility Industry. [EMD-81-35] Washington, D.C.: U.S. General Accounting Office, March 2, 1981.

Council on Economic Priorities and White, Ronald. The Price of Power Update: Electric Utilities and the Environment. New York: Council on Economic Priorities, 1977.

Dickson, Charles; Eichen, Marc; and Feldman, Stephen. "Solar Energy and U.S. Public Utilities: The Impact on Rate Structure and Utilization." Energy Policy, 5 (September 1977), 195-210.

Doyle, Jack. Lines Across the Land. Washington, D.C.: Environmental Policy Institute, 1979.

Edison Electric Institute. 1975 Year-End Summary of the Electric Power Situation in the United States. New York: Edison Electric Institute, 1975.

_____. Statistical Yearbook of the Electric Utility Industry. New York: Edison Electric Institute, 1975.

_____. The Transitional Storm. New York: Edison Electric Institute, 1977.

Funigiello, Philip J. Toward a National Power Policy: The New Deal and the Electric Utility Industry. Pittsburgh, Pennsylvania: University of Pittsburgh Press, 1973.

Gandara, Arturo. Utility Decision-Making and the Nuclear Option. [R-2148-NSF] Santa Monica, California: RAND Corporation, June 1977.

Garfield, Paul J. and Lovejoy, Wallace F. Public Utility Economics. Englewood Cliffs, New Jersey: Prentice-Hall, 1964.

Gordon, Richard L. U.S. Coal and the Electric Power Industry. Baltimore, Maryland: The Johns Hopkins University Press, 1975.

Hamlen, William and Tschirhart, John. "Solar Energy, Public Utilities, and Economic Efficiency." Southern Economic Journal, 47 (October 1980), 348-365.

Hounshell, D. A. "Edison and the Pure Science Ideal in 19th Century America." Science, 207 (February 8, 1980), 612-617.

Howard, Peter R. "Electrical Transmission of Energy: Current Trends." Energy Policy, 1 (September 1973), 154-160.

Hub, K., et al. A Study of Social Costs for Alternate Means of Electric Power Generation for 1980 and 1990. Argonne, Illinois: Argonne National Laboratory, 1973.

Institute of Gas Technology. The Role of Utility Companies in Solar Energy. Chicago, Illinois: Institute of Gas Technolgy, 1978.

Johnson, Charles J. Coal Demand in the Electric Utility Industry, Nineteen Forty-Six to Nineteen Ninety. Ed. by Stuart Bruchey. New York: Arno Press, 1979.

Lagassa, George K. "Implementing the Soft Path in a Hard World: Decentralization and the Problem of Electric Power Grids." In Gregory Daneke and George Lagassa, eds. Energy Policy and Public Administration. Lexington, Massachusetts: Lexington Books, 1980, pp. 167-189.

Levy, Paul F. "The Politics of Rate Reform." Technology Review, 80 (February 1978), 37-43.

Lowe, William E. "Creating Power Plants: The Costs of Controlling Technology." Technology Review, 74 (January 1972), 22-30.

McCraw, Thomas K. TVA and the Public Power Fight, 1933-1939. Philadelphia, Pennsylvania: J. B. Lippincott & Company, 1971.

Maidique, Modesto and Woo, Benson. "Solar Heating and the Electric Utilities." Technology Review, 82 (May 1980), 24-33.

Marsh, W. D. The Economics of Electric Utility Power Generation. New York: Oxford, University Press, 1980.

Marts, M. E. Electric Power and the Future of the Pacific Northwest. Seattle, Washington: University of Washington Press, 1980.

Meadows, Dennis. "Ten Facts Hindering Utility Support for Solar Energy." In Community Energy Self-Reliance. [SERI/CP-354-421] Washington, D.C.: U.S. Government Printing Office, July 1980, 456-459.

Messing, Mark; Friesema, Paul H.; and Morrell, David. Centralized Power. Washington, D.C.: Environmental Policy Institute, March 1979.

Miller, John T., Jr. "A Needed Reform of the Organization and Regulation of the Interstate Electric Power Industry." Fordham Law Review, 38 (May 1970), 635-673.

Mitchell, Bridger M. and Manning, Willard G. Peak-Load Pricing: European Lessons for U.S. Energy Policy. Cambridge, Massachusetts: Ballinger Publishing Company, 1978.

Mitzman, Barry. "Abusing Public Power." Environmental Action, 7 (April 24, 1976), 7-11.

Novick, Sheldon. "The Electric Power Industry." Environment, 17 (November 1975), 7-13, 32-39.

Quarles, John R., Jr. "The Electric Power Industry and the Environment." In Harold Wolozin, ed. Energy and the Environment. Morristown, New Jersey: General Learning Press, 1974, pp. 127-133.

Ramsay, William. Unpaid Costs of Electrical Energy: Health and Environmental Impacts of Coal and Nuclear Power. Baltimore, Maryland: The Johns Hopkins University Press, 1979.

Robert, Marc J. and Bluhm, Jeremy S. The Choices of Power: Utilities Face the Environmental Challenge. Cambridge, Massachusetts: Harvard University Press, 1981.

Rodgers, William. Brown-Out: The Power Crisis in America. New York: Stein and Day, 1972.

Sayre, Kenneth, ed. Values in the Electric Power Industry. Notre Dame, Indiana: University of Notre Dame Press, 1977.

Shapley, Deborah. "TVA Today: Former Reformers in an Era of Expensive Electricity." Science, 194 (November 19, 1976), 814-818.

Sichel, Werner. Public Utility Rate Making in an Energy Conscious Environment. Boulder, Colorado: Westview Press, 1978.

Smackey, Bruce M. "Should Electric Utilities Market Solar Energy?" Public Utilities Fortnightly, 102 (September 28, 1978), 37-43.

Smith, Bruce A. Technological Innovation in Electric Power Generation, 1950-1970. East Lansing, Michigan: Graduate School of Business Administration, Michigan State University, 1977.

Sporn, Philip. The Social Organization of Electric Power Supply in Modern Societies. Cambridge, Massachusetts: M.I.T. Press, 1971.

Stelzer, Irwin M. "The Electric Utilities Face the Next Twenty Years." In Hans Landsberg. ed. Selected Studies on Energy: Background Papers for Energy: The Next Twenty Years. Cambridge, Massachusetts: Ballinger Publishing Company, 1980, 127-144.

Troutman, Michael and Morgan, Richard. "Electric Utilities: Private Rule vs. Public Control." Environmental Action, 7 (April 24, 1976), 4-6.

United States. Department of Energy. Economic Regulatory Commission. Division of Power Supply and Reliability. Additions to Generating Capacity 1978-1987 for the Contiguous United States. Washington, D.C.: U.S. Government Printing Office. [Annual Update]

_____. Bulk Electric Power Load and Supply Projections 1988-1997 for the Contiguous United States. Washington, D.C.: U.S. Government Printing Office. [Annual Update]

_____. Electric Power Supply and Demand 1978-1987 for the Contiguous United States. Washington, D.C.: U.S. Government Printing Office. [Annual Update]

United States. Department of Energy. Federal Energy Regulatory Commission. Office of Electric Power Regulation. Annual Summary of Cost and Quality of Electric Utility Plant Fuels. Washington, D.C.: U.S. Government Printing Office. [Annual]

United States. Library of Congress. Congressional Research Service. Centralized vs. Decentralized Energy Systems: Diverging or Parallel Roads? [A report prepared for use by the Subcommittee on Energy and Power, Committee on Interstate and Foreign Commerce, House of Representatives.] Washington, D.C.: U.S. Government Printing Office, May 1979.

Walsh, John. "Electric Power Research Institute: A New Formula for Industry R&D." Science, 182 (October 19, 1973), 263-265.

Waltrip, John R. Public Power During the Truman Administration. Ed. by Stuart Bruchey. New York: Arno Press, 1979.

Westley, Steven, ed. Energy Efficiency and the Utilities: New Directions. San Francisco, California: California Public Utilities Commission, July 1980.

Young, Louise B. Power Over People. New York: Oxford University Press, 1973.

3. Energy and Social Structure

Adams, Richard Newbold. Energy and Structure: A Theory of Social Power. Austin, Texas: University of Texas Press, 1975.

Althouse, Ronald. Work, Safety, and Lifestyle Among Southern Appalachian Coal Miners: A Survey of the Men of Standard Mines. Morgantown, West Virginia: Office of Research and Development, Division of Social and Economic Development, Appalachian Center, West Virginia University, 1974.

Appleby, A. J. "Energy Costs and Society: The High Price of Future Energy." Energy Policy, 4 (1976), 87-97.

Arble, Mead. The Long Tunnel: A Coal Miner's Journal. New York: Atheneum Publishers, 1976.

Balliet, Lee. 'A Pleasing Tho' Dreadful Sight': Social and Economic Impacts of Coal Production in the Eastern Coalfields. [A report to the Office of Technology Assessment, Congress of the United States] Washington, D.C.: Office of Technology Assessment, 1978.

Bender, Frederick L. "Energy and USA Social Policy: The Opportunity for the Left." Social Praxis, 6 (1979), 161-176.

Boulding, Kenneth E. "The Social System and the Energy Crisis." Science, 184 (April 19, 1974), 255-257.

Bowman, Mary Jean and Haynes, Warren W. Resources and People in East Kentucky: Problems and Potentials of a Lagging Economy. Baltimore, Maryland: Published for Resources for the Future by the Johns Hopkins University Press, 1963: New York: AMS Press, 1977.

Brown, George H. "Suburban Sprawl and the Energy Situation." Conference Board Record, 11 (November 1974), 35-38.

Bryant, F. Carleen. The Social Impact of Surface Mining in a Rural Appalachian Community. Knoxville, Tennessee: Appalachian Resources Project, University of Tennessee, 1976.

Budnitz, Robert J. and Holdren, John P. "Social and Environmental Costs of Energy Systems." In Jack M. Hollander and Melvin K. Simmons, eds. Annual Review of Energy. Volume 1. Palo Alto, California: Annual Reviews, Inc., 1976, pp. 553-580.

Bulmer, M.I.A. "Sociological Models of the Mining Community." The Sociological Review, 23 (February 1975), 61-92.

Burpy, Raymond and Bell, A. Fleming. Energy and the Community. Cambridge, Massachusetts: Ballinger Publishing Company, 1978.

Caldwell, Lynton K. "Energy and the Structure of Social Institutions." Human Ecology, 4 (1976), 31-46.

Callaway, Donald G.; Levy, Jerrold E.; and Henderson, Eric B. The Effects of Power Production and Strip Mining on Local Navaho Populations. Los Angeles, California: Institute of Geophysics and Planetary Physics, University of California, 1976.

Cameron, J. and Wood, G. "Energy Organizing in the 1980's." Socialist Review, No. 52 (July-August 1980), 120-129.

Caudill, Harry M. My Land Is Dying. New York: E. P. Dutton & Company, 1971.

_____ . Night Comes to the Cumberlands. Boston, Massachusetts: Atlantic Monthly Press/Little Brown and Company, 1962.

Centaur Management Consultants, Inc. Managing the Social and Economic Impacts of Energy Developments. [Report prepared for the U.S. Energy Research and Development Administration] Washington, D.C.: Energy Research and Development Administration, 1978.

Coleman, McAlister. Men and Coal. New York: Farrar & Rinehart, 1943; New York: Arno and the New York Times, 1969.

Cottrell, Fred. "Energy and Sociology." Humanity and Society, 4 (1978), 237-249.

Curran, Samuel C. and Curran, John S. Energy and Human Needs. New York: Halsted Press, 1980.

Cushing, George H. The Human Story of Coal. New York: W. E. Rudge, 1924.

Densmore, Raymond E. The Coal Miner of Appalachia. Parsons, West Virginia: McClain Printing Company, 1977.

Diesendorf, Mark, ed. Energy and People: Social Implications of Different Energy Futures. Forest Grove, Oregon: International Scholarly Book Service, 1980.

Ergood, Bruce and Kuhre, Bruce E., eds. Appalachia: Social Context, Past and Present. Dubuque, Iowa: Kendall/Hunt Publishing Company, 1976.

Erickson, Kai T. Everything in its Path: Destruction of a Community in the Buffalo Creek Flood. New York: Simon and Schuster, 1976.

Frankena, Frederick. "Regional Socioeconomic Impacts of Declining Net Energy." Urban Ecology, 3 (1978), 101-110.

Gaines, Linda and Berry, K. Stephen. TOSCA: The Total Social Cost of Coal and Nuclear Power. Cambridge, Massachusetts: Ballinger Publishing Company, 1979.

Garrett, P.; Webb, C.; Peck, J.; Crane, D.;and Sweeney, K. Energy Impact: A Community in Action. Frisco, Colorado: Northwest Colorado Council of Governments, February 1977.

Garvin, D. F. "Social, Economic, and Utility Growth." Public Utilities Fortnightly, 91 (February 15, 1973), 23-38.

Gaskin, M. "Energy and Social Economics." In I. M. Blair, B. D. Jones and A. J. van Horn, eds. Aspects of Energy Conversion. Elmsford, New York: Pergamon Press, 1976, pp. 689-707.

Hannon, Bruce. "Energy, Labor, and the Conserver Society." Technology Review, 79 (1977), 47-53.

Healy, Timothy J. Energy and Society. San Francisco, California: Boyd & Fraser Publishing Company, 1976.

Hub, K., et al. A Study of Social Costs for Alternate Means of Electric Power Generation for 1980 and 1990. Argonne, Illinois: Argonne National Laboratory, 1973.

Klineberg, Otto, ed. Social Implications of the Peaceful Uses of Nuclear Energy. New York: Unipub, 1964.

Kolbash, Ronald Lee. A Study of Appalachia's Coal Mining Communities and Associated Environmental Problems. Ann Arbor, Michigan: University Microfilms International, 1978.

Lantz, Herman R. People of Coal Town. Carbondale, Illinois: Southern Illinois University Press, 1958.

Laramey, Tom A., Jr. "The National Response to Energy-Related Needs of the Poor." The Urban Lawyer, 12 (Summer 1980), 526-533.

Leistritz, F. Larry and Murdock, Steven H. The Socioeconomic Impact of Resource Development: Methods of Assessment. Boulder, Colorado: Westview Press, 1981.

Lewis, Helen M.; Johnson, Linda; and Askins, Donald, eds. Colonialism in Modern America: The Appalachian Case. Boone, North Carolina: Appalachian Consortium Press, 1978.

Matthews, J. "Marxism, Energy, and Technological Change." In Adlam, Diana, et al., eds. Politics & Power One. London, England: Kegan Paul, 1981, pp. 19-37.

Mazur, Allan and Rosa, Eugene. "Energy and Life-Style." Science, 186 (1974), 607-610.

Morgan, M. Granger, ed. Energy and Man: Technical and Social Aspects of Energy. New York: Institute of Electrical and Electronics Engineers Press, 1975.

Morris, Deane N. Effects of Energy Shortages on the Way We Live. Santa Monica, California: RAND Corporation, December 1974.

National Research Council. Energy Choices in a Democratic Society. Washington, D.C.: National Academy of Sciences, 1980.

Nelkin, Dorothy. "Native Americans and Nuclear Power." Science, Technology and Human Values, No. 35 (Spring 1981), 2-13.

_____. "Some Social and Political Dimensions of Nuclear Power: Examples from Three Mile Island." American Political Science Review, 75 (March 1981), 131-142.

Nevins, A.; Dunlop, R. G.; Teller E.; Mason, E. S.; and Hoover, H., Jr. Energy and Man: A Symposium. New York: Appleton-Century-Crofts, 1960.

Newman, Dorothy, K. and Day, Dawn. The American Energy Consumer. Cambridge, Massachusetts: Ballinger Publishing Company, 1975.

Nore, Peter and Turner, Terisa, eds. Oil and Class Struggle. Westport, Connecticut: Lawrence Hill, 1980.

Odum, H. T. and Odum, E. C. Energy Basis for Man and Nature. New York: McGraw-Hill, 1976.

O'Toole, James. Energy and Social Change. Cambridge, Massachusetts: M.I.T. Press, 1976.

Perlman, Robert and Warren, Roland, L. Families in the Energy Crisis: Impacts and Implications for Theory and Policy. Cambridge, Massachusetts: Ballinger Publishing Company, 1977.

Peterson, Bill. Coaltown Revisited: An Appalachian Notebook. Chicago, Illinois: Henry Regnery Company, 1972.

Pfuhl, John J. Oil and Its Impact: A Case Study of Community Change. Washington, D.C.: University Press of America, 1980.

Plunkett, H. D. and Bowman, M. J. Elites and Change in the Kentucky Mountains. Lexington, Kentucky: University of Kentucky Press, 1973.

Reiff, I. S. Managing the Social and Economic Impacts of Energy Development. Washington, D.C.: Energy Research and Development Administration, July 1976.

Riddel, Frank S., ed. Appalachia: Its People, Heritage, and Problems. Dubuque, Iowa: Kendall/Hunt Publishing Company, 1974.

Sagan, Leonard A., ed. Human and Ecologic Effects of Nuclear Power Plants. Springfield, Illinois: Charles C. Thomas, Publisher, 1974.

_____. "Human Costs of Nuclear Power." Science, 177 (August 11, 1972), 487-493.

Sheppard, Muriel. Cloud By Day: The Story of Coal and Coke and People. Chapel Hill, North Carolina: University of North Carolina Press, 1947.

Sherfield, Lord, ed. Economic and Social Consequences of Nuclear Energy. London, England: Oxford University Press, 1972.

Sporn, Philip. The Social Organization of Electric Power Supply in Modern Societies. Cambridge, Massachusetts: M.I.T. Press, 1971.

Stenehjem, E. J.; Hoover, L. J.; and Krohm, G. C. Empirical Investigation of the Factors Affecting Socioeconomic Impacts from Energy Development. Argonne, Illinois: Argonne National Laboratory, September 1977.

Thomas, Jerry Bruce. Coal Country: The Rise of the Southern Smokeless Coal Industry and Its Effect on Area Development, 1872-1910. [Thesis, University of North Carolina at Chapel Hill, 1971] Ann Arbor, Michigan: University Microfilms International, 1972.

Thring, M. W. Energy and Humanity. Stevenage, Hartfordshire, England: Peter Peregrinus, 1974.

Toole, K. Ross. The Rape of the Great Plains: Northwest America, Cattle, and Coal. Boston, Massachusetts: Little, Brown and Company, 1976.

Unseld, Charles. T., et al., eds. Sociopolitical Effects of Energy Use and Policy. [Supporting Paper No. 5 to the Committee on Nuclear and Alternative Energy Systems] Washington, D.C.: National Academy of Sciences, 1979.

Warkov, Seymour, ed. <u>Energy Policy in the United States: Social and Behavioral Dimensions</u>. New York: Praeger Publishers, 1978.

Weinberg, Alvin M. "Social Institutions and Nuclear Energy." <u>Science</u>, 177 (July 7, 1972), 27-34.

Whipple, D. Sawyer. "The Social Costs of Coal." <u>Environmental Action</u>, 8 (September 11, 1976), 3-7.

4. Energy and the Environment

Abrahamson, Dean E. Environmental Costs of Electric Power. New York: Scientists' Institute for Public Information, 1970.

Ackerman, Bruce A. and Hassler, William T. Clean Coal/Dirty Air. New Haven, Connecticut: Yale University Press, 1981.

Alexander, Tom. "A Promising Try at Environmental Detente for Coal." Fortune, February 13, 1978, 94-102.

Appalachian Regional Commission. Challenges for Appalachia Energy, Environment, and Natural Resources. Washington, D.C.: Appalachian Regional Commission, 1976.

Ashley, Holt, et al., eds. Energy and the Environment. Elmsford, New York: Pergamon Press, 1977.

Atwood, Genevieve. "The Strip Mining of Western Coal." Scientific American, 233 (December 1975), 23-29.

Austin, Richard Cartwright. Spoil: A Moral Study of Strip Mining for Coal. New York: National Division, Board of Global Ministries, Methodist Church, 1973.

_____ and Borrelli, Peter. The Strip Mining of America: An Analysis of Surface Coal Mining and the Environment. New York: Sierra Club Books, 1971.

Axelrod, Regina, ed. Environment, Energy, and Public Policy. Lexington, Massachusetts: Lexington Books, 1980.

Ballard, Steven C.; Devine, Michael; and Associates. Water and Western Energy: Impacts, Issues, and Choices. Boulder, Colorado: Westview Press, 1982.

Base, C. F.; Goeller, H. E.; Olson, J. S.; and Rotty, R. M. The Global Carbon Dioxide Problem. [ORNL-5194] Oak Ridge, Tennessee: Oak Ridge National Laboratory, 1976.

Bernard, Harold W., Jr. The Greenhouse Effect. Cambridge, Massachusetts: Ballinger Publishing Company, 1980.

Berry, Wendell. "Strip Mining Morality: The Landscaping of Hell." Nation, 202 (January 24, 1966), 96-100.

Bishop, A. S. "Environmental Aspects of Energy Alternatives." In N. Polunin, ed. Growth Without Ecodisaster? New York: John Wiley & Sons, 1980, pp. 343-378.

Box, Thadis, et al. Rehabilitation Potential of Western Coal Lands. Cambridge, Massachusetts: Ballinger Publishing Company, 1974.

Budnitz, Robert J. and Holdren, John P. "Social and Environmental Costs of Energy Systems." In Jack M. Hollander and Melvin K. Simmons, eds. Annual Review of Energy. Volume 1. Palo Alto, California: Annual Reviews, Inc., 1976, pp. 553-580.

Caldwell, Lynton, et al. Citizens and the Environment. Bloomington, Indiana: Indiana University Press, 1976.

Camplin, Paul, ed. Strip Mining in Kentucky. Frankfort, Kentucky: Kentucky Division of Strip Mining and Reclamation, 1965.

Carter, Luther J. "Coal: Invoking 'The Rule of Reason' in an Energy-Environment Conflict." Science, 198 (October 21, 1977), 276-280.

Caudill, Harry M. My Land Is Dying. New York: E. P. Dutton & Company, 1971.

Coffin, David L. and Knelson, John H. "Acid Precipitation: Effects of Sulfur Dioxide and Sulfate Aerosol Particles on Human Health." Ambio, 5 (1976), 239-242.

Cohen, Alan, et al. Residential Fuel Policy and the Environment. Cambridge, Massachusetts: Ballinger Publishing Company, 1974.

Congressional Quarterly Editorial Research Reports Staff. Editorial Research Reports on Earth, Energy, and Environment. Washington, D.C.: Congressional Quarterly, Inc., 1977.

Cook, Constance E. Nuclear Power and Legal Advocacy: The Environmentalists and the Courts. Lexington, Massachusetts: Lexington Books, 1980.

Council on Economic Priorities and White, Ronald. The Price of Power Update: Electric Utilities and the Environment. New York: Council on Economic Priorities, 1977.

Council on Environmental Quality. The Good News About Energy. Washington, D.C.: U.S. Government Printing Office, 1979.

Daly, Herman E. and Umana, Alvaro, eds. Energy, Economics, and the Environment: Conflicting Views of an Essential Interrelationship. Boulder, Colorado: Westview Press, 1981.

Damon, Paul E. and Kunen, Steven M. "Global Cooling?" Science, 193 (August 6, 1976), 447-453.

Doyle, W. S. Strip Mining of Coal - Environmental Solutions. Park Ridge, New Jersey: Noyes Data Corporation, 1976.

Eicholz, Geoffrey G. Environmental Aspects of Nuclear Power. Ann Arbor, Michigan: Ann Arbor Science Publishers, 1976.

El-Hinnawi, Essam E. Energy and the Environment. New York: American-Elsevier, 1980a.

_____ , ed. Nuclear Energy and the Environment. Elmsford, New York: Pergamon Press, 1980b.

Eliassen, Rolf. "Power Generation and the Environment." Bulletin of the Atomic Scientists, 27 (September 1971), 37-42.

Ellickson, Phyllis, et al. Balanced Energy and the Environment: The Case for Geothermal Development. [R-2274-DOE] Santa Monica, California: RAND Corporation, 1978.

"Energy and the Environment." [Special Issue, edited by Manes L. Regens] American Behavioral Scientist, 22 (November-December 1978).

"Energy Law and the Environment." [Special Section] Ecology Law Quarterly, 8 (1980), 725-830.

Environmental Studies Board. Study Committee on the Potential for Rehabilitating Lands Surface-Mined for Coal in the Western United States. Rehabilitation Potential of Western Coal Lands. [A Report to the Energy Policy Project of the Ford Foundation] Cambridge, Massachusetts: Ballinger Publishing Company, 1974.

Evans, Allen R. Energy and Environment. Washington, D.C.: Communications Press, 1980.

Finkel, Asher J., ed. Energy, the Environment, and Human Health. Acton, Massachusetts: Publishing Sciences Group, 1973.

Foell, Wesley K. Management of Energy-Environment Systems: Methods and Case Studies. New York: John Wiley & Sons, 1979.

Ford, A. "Environmental Policies for Electricity Generation: A Study of the Long-Term Dynamics of the SO_2 Problem." Energy Systems and Policy, 1 (1975), 287-304.

Fowler, John M. Energy and the Environment. New York: McGraw-Hill, 1975.

Freeman, A. Myrick, III; Haveman, Robert H.; and Kneese, Allen V. The Economics of Environmental Policy. New York: John Wiley & Sons, 1973.

Garvey, Gerald. Energy, Ecology, Economy: A Framework for Environmental Policy. New York: W. W. Norton & Company, 1972.

Glasstone, Samuel and Jordan, Walter H. Nuclear Power and Its Environmental Effects. LaGrange Park, Illinois: American Nuclear Society, 1980.

Goodwin, Irwin, ed. Energy and Environment: A Collision of Crises. Acton, Massachusetts: Publishing Sciences Group, 1974.

Greenburg, William. "Chewing It Up at 200 Tons a Bite: Strip Mining." Technology Review, 75 (February 1973), 46-55.

Hardt, Jerry. Harlan County Flood Report. Corbin, Kentucky: Appalachia-Science in the Public Interest, 1978.

Harte, John and El-Gasseir, Mohamed. "Energy and Water." Science, 199 (February 10, 1978), 623-634.

Holdren, John P.; Morris, Gregory; and Mintzer, Irving. "Environmental Aspects of Renewable Energy Sources." In Jack M. Hollander, et al., eds. Annual Review of Energy. Volume 5. Palo Alto, California: Annual Reviews, Inc., 1980, pp. 241-292.

International Atomic Energy Agency. Environmental Effects of Cooling Systems at Nuclear Power Plants. Vienna, Austria: International Atomic Energy Agency, 1975.

Jimeson, Robert M. and Spindt, Roderick S. Pollution Control and Energy Needs. Washington, D.C.: American Chemical Society, 1973.

Karam, R. A. and Morgan, Karl J., eds. Energy and the Environment: Cost Benefit Analysis. Elmsford, New York: Pergamon Press, 1976.

_____ , et al., eds. Environmental Impact of Nuclear Power Plants. Elmsford, New York: Pergamon Press, 1977.

Kaufman, Alvin. "Beauty and the Beast: The Siting Dilemma in New York State." Energy Policy, 1 (December 1973), 243-253.

Keating, William. Politics, Technology, and the Environment: Technology Assessment and Nuclear Energy. Stuart Bruchey, ed. New York: Arno Press, 1979.

Kelley, Donald R., ed. The Energy Crisis and the Environment: An International Perspective. New York: Praeger Publishers, 1977.

Kellogg, William W. "Is Mankind Warming the Earth?" Bulletin of the Atomic Scientists, 34 (February 1978), 10-19.

Keyes, Dale L. "Energy and Land Use: An Instrument of U.S. Conservation Policy?" Energy Policy, 4 (September 1976), 225-236.

Kolbash, Ronald Lee. A Study of Appalachia's Coal Mining Communities and Associated Environmental Problems. Ann Arbor, Michigan: University Microfilms International, 1978.

Kwee, S. L. and Mullender, J. S. R. Growing Against Ourselves - The Energy Environment Tangle. Lexington, Massachusetts: Lexington Books, 1972.

Landy, Marc Karnis. The Politics of Environmental Reform: Controlling Kentucky Strip Mining. Washington, D.C.: Resources for the Future; Baltimore, Maryland; Distributed by the Johns Hopkins University Press, 1976.

Lenzer, C.; Phipps, C.; Valleix, J.; and Surrey, J., eds. Energy and the Environment: Democratic Decision-Making. New York: Macmillan, 1978.

Lowe, Julian and Lewis, David. "Energy Usage and the Economics of Environmental Control." Energy Policy, 9 (March 1981), 25-31.

Miller, G. Tyler, Jr. Energy and Environment: The Four Energy Crises. 2nd ed. Belmont, California: Wadsworth Publishing Company, 1980.

Mountain Community Union. Land Use and Environmental Rights Committee. You Can't Put It Back: A West Virginia Guide to Strip Mine Opposition. Fairmont, West Virginia: Mountain Community Union, 1976.

Muntzing, Manning. "Siting and Environment: Essentials in an Effective Nuclear Siting Policy." Energy Policy, 4 (March 1976), 3-11.

Murphy, Earl F. Energy and Environmental Balance. Elmsford, New York: Pergamon Press, 1980.

Muschett, F. Douglas. Coal Development in Montana: Economic and Environmental Impacts. Ann Arbor, Michigan: Department of Geography, University of Michigan, 1977.

National Research Council. Energy and Climate: Outer Limits to Growth? Washington, D.C.: National Academy of Sciences, Geophysics Board, 1977.

National Research Council. Committee on Nuclear and Alternative Energy Systems. Risk and Impact Panel. Ecosystems Resource Group. Energy and the Fate of Ecosystems. Washington, D.C.: National Academy of Sciences. [In preparation]

_____ . Environmental Studies Board. Underground Disposal of Coal Mine Wastes. Washington, D.C.: National Academy of Sciences, 1975.

Navarro, Peter. "The 1977 Clean Air Act Amendments: Energy, Environmental, Economic, and Distributional Impacts." Public Policy, 29 (Spring 1981), 121-146.

Neyman, Jerzy. "Public Health Hazards from Electricity-Producing Plants." Science, 195 (February 25, 1977), 754-758.

Odum, Howard T. Environment, Power, Society. New York: Wiley Interscience, 1971.

Organization for Economic Cooperation and Development. Energy and the Environment: Methods to Analyze the Long-Term Relationship. Paris, France: Organization for Economic Cooperation and Development, 1974.

Pentreath, R. J. Nuclear Power, Man, and the Environment. New York: Crane-Russak Company, 1980.

Rall, David P., Director. Report of the Committee on Health and Environmental Effects of Increased Coal Utilization. Washington, D.C.: U.S. Department of Health, Education, and Welfare, December 27, 1977.

Ramsay, William. Unpaid Costs of Electrical Energy: Health and Environmental Impacts of Coal and Nuclear Power. Baltimore, Maryland: The Johns Hopkins University Press, 1979.

Randall, Alan, et al. Estimating Environmental Damages from Surface Mining of Coal in Appalachia: A Case Study. Cincinnati, Ohio: Office of Research and Development, U.S. Environmental Protection Agency, Janaury 1978.

Reitze, Arnold W., Jr. "Old King Coal and the Merry Rapists of Appalachia." Case Western Reserve Law Review, 22 (1971), 650-737.

Rowe, James E., ed. Coal Surface Mining: The Impacts of Reclamation. Boulder, Colorado: Westview Press, 1979.

Ruedisili, Lon C. and Firebaugh, Morris W., eds. Perspectives on Energy: Issues, Ideas, and Environmental Dilemmas. 2nd ed. New York: Oxford University Press, 1978.

Schlottman, Alan M. Environmental Regulation and the Allocation of Coal: A Regional Analysis. New York: Praeger Publishers, 1977.

Schlottman, Alan M. and Spore, Robert L. Economic Impacts of Surface Mine Reclamation. Knoxville, Tennessee: Appalachian Resources Project, University of Tennessee, 1976.

Schneider, S. H. and Dennett, R. G. "Climatic Barriers to Long-Term Energy Growth." Ambio, 4 (1975), 65-74.

Schurr, Sam H., ed. Energy, Economic Growth, and the Environment. Baltimore, Maryland: The Johns Hopkins University Press, 1972.

Seale, Robert L. and Sierka, Raymond eds. Energy Needs and the Environment. Tucson, Arizona: University of Arizona Press, 1973.

Spangler, Miller B. "Environmental and Social Issues of Site Choice for Nuclear Power Plants." Energy Policy, 2 (March 1974), 18-32.

Stacks, John F. Stripping. New York: Sierra Club Books, 1972.

Steinman, Michael. Energy and Environmental Issues: The Making and Implementation of Public Policy Issues. Lexington, Massachusetts: Lexington Books, 1979.

Stephenson, Lee; Krammer, Jackie; and Lahn, Dick. "Energy, Exploitation, and Public Lands: An Overview." Environmental Action, 8 (October 23, 1976), 4-7.

The Strip Mine Handbook. Washington, D.C.: Environmental Policy Institute and Center for Law and Social Policy, 1978.

Talbot, Albert R. Power Along the Hudson: The Storm King Case and the Birth of Environmentalism. New York: E. P. Dutton & Company, 1972.

Thames, John L., ed. Reclamation and Use of Disturbed Land in the Southwest. Tuscon, Arizona: University of Arizona Press, 1977.

Theodore, Louis, et al. Energy and the Environment: Interactions. Vol. I, Pts. A and B. Boca Raton, Florida: CRC Press, 1980.

Toole, K. Ross. The Rape of the Great Plains: Northwest America, Cattle, and Coal. Boston, Massachusetts: Little, Brown and Company, 1976.

Train, Russell E. "Energy Problems and Environmental Concern." Bulletin of the Atomic Scientists, 29 (November 1973), 43-47.

Tuve, George L. Energy, Environment, Population, and Food. New York: John Wiley & Sons, Inc., 1976.

United States. Congress. Senate. Committee on Interior and Insular Affairs. Coal Surface Mining and Reclamation. Washington, D.C.: U.S. Government Printing Office, 1973.

United States. Council on Environmental Quality. Coal Surface Mining and Reclamation: An Environmental and Economic Assessment of Alternatives. Washington, D.C.: U.S. Government Printing Office, 1973.

United States. Department of the Interior. Surface Mining and Our Environment: A Special Report to the Nation. Washington, D.C.: U.S. Government Printing Office, 1967.

Van Tassel, Alfred J. The Environmental Price of Energy. Lexington, Massachusetts: Lexington Books, 1975.

Vietor, Richard H. K. Environmental Politics and the Coal Coalition. College Station, Texas: Texas A&M University Press, 1980.

Wali, Mohan K., ed. Practices and Problems of Land Reclamation in Western North America. Grand Forks, North Dakota: University of North Dakota Press, 1975.

Weinberg, Alvin M. "Nuclear Energy and the Environment." Bulletin of the Atomic Scientists, 26 (June 1970), 69-74.

_____ and Hammond R. Philip. "Global Effects of Increased Use of Energy." Bulletin of the Atomic Scientists, 28 (March 1972), 5-8.

White, David C. "The Energy-Environment-Economy Triangle." Technology Review, 76 (December 1973), 10-19.

_____ . "Energy, the Economy, and the Environment." Technology Review, 74 (October/November 1971), 18-31.

Wiener, Daniel Philip. Reclaiming the West. New York: Inform, Inc., 1980.

Wilson, Richard and Jones, William. Energy, Ecology, and the Environment. New York: Academic Press, 1974.

Woodwell, George M. "The Carbon Dioxide Question." Scientific American, 38 (January 1978), 34-43.

_____ ; MacDonald, G. J.; Revelle, R.; and Keeling, C. D. The Carbon Dioxide Problem: Implications for Policy in the Management of Energy and Other Resources. [Report to the Council on Environmental Quality] Washington, D.C.: National Academy of Sciences, July 1979.

Wright, Robert A., ed. The Reclamation of Disturbed Arid Lands. Albuquerque, New Mexico: University of New Mexico Press, 1978.

Yanarella, Ernest J. and Yanarella, Ann-Marie, eds. Surface Mining and the Limits of Reclamation: What Is Being Reclaimed? What Is Not? [Proceedings of a Symposium, University of Kentucky, May 3, 1979] Lexington, Kentucky: Social Science/Technology Development Group, University of Kentucky, June 1979.

Young, Louis B. and Young, H. Peyton. "Pollution by Electrical Transmission." <u>Bulletin of the Atomic Scientists</u>, 30 (December 1974), 34-38.

5. Energy and Ethics

Abbate, Fred J. "Kilowatts and Morality: The New Criticism of Utility Decision-Making." Electric Perspectives, 1 (1978), 2-7.

Austin, Richard Cartwright. Spoil: A Moral Study of Strip Mining for Coal. New York: National Division, Board of Global Ministries, Methodist Church, 1973.

Berry, Wendell. "Strip Mining Morality: The Landscaping of Hell." Nation, 202 (January 24, 1977), 96-100.

Boffey, Philip M. "Plutonium: Its Morality Questioned by National Council of Churches." Science, 192, (April 23, 1976), 356-359.

Cesaretti, C. A., ed. The Prometheus Question: A Moral and Theological Perspective on the Energy Crisis. New York: Seabury Press, 1980.

Cose, Ellise, ed. Energy and Equity: Some Social Concerns. Washington, D.C.: Joint Center for Political Studies, 1979.

Fritsch, Albert J. and the Science Action Coalition. Environmental Ethics: Choices for Concerned Citizens. Garden City, New York: Doubleday & Company, 1980.

Gremillion, Joseph B. Food/Energy and the Major Faiths. Maryknoll, New York: Orbis Books, 1978.

Harnik, Peter. "The Ethics of Energy Production and Use: Debate Within the National Council of Churches." Bulletin of the Atomic Scientists, 35 (February 1979), 5-9.

Hessel, Dieter T., ed. Energy Ethics: A Christian Response. New York: Friendship Press, 1979.

Illich, Ivan D. Energy and Equity. London, England: Calder and Boyars, 1973.

Kannan, Narishmhan P. Energy, Economic Growth, and Equity in the U.S. New York: Praeger Publishers, 1979.

Lovins, Amory B. and Price, John. Non-Nuclear Futures: The Case for an Ethical Energy Strategy. San Francisco, California: Friends of the Earth, 1975.

McGinn, Robert E. "Energy and Ethics." Stanford Journal of International Studies, 9 (Spring 1974), 246-250.

Morrison, Denton. "Equity Impacts of Some Major Energy Alternatives." In Seymour Warkov, ed. Energy Policy in the United States: Social and Behavioral Dimensions. New York: Praeger Publishers, 1978, pp. 164-219.

Myers, Desaix. The Nuclear Power Debate: Moral, Economic, Technical, and Political Issues. New York: Praeger Publishers, 1977.

National Council of Churches. Energy and Ethics: The Ethical Implications of Energy Production and Use. New York: Energy Education Project, Division of Church and Society, National Council of Churches, March 1979.

_____. Committee of Inquiry. The Plutonium Economy: A Statement of Concern. New York: National Council of Churches, 1975.

Partridge, Ernest, ed. Responsibilities to Future Generations: Environmental Ethics. Buffalo, New York: Prometheus Books, 1981.

Reader, Mark, ed. Energy: The Human Dimension. [Research Paper No. 5] Tempe, Arizona: Center for Environmental Studies, Arizona State University, February 1977.

Routley, R. and Routley, V. "Nuclear Energy and Obligations to the Future." Inquiry, 21 (Summer 1978), 133-179.

Ryan, C. J. "The Choices in the Next Energy and Social Revolution." Technological Forecasting and Social Change, 16 (March 1980), 191-208.

Sayre, Kenneth. "Morality, Energy, and the Environment." Environmental Ethics, 3 (Spring 1981), 5-18.

_____, ed. Values in the Electric Power Industry. Notre Dame, Indiana: University of Notre Dame Press, 1977.

Shrader-Frechette, K. S. Nuclear Power and Public Policy: The Social and Ethical Problems of Fission Technology. Hingham, Massachusetts: Klewer Boston, 1980.

Thring, M. W. and Crooks, R. J., eds. Energy and Humanity. Forest Grove, Oregon: International Scholarly Books Services, 1974.

Part IV
Energy Future

It is a matter of some significance for the energy debate in its national and international dimensions that the years following the Arab oil embargo have witnessed the publication and proliferation of numerous energy blueprints. Designed as guides through a transition period from an energy economy based primarily upon depletable energy sources to one founded upon renewable energy sources, these scenarios hold forth the promise for certain nations individually – or for the international community collectively -- of taking command of their energy future and constructing it on a basis compatible with a set of social, political, economic, and ethical assumptions and values upon which a consensus has been reached. The appearance of these examples of energy futuristics has coincided with, and indeed been advanced by, the development and incorporation into energy analysis of a host of methodological techniques within the natural science and social science communities for projecting and evaluating alternative energy futures. Meanwhile, high-technologists and garage shop inventors have been busy exploring possible applications of scientific ideas for expanding the supply of energy or reducing its waste.

A wide-ranging sample of the steady stream of studies on energy and the future is featured in the first chapter of this part of the bibliography. These illustrations of energy futuristics differ in terms of the technological paths they advance (soft, hard, or mixed) as well as the degree of methodological rigor, the explicitness of the social image of the future, and the degree of sensitivity to issues of political economy informing the energy paths charted. Falling into the soft energy path category are those energy scenarios projected by Clark (1975), Commoner (1979), Kendall, et al. (1980), Steinhart (1979), and Stobaugh and Yergin (1979). Wilson Clark's book (1975) is in many ways the forerunner of the energy strategy outlined more explicitly and with greater conceptual refinement by Amory Lovins (1976). The distinctiveness of Commoner's work lies in his careful elaboration of a viable transition to a solar future -- one which is keenly aware of the salient matters relating to the political economy of such an energy transition. Least sensitive to questions of political strategy or to issues of political economy is the Union of Concerned Scientists' study directed by Henry Kendall. Its chief merit is its impressive demonstration of the technical feasibility of phasing in energy efficiency and solar technologies now or in the near future in order to hasten the institutionalization of solar/conservation options into our energy economy. Finally, the Harvard Energy Study, coordinated by Robert Stobaugh and Daniel Yergin, offers a valuable contribution to the popularizing of soft energy technologies,

99

despite the fact that it derides the social image of soft path advocates as "romantic" and merely assumes the compatibility of soft path technologies with the present corporate structure of our political economy.

Hard energy path scenarios or analyses are represented in this section by the works of Herman and Cannon (1977), Kash (1977), the National Academy of Engineering (1974), and U.S. ERDA (1976). Actually, the ERDA-76 plan is probably the only genuine energy scenario listed in this section, although the National Academy of Engineering study, due to its explicit technological focus, does exemplify some of the key characteristics of an energy futures study guided by hard path assumptions. Don Kash's study, on the other hand, is more concerned with analyzing various energy technologies in terms of technical feasibility and social and environmental impacts. The INFORM, Inc. study, led by Stewart Herman and James Cannon, sketches the energy futures being prepared for us through the energy technologies under development by the leading American corporations in their research laboratories and testing facilities. Two serious deficiencies in the corpus of writings on energy futures by the hard technology proponents are the absence of any real clarification of the contours of future society implied by the energy scenarios outlined and, consequently, the lack of overt attention to the ramifications of dependence on capital-intensive, high-technology energy projects upon the structure of our political economy. The obvious inference to be drawn from this dual silence is that the shape of society and political economy will remain unaltered; but, as soft path advocates have been vocal in arguing, such an inference is undoubtedly incorrect and in any case is subject to considerable question.

A third category of futuristic studies on energy presents more pluralistic or multi-plex energy futures based upon a mix of soft and hard energy technologies, often coupled with energy conservation measures. These studies (Landesberg, 1979; National Research Council, 1980; and Schurr, et al., 1980) are notable both for the respectability and expertise of their contributors and for the technical sophistication and statistical grounding of their analysis and conclusions. Required reading for any serious energy policy analyst, these works are directed to key participants in the policy-making arena. While offering policy recommendations for grappling with the long-term energy crisis, they eschew proposing any single path or technological fix for securing for America energy independence. Instead, they tend as a rule to acknowledge the irreducible social and political dimensions of the energy crisis and to articulate energy policies embracing a technological pluralism only somewhat undercut by certain latent hard energy path assumptions and/or political assumptions tied to interest-group liberalism. [For further comments on these works, see Part IX]

Less tied to the American context are a number of books which either examine possible energy futures of many nations or deal with energy prospects in a global frame of reference. Landesberg (1980), Lonnroth and Steen (1980), Lonnroth, et al. (1980), and W.A.E.S. (1977) all broaden the scope of analysis of energy futuristics to include consideration of the energy crisis in an international context. The companion volume to Landesberg's above-mentioned work is especially valuable for its speculative essays by noted specialists on energy prospects for Western Europe, the Soviet Union and China, Japan, the OPEC countries, and the non-OPEC developing nations. Lonnroth's two works highlight the energy transition generally and the different energy futures unfolding from the choice of the nuclear or solar option as the foundation of a national energy policy. Equally comprehensive is the study of global energy prospects which was carried out by the Workshop on Alternative Energy Strategies (W.A.E.S.) under the able direction of Carroll Wilson.

Beyond these contending, but useful, energy scenarios and studies, a host of other writings may be found in this section – including high-quality journalistic and scholarly surveys (Abelson, 1975; Hammond, et al., 1973; Hayes, 1977; Oak Ridge, 1977; and Rothman, 1970); valuable scholarly compilations of varying points of view (Bergman, et al., 1978; Diesendorf, 1980; and U.S. Congress, 1977); reviews and interpretive analysis of important scenarios for future U.S. energy use (Brooks and Hollander, 1979; Carlson, et al., 1981; and Just and Lave, 1979); philosophical commentaries on energy futuristics (Cook, 1980; Daly, 1979; Demand and Conservation Panel, 1978; and Leach, 1976); and, finally, source-specific projections (Casper, 1976; McDaniels, 1979; Seaborg, 1970; Spaid, 1975; Squires, 1974; Surrey, 1973a and b; and United Nations, 1978). In general, these writings both supplement and deepen our understanding of the promise and limits of energy futures studies.

The economic jolt of the Arab oil embargo to the national economies of virtually all non-OPEC societies and the accompanying recognition of the existence of at least a long-term energy crisis not only elevated energy to a high-level policy position for governmental decision-making; it also led to the proliferation of techniques of energy analysis and energy policy-making by scholars and practitioners who often borrowed methodological approaches and hardware from the fields of economics, game theory, and statistics, systems analysis, and operations research. Chief among the approaches and techniques appropriated by energy analysts were: energy modeling and forecasting, decision analysis, the Delphi technique, sensitivity analysis, operational gaming, and input-output analysis.

The books and essays listed in the section on energy analysis and modeling may be conveniently broken down into three areas: methodological works, applications, and assessments. The writings devoted to methodological considerations are clearly the most numerous, and the most important are those works which discuss energy analysis generally (Common, 1976; Gilliland, 1978; Maddala, et al., 1978; and Webb and Pearce, December 1975, June 1977); varieties of energy modeling (Energy Policy, 1974; Hartman, 1979; Hitch, 1977; Hoffman and Wood, 1976; Limaye, 1974; Ormerod, December 1980; Samouilidis, 1980; Searl, 1974; and U.S. Congress, 1976); forecasting (DeSouza, 1980; and Hoffman and Wood, 1976); gaming (Saaty, et al., May 1977); and the Delphi method (O'Toole, 1976). Some of these methods have been applied to specific energy problems (e.g., Carter, 1974; Hill and Walford, 1975; Keeney and Nair, 1977; Konno and Srinivasan, 1975; and Manne and Richels, 1980) and to national energy policy-making (Greenburg and Murphy, 1980; Tietenberg, 1976; U.S. F.E.A., 1977; and Weyant, 1980). Because of the relative infancy of energy analysis and the omnipresent danger of it serving to rationalize dominant political preferences, the critical perspective on energy modeling provided by Commoner (1979) and the assessments offered by Gass (1980), Greenberger and Richels (1979), and Koreisha (1980) are particularly valuable.

No form of energy analysis is more controversial within the scholarly community or within elements of the attentive public than risk analysis. For this reason, it has been singled out for special consideration in a separate section. For those unfamiliar with this specialty, risk analysis is a developing field of expertise growing out of the efforts of decision-makers interested in long-range planning or policy-making to deal with risk, uncertainty, and decision in a variety of spheres (see, generally, Rowe, 1977; Schwing, 1980; and Starr, et al., 1976). In the energy domain, it has served as a method for assessing the comparative risks of different sources of energy production by measuring these sources against a set of factors or criteria (Goodman and Rowe,

1979; National Academy of Sciences, 1974; and National Research Council, 1979). Perhaps the two most noted examples of this type of energy analysis are the comparative energy risk study by Dr. Herbert Inhaber (1978; 1979) and the WASH-1400 Nuclear Reactor Safety Study (U.S., A.E.C., August 1974), directed by Norman Rasmussen for the Atomic Energy Commission. The former study, which compared conventional and alternative energy sources on the basis of their relative dangers along certain dimensions, has been the object of broad-ranging and detailed criticism by environmentalists and soft-energy path proponents (Ehrlich and Ehrlich, 1979; Holdren, 1979); the validity of the latter AEC-commissioned study has been put into question by a government panel (Lewis, et al., 1978), as well as by other acknowledged experts (Green, 1976; and Lovins, 1977).

On the positive side, used sensitively and with humility, risk analysis studies compel citizens and policy-makers alike to realize that no energy source is without hazard or risk and that energy policy decisions may involve engaging in trade-offs with other policy areas or values which we ideally would like to maximize. On the negative side, risk analysis at best can offer us only the first word (if that), not the last word, on choosing energy options. For, this type of analysis is as prone as any other form of technical analysis to trying to quantify the unquantifiable or to smuggling interpretive or political judgments into calculations under the guise of technical estimates. Consequently, while risk analyses of nuclear reactors (Lewis, et al., 1978; MITRE Corporation, 1976; National Research Council, 1979; and U.S. A.E.C., 1974) and of alternative energy systems (National Research Council, in prep.) can provide some assistance in making political choices regarding energy alternatives, they can never overcome the irreducible and ineliminable political and ethical dimensions of public decision on energy alternatives and their varying risks (see, especially, Holdren, 1979; Lovins, 1977; Lowrence, 1976; Sorensen, 1979; and Weatherwax, September 1975). And, as Commoner (1970) has argued, while many citizens seem evidently prepared to accept a higher level of risk associated with individual, voluntary choices relating to travel, drinking alcoholic beverages, and sports activities, they are less and less willing to permit a similar level of risk when it flows from arbitrary or bureaucratic decision foisted upon them by governmental agencies or industries. The significance of this growing shift in public perception and attitude toward radioactivity, toxic chemicals, and environmental pollutants is that it destroys the technocratic foundations of the risk/benefit calculus that some scientists, bureaucrats, and corporate leaders have tried to sell to the American public in considering the relative value of nuclear power vis-a-vis other energy sources and the comparative costs and benefits of new petrochemical products.

The organization and direction of technological development in the energy realm in the United States, in other countries, and globally will greatly influence the design of the energy future which succeeding generations will inherit from the present. The organizational structure, funding levels, and program priorities in energy R&D (research and development) highlighted in the last chapter of this part are critical aspects of any nation's energy policies (Committee for Scientific and Technological Policy, 1975; Hammond, et al., 1973; and Oak Ridge, 1977). Historically, as some of these writings bear out, these dimensions of innovation in energy technologies in the United States have never been neutral or unbiased in their emphases; rather, the inequalities of a market dominated since the early twentieth century by major oil corporations and/or the consensus among elites over budgetary and program priorities structured into national energy policy have decisively skewed American energy R&D policy toward capital-intensive, high-technology energy projects (Comptroller General, 1978; Holloman and Grenon, 1975; Teece, 1977; Tilton, 1974; U.S. Congress, 1973; and U.S. Energy Study Group, 1965, 1978). As a consequence of these inequalities and biases, the organizational pursuit of some energy technologies (e.g., nuclear power, breeder reactors, fusion power) has involved highly-centralized, generously-funded undertakings (Dale, 1979; Schmandt, 1972; and Herman and Cannon, 1978), while other energy technologies (e.g., coal) have been greatly dispersed, meagerly supported programs (Hammond, 1976a, b, c, d). Still others, like solar technologies and applications, have been woefully neglected or actively suppressed within government and industry (Bobrow and Kudre, 1975; Hammond, 1971; Maize, 1977; and, especially, Reece, 1979) because these alternative avenues of technological development do not easily fit into the "technological breakthrough" strategy patterned after the model of technological in-novation characterizing the development of nuclear power (see, generally, Yanarella and Ihara, 1978). From this perspective, the problems of energy R&D reduplicate the unresolved problems and contested issues of energy policy generally, and this observa-tion implies that the questions relating to the organization, funding, and orientation of energy R&D will remain a focus of the politics of energy for some time to come.

1. Energy and the Future

Abelson, Philip H. Energy for Tomorrow. Seattle, Washington: University of Washington Press, 1975.

Bacher, Robert. "Nuclear Energy and Our Future." Bulletin of the Atomic Scientists, 33 (March 1977), 63-65.

Bergman, Elihu, et al., eds. American Energy Choices for the Year Two Thousand. Lexington, Massachusetts: Lexington Books, 1978.

Beyond Today's Energy Crisis. Austin, Texas: Lyndon B. Johnson School of Public Affairs, University of Texas at Austin, 1975.

Brooks, Harvey and Hollander, Jack M. "United States Energy Alternatives to 2010 and Beyond: The CONAES Study." In Jack M. Hollander, et al., eds. Annual Review of Energy. Volume 4. Palo Alto, California: Annual Reviews, Inc., 1979, 1-70.

Brown, Harrison. "Energy in Our Future." In Jack M. Hollander and Melvin K. Simmons, eds. Annual Review of Energy. Volume 1. Palo Alto, California: Annual Reviews, Inc., 1976, pp. 1-36.

Carlson, Richard C.; Harman, Willis W.; and Schwartz, Peter and Associates. Energy Futures, Human Values, and Lifestyles. Boulder, Colorado: Westview Press, 1981.

Casper, David A. "A Less Electric Future?" Energy Policy, 4 (September 1976), 191-211.

Clark, Wilson. Energy for Survival: An Alternative to Extinction. New York: Anchor/Doubleday, 1975.

Commoner, Barry. The Politics of Energy. New York: Random House, 1979.

Cook, Earl. "Charting Our Energy Future: Progress or Prudence?" The Futurist, 14 (April 1980), 64-69.

_____. "Energy for Millenium Three." Technology Review, 75 (December 1972), 16-23.

Daly, Herman E. "On Thinking About Future Energy Requirements." In Charles T. Unseld, et al., eds. Sociopolitical Effects of Energy Use and Policy. Washington, D.C.: National Academy of Sciences, 1979, pp. 229-241.

Demand and Conservation Panel of the Committee on Nuclear and Alternative Energy Systems. "U.S. Energy Demand: Some Low Energy Futures." Science, 200 (April 14, 1978), 142-152.

Diesendorf, Mark. Energy and People: Social Implications of Different Energy Futures. Forest Grove, Oregon: International Scholarly Book Service, 1980.

"The Energy Potential - More Than We Think." [Special Issue] Impact of Science on Society, 29 (October-December 1979).

Hammond Allen L.; Metz, William D.; and Maugh, Thomas H., II. Energy and the Future. Washington, D.C.: American Association for the Advancement of Science, 1973.

Harsany, Peter. Energy Tomorrow. New York: James H. Heineman, Publishers, 1980.

Hartnett, James P., ed. Alternative Energy Sources. New York: Academic Press, 1976.

Hayes, Denis. Rays of Hope: The Transition to a Post-Petroleum World. New York: Published for the Worldwatch Institute by W. W. Norton and Company, 1977,

Hayes, Earl T. "Energy Resources Available to the United States, 1985-2000." Science, 203 (January 19, 1979), 223-239.

Herman, Stewart W. and Cannon, James C., with INFORM, Inc. Energy Futures: Industry and the New Technologies. New York: INFORM, Inc., 1976; Cambridge, Massachusetts: Ballinger Publishing Company, 1977.

Hirst, Eric. "Residential Energy Use Alternatives: 1976 to 2000." Science, 194 (December 17, 1976), 1247-1252.

Just, James and Lave, Lester. "Review of Scenarios of Future U.S. Energy Use." In Jack M. Hollander, et al., eds. Annual Review of Energy. Volume 4. Palo Alto, California: Annual Reviews, Inc., 1979, 501-536.

Kash, Don E., et al. Our Energy Future: The Role of Research, Development, and Demonstration in Reaching a National Consensus on Energy Supply. Norman, Oklahoma: University of Oklahoma Press, 1977.

Kendall, Henry, et al. Energy Strategies: Toward a Solar Future. Cambridge, Massachusetts: Ballinger Publishing Company, 1980.

Kent, Peter, ed. Energy in the 1980's. London, England: The Royal Society, 1974.

Landsberg, Hans, ed. Energy: The Next Twenty Years. Cambridge, Massachusetts: Ballinger Publishing Company, 1979.

_____. Selected Studies on Energy: Background Papers for Energy: The Next Twenty Years. Cambridge, Massachusetts: Ballinger Publishing Company, 1980.

Leach, Gerald. "Energy Futures - Wide Open to Change and Choice." Ambio, 5 (1976), 108-116.

Little, Dennis L.; Pils, Robert E.; and Gray, John. Renewable Natural Resources: A Management Handbook for the Eighties. Boulder, Colorado: Westview Press, 1981.

Lonnroth, Mans and Steen, Peter. Energy in Transition: A Report on Energy Policy and Future Options. Berkeley, California: University of California Press, 1980.

_____, et al. Solar versus Nuclear: Choosing Energy Futures. Elmsford, New York: Pergamon Press, 1980.

McDaniels, D. K. The Sun: Our Future Energy Source. New York: John Wiley & Sons, 1979.

Marshall, Eliot. "Energy Forecasts: Sinking to New Lows." Science, 208 (June 20, 1980), 1353-1354, 1356.

Moore, T. G. "Energy Options." In P. Doignan and A. Rabushka, eds. United States in the 1980's. Stanford, California: Hoover Institution Press, 1980, pp. 221-252.

National Academy of Engineering. Task Force on Energy. U.S. Energy Prospects: An Engineering Viewpoint. Washington, D.C.: National Academy of Engineering, 1974.

National Academy of Sciences. Energy: Future Alternatives and Risks. [Academy Forum] Cambridge, Massachusetts: Ballinger Publishing Company, 1974.

National Research Council. Energy in Transition, 1095-2010: Final Report of the Committee on Nuclear and Alternative Energy Systems. San Francisco, California, W. H. Freeman Company, 1980.

Oak Ridge Associated Universities. Future Strategies for Energy Development: A Question of Scale. Oak Ridge, Tennessee: Oak Ridge Associated Universities, 1977.

O'Neill, Gerard K. "Space Colonies and Energy Supply to Earth." Science, 190 (December 5, 1975), 943-947.

Perelman, Lewis J.; Giebelhaus, August W.;and Yokell, Michael D. Energy Transitions: Long-Term Perspectives. Boulder, Colorado: Westview Press, 1980.

Rothman, Milton. Energy and the Future. New York: Franklin Watts, 1975.

Sant, Roger, Director. Eight Great Energy Myths: The Least-Cost Energy Strategy – 1978-2000. Arlington, Virginia: Mellon Institute, 1981.

Schurr, Sam H., et al., eds. Energy in America's Future: The Choices Before Us. Baltimore, Maryland: The Johns Hopkins University Press, 1980.

Seaborg, Glenn T. "Our Nuclear Future - 1995." Bulletin of the Atomic Scientists, 26 (June 1970), 7-14.

"Searching for Energy in the 1980's: Challenges and Opportunities for Business." [Special Issue] Journal of Contemporary Business, 9 (1980).

Spaid, Ora. "Forecast: Doubled Coal Production in Appalachia." Appalachia (June/ July 1975), 1-10.

Squires, Arthur M. "Coal: A Past and Future King." Ambio, 3 (1974), 1-14.

Steinhart, John, et al. Pathway to Sustainable Energy: The 2050 Study. San Francisco, California: Friends of the Earth, 1979.

Stobaugh, Robert and Yergin, Daniel, eds. Energy Future: Report of the Energy Project at the Harvard Business School. New York: Random House, 1979.

Surrey, A. J. "The Future Growth of Nuclear Power. Part I. Demand and Supply." Energy Policy, 1 (September 1973), 107-129.

_____ . "The Future Growth of Nuclear Power. Part II. Choices and Obstacles." Energy Policy, 1 (December 1973), 208-224.

Thomas, John A. G. "9th World Energy Conference: The Economic and Environmental Challenge of Future Energy Requirements." Energy Policy, 2 (December 1974), 330-339.

United Nations. Coal: 1985 and Beyond: A Prospective Study of the Coal Industry in Europe and North America. Elmsford, New York: Pergamon Press, 1978.

United States. Congress. Senate. Committee on Energy and Natural Resources. Energy: An Uncertain Future - An Analysis of U.S. and World Energy Projections Through 1990. [95th Cong., 2nd sess.] Washington, D.C.: U.S. Government Printing Office, December 1978.

United States. Congress. Senate. Select Committee on Small Business and the Committee on Interior and Insular Affairs. Alternative Long-Range Energy Strategies. 2 Volumes. Washington, D.C.: U.S. Government Printing Office, 1977.

United States. Department of Energy. Assistant Secretary for Policy and Evaluation. Deputy Assistant Secretary for Conservation and Renewable Resources. Low Energy Futures for the United States. [DOE/PE-0020] Washington, D.C.: U.S. Government Printing Office, June 1980.

United States. Department of the Interior. Bureau of Mines. United States Energy Through the Year 2000 (Revised). Washington, D.C.: U.S. Government Printing Office, 1975.

United States. Energy Research and Development Administration. A National Plan for Energy Research, Development, and Demonstration: Creating Energy Choices for the Future. Volumes I and II. Washington, D.C.: U.S. Government Printing Office, 1976.

Voegeli, Henry E. and Tarrant, John J. Survival Two Thousand One: Scenarios From the Future. New York: Van Nostrand Reinhold Company, 1975.

White, David C. "Energy Choices for the 1980's." Technology Review, 82 (August/September 1980), 30-41.

Wilson, Carroll, ed. Energy Demand to the Year 2000 and Energy Supply-Demand Integrations. Cambridge, Massachusetts: M.I.T. Press, 1977.

Wilson, Richard, ed. Energy for the Year Two Thousand. New York: Plenum Publishers, 1980.

Workshop on Alternative Energy Strategies [WAES]. Energy: Global Prospects, 1985-2000. New York: McGraw-Hill, 1977.

Yanarella, Ernest J. and Yanarella, Ann-Marie, eds. The Role of Coal in the Energy Picture to the Year 2000: Economic and Political Perspectives. [Proceedings of a Symposium, University of Kentucky, February 9, 1979] Lexington, Kentucky: Social Science/Technology Development Group, University of Kentucky, March 1979.

Young, Gordon. "Will Coal Be Tomorrow's 'Black Gold'?" National Geographic Magazine, 148 (August 1975), 234-259.

2. Energy Modeling

Carter, Anne P. "Applications of Input-Output Analysis to Energy Problems." Science, 184 (April 19, 1974), 325-329.

Common, Michael. "The Economics of Energy Analysis Reconsidered." Energy Policy, 4 (June 1976), 158-165.

Commoner, Barry. The Politics of Energy. New York: Alfred A. Knopf, 1979.

DeSouza, Glenn R. Energy Policy and Forecasting: Economic, Financial, and Technological Dimensions. Lexington, Massachusetts: Lexington Books, 1980.

Energy Policy, eds. Energy Modelling. Guildford, England: IPC Science and Technology Press, 1974.

Gass, S. I. Validation and Assessment Issues of Energy Models. [Sponsored by the National Bureau of Standards] Washington, D.C.: U.S. Government Printing Office, 1980.

Gilliland, Martha W., ed. Energy Analysis: A New Public Policy Tool. Boulder, Colorado: Westview Press, 1978.

Greenberg, H. J. and Murphy, F. H. "Modeling the National Energy Plan." The Journal of the Operational Research Society, 31 (November 1980), 965-974.

Greenberger, Martin and Richels, Richard. "Assessing Energy Policy Models: Current State and Future Directions." In Jack M. Hollander, et al., eds. Annual Review of Energy. Volume 4. Palo Alto, California: Annual Reviews, Inc., 1979, 467-500.

Hartman, Raymond S. "Frontiers in Energy Demand Modeling." In Jack M. Hollander, et al., eds. Annual Review of Energy. Volume 4. Palo Alto, California: Annual Reviews, Inc., 1979, 443-446.

Hill, K. M. and Walford, F. J. "Energy Analysis of a Power Generating System." Energy Policy, 3 (December 1975), 306-317.

Hitch, Charles J., ed. Modeling Energy-Economy Interactions: Five Approaches. Baltimore, Maryland: The Johns Hopkins University Press, 1977.

Hoffman, Kenneth C. and Wood, David O. "Energy System Modeling and Forecasting." In Jack M. Hollander and Melvin K. Simmons, eds. Annual Review of Energy. Volume 1. Palo Alto, California: Annual Reviews, Inc., 1976, pp. 423-453.

Hughes, Barry. Setups: Public Policy: U.S. Energy, Environment, and Economic Problems. Washington, D.C.: American Political Science Association, 1975.

Jorgenson, D. W. Economic Studies of U.S. Energy Policy. New York: Elsevier-North Holland Publishing Company, 1976.

Keeney, Ralph L. and Nair, Keshaven. "Nuclear Siting Using Decision Analysis." Energy Policy, 5 (September 1977), 223-231.

Konno, Hiroshi and Srinivasan, T. N. "Nuclear Reactor Strategies: Sensitivity Analysis of the Hafele-Manne Model." Energy Policy, 3 (September 1975), 211-222.

Koreisha, S. "The Limitations of Energy Policy Models." Energy Economics, 2 (April 1980), 96-110.

Lawrence Livermore Laboratory. An Assessment of U.S. Energy Options for Project Independence. [UCRL-51638] Springfield, Virginia: National Technical Information Service, U.S. Department of Commerce, 1974.

Leach, Gerald. "Net Energy Analysis - Is It Any Use?" Energy Policy, 3 (December 1975), 332-344.

Limaye, Dilip R. Energy Policy Evaluation: Modeling and Simulation Approaches. Lexington, Massachusetts: Lexington Books, 1974.

_____ and Sharko, John R. "U.S. Energy Policy Evaluation: Some Analytical Approaches." Energy Policy, 2 (March 1974), 3-17.

Maddala, G. S., et. al. Economic Studies in Energy Demand and Supply. New York: Praeger Publishers, 1978.

Manne, Alan S. and Richels, Richard G. "Evaluating Nuclear Fuel Cycles: Decision Analysis and Probability Assessments." Energy Policy, 8 (March 1980), 3-16.

Ormerod, R. J. "Energy Models for Decision-Making." European Journal of Operational Research, 5 (December 1980), 366-377.

O'Toole, James. Energy and Social Change. Cambridge, Massachusetts: M.I.T. Press, 1976.

Saaty, Thomas; Ma, Fred; and Blair, Peter. "Operational Gaming for Energy Policy Analysis." Energy Policy, 5 (March 1977), 63-75.

Samouilidis, J. E. "Energy Modeling - A New Challenge for Management Science." Omega, 8 (1980), 609-622.

Searl, Milton F., ed. Energy Modeling: Art, Science, Practice. Baltimore, Maryland: The Johns Hopkins University Press, 1974.

Tietenberg, Thomas H. Energy Planning and Policy: The Political Economy of Project Independence. Lexington, Massachusetts: Lexington Books, 1976.

United States. Congress. House of Representatives. Committeee on Interstate and Foreign Commerce. "Energy Demand Studies - An Analysis and Comparison." Middle and Long-Term Energy Policies and Alternatives. Washington, D.C.: U.S. Government Printing Office, 1976.

United States. Federal Energy Administration. Office of Energy Information and Analysis. Project Independence Evaluation System (PIES) Documentation: Volume XIV, A User's Guide. Springfield, Virginia: National Technical Information Service, June 1977.

Waverman, Leonard. "Estimating the Demand for Energy: Heat Without Light." Energy Policy, 5 (March 1977), 2-11.

Webb, Michael and Pearce, David. "The Economics of Energy Analysis." Energy Policy, 3 (December 1975), 318-331.

_____. "The Economics of Energy Analysis Revisited." Energy Policy, 5 (June 1977), 158-159.

Weyant, John P. "Quantitative Models in Energy Policy." Policy Analysis, 6 (Spring 1980), 211-234.

Ziemba, W. T. and Schwartz, S. L. Energy Policy Modeling. Volume 2: United States and Canadian Experiences. The Hague, The Netherlands: Martins Nijhoff, 1980.

3. Risk Analysis

Bazelon, D. L. "Risk and Responsibility." Science, 205 (July 20, 1979), 277-280.

Commoner, Barry. "Nuclear Power: Benefits and Risks." In Harry Foreman, ed. Nuclear Power and the Public. Minneapolis, Minnesota: University of Minnesota Press, 1970, pp. 224-239.

Ehrlich, Anne and Ehrlich, Paul. "The 'Inhaber Report', Parts I and II." The Mother Earth News, July/August and September/October, 1979, 115-116 and 115-116, respectively.

Gehrs, C. W., et al. "Environmental Health and Safety Implications of Increased Coal Utilization." In Elliott, M. A., ed. Chemistry of Coal Utilization, Supplement, Volume 2. New York: Wiley Interscience [In press].

Goodman, G. T. and Rowe, W. D., eds. Energy Risk Management. London, England: Academic Press, 1979.

Green, Harold P. "The Risk-Benefit Calculus in Safety Determinations." George Washington Law Review, 43 (1975), 791-803.

Holdren, J. P.; Smith, K. R.; and Morris, G. "Energy: Calculating the Risks (II)." Science, 204 (May 11, 1979), 564-568.

_____, et al. Risk of Renewable Energy Sources: A Critique of the Inhaber Report. [Energy and Resources Group] Berkeley, California: University of California, June 1979.

Inhaber, Herbert. Risk of Energy Production. Ottowa, Ontario: Atomic Energy Control Board, 1978.

_____. "Risk with Energy from Conventional and Nonconventional Sources." Science, 203 (February 23, 1979), 718-723.

Keeney, Ralph L.; Kulkarni, Ram B.; and Nau, Keshaven. "Assessing the Risk of an LNG Terminal." Technology Review, 81 (October 1978), 64-72.

Lewis, H. W., et al. Risk Assessment Review Group Report to the U.S. Nuclear Regulatory Commission. [NUREG/CR-0400] Washington, D.C.: U.S. Nuclear Regulatory Commission, 1978.

Lovins, Amory B. "Cost-Risk-Benefit Assessments in Energy Policy." George Washington Law Review, 45 (August 1977), 911-943.

Lowrance, William W. Of Acceptable Risk: Science and the Determination of Safety. Los Altos, California: William Kaufmann, 1976.

MITRE Corporation. Metreck Division. Accidents and Unscheduled Events Associated with Non-Nuclear Energy Resources and Technology. [M76-68] McLean, Virginia: MITRE Corporation, December 1976.

National Academy of Sciences. Energy: Future Alternatives and Risks. [Academy Forum] Cambridge, Massachusetts: Ballinger Publishing Company, 1974.

National Research Council. "Risks of Energy Systems." In Energy in Transition 1985-2010: Final Report of the Committee on Nuclear and Alternative Energy Systems. San Francisco, California: W. H. Freeman and Company, 1979, 422-499.

_____. Committee on Nuclear and Alternative Energy Systems. Risk and Impact Panel. Risks and Impacts of Alternative Energy Systems. Washington, D.C.: National Academy of Sciences. [In preparation]

_____. Committee on Science and Public Policy. Committee on Literature Survey of Risks Associated with Nuclear Power. Risks Associated with Nuclear Power: A Critical Review of the Literature. Washington, D.C.: National Academy of Sciences, 1979.

Rowe, William D. An Anatomy of Risk. New York: John Wiley & Sons, 1977.

Schwing, Richard and Albers, Walter A., Jr., eds. Societal Risk Assessment: How Safe Is Safe Enough? New York: Plenum Press, 1980.

Sorensen, Bent. "Nuclear Power: The Answer that Became a Question: An Assessment of Accident Risks." Ambio, 8 (1979), 10-17.

Starr, Chauncey; Rudman, Richard; and Whipple, Chris. "Philosophical Basis for Risk Analysis." In Jack M. Hollander and Melvin K. Simmons, eds. Annual Review of Energy. Volume 1. Palo Alto, California: Annual Reviews, Inc., 1976, pp. 629-662.

United States. Atomic Energy Commission. Reactor Safety Study: An Assessment of Accident Risks in U.S. Commerical Nuclear Power Plants. [WASH-1400] Norman Rasmussen, Chairman. Washington, D.C.: Atomic Energy Commission, August 1974 [Draft] ; Nuclear Regulatory Commission, October 1975 [Final].

United States. Nuclear Regulatory Commission. Overview of the Reactor Safety Study Consequence Model. [NUREG-0340] Washington, D.C.: U.S. Nuclear Regulatory Commission, 1977.

Weatherwax, Robert K. "Virtues and Limitations of Risk Analysis." Bulletin of the Atomic Scientists, 31 (September 1975), 29-32.

4. Energy R&D

Auer, Peter L. "An Integrated National Energy Research and Development Program." Science, 184 (April 19, 1974), 295-301.

Bobrow, Davis B. and Kudre, Robert T. "Energy R&D: In Tepid Pursuit of Collective Goods." International Organization, 33 (Spring 1979), 149-175.

Boffey, Philip M. "Energy Research: A Harsh Critique Says Federal Effort May Backfire." Science, 190 (November 7, 1975), 535-537.

Chen, K. "Optimizing Budgetary Allocation to Energy R&D Programs." Policy Studies Journal, 8 (1980), 1060-1069.

Committee for Scientific and Technological Policy. Energy R&D: Problems and Perspectives. Washington, D.C.: Organization for Economic Cooperation and Development, 1975.

Comptroller General. The Multiprogram Laboratories: A National Resource for Non-nuclear Energy Research, Development, and Demonstration. Washington, D.C.: General Accounting Office, May 22, 1978.

Dale, Alfred. Nuclear Power Development in the U.S. to Nineteen Sixty: A New Pattern in Innovation and Technological Change. Stuart Bruchey, ed. New York: Arno Press, 1979.

DeWinter, F. Description of the Solar Energy R&D Programs of Many Nations. J. W. deWinter, ed. San Mateo, California: Solar Energy Information Services, 1979.

Hammond, Allen L. "Coal Research I: Is the Program Moving Ahead?" Science, 193 (August 20, 1976), 665-667, 704.

_____. "Coal Research II: Gasification Faces an Uncertain Future." Science, 193 (August 27, 1976), 750-753.

Hammond, Allen L."Coal Research III: Liquefaction Has Far to Go." Science, 193 (September 3, 1976), 873-875.

_____ . Coal Research IV: Direct Combustion Lags Its Potential." Science, 194 (October 8, 1976), 172-173, 218-220.

_____ . "Federal R&D: Domestic Problems Get New Efforts But Little Money." Science, 171 (February 19, 1971), 657-661.

_____ ; Metz, William D.; and Maugh, Thomas H., II. Energy and the Future. Washington, D.C.: American Association for the Advancement of Science, 1973.

Herman, Stewart W. and Cannon, James S. Energy Futures: Industry and the New Technologies. Cambridge, Massachusetts: Ballinger Publishing Company, 1978.

Holloman, J. Herbert and Grenon, Michele. U.S. Energy Research and Development Policy. Cambridge, Massachussetts: Ballinger Publishing Company, 1975.

An Inventory of Energy Research. 2 Volumes. Washington, D.C.: U.S. Government Printing Office, 1972.

Kash, Don E., et al. Our Energy Future: The Role of Research, Development, and Demonstration in Reaching a National Consensus on Energy Supply. Norman, Oklahoma: University of Oklahoma Press, 1977.

Kenward, Michael. Potential Energy. Cambridge, England: Cambridge University Press, 1976.

Maize, Kennedy P. "Government R&D Programs: A Look at the Sunny Side." Environmental Action, 8 (March 12, 1977), 4-8.

Mansfield, E. "Firm Size and Technological Change in the Petroleum and Bituminous Coal Industries." In Thomas D. Duchesneau, ed. Competition in the U.S. Energy Industry. Cambridge, Massachusetts: Ballinger Publishing Company, 1975, pp. 317-345.

Nichols, Rodney W. "Mission-Oriented R&D." Science, 172 (April 2, 1971), 29-37.

Oak Ridge Associated Universities. Future Strategies for Energy Development: A Question of Scale. Oak Ridge, Tennessee: Oak Ridge Associated Universities, 1977.

Reece, Ray. The Sun Betrayed. A Study of the Corporate Seizure of Solar Energy Development. Boston, Massachusetts: South End Press, 1979.

Schmandt, Jurgen. One Aspect of the Energy Crisis: The Unbalanced State of Energy R&D. [Research Report No. 1] Austin, Texas: Occasional Papers, Lyndon B. Johnson School of Public Affairs, 1972.

Teece, David J., ed. R&D in Energy: Implications of Petroleum Industry Reorganization. Stanford, California: Institute for Energy Studies, Stanford University, 1977.

Tilton, John E. U.S. Energy R&D Policy: The Role of Economics. Washington, D.C.: Resources for the Future, 1974.

United States. Congress. Senate. Energy Research and Development: Problems and Prospects. [Prepared at the request of Henry M. Jackson, Chairman, Committee on Interior and Insular Affairs; Serial No. 93-21 (92-56)] Washington, D.C.: U.S. Government Printing Office, 1973.

United States. Energy Research and Development Administration. A National Plan for Energy Research, Development, and Demonstration: Creating Energy Choices for the Future. Volumes I and II. Washington, D.C.: U.S. Government Printing Office, 1976.

United States Energy Study Group. Energy R&D and National Progress. Ali B. Cambel, ed. Westport, Connecticut: Greenwood Press, 1965, 1978.

Walsh, John. "Electric Power Research Institute: A New Formula for Industry R&D." Science, 182 (October 19, 1973), 263-265.

Yanarella, Ernest J. and Ihara, Randal H. "The Military/Energy Connection: The Institutionalization of the 'Technological Breakthrough' Approach to Energy R&D." Northeast Peace Science Review, 1 (1978), 187-207.

Part V
Conventional and Swing Fuels

The emerging age of ecological scarcity poses a number of serious problems and dilemmas in the realm of energy for all nations of the world community. Insofar as the long-term energy crisis is a genuine global problem and not simply a contrivance of powerful political forces in the international political economy, the remaining reserves of petroleum, natural gas, and coal will have to be shepherded so as to stretch out their availability for use as fuels (and in some cases as valuable raw materials for drugs and synthetics) in the transition to an energy future based upon alternative energy sources. Less widely recognized is the fact that, in the absence of the full-scale development and exploitation of breeder reactor technology or of the closing off of the nuclear option, the world's supplies of uranium, too, will have to be treated as a depletable and a diminishing energy resource. Moreover, in virtually all energy scenarios or blueprints for getting from here (the era of resource depletion) to there (the new era of alternative energy sources), one or a combination of exhaustible conventional fuels must serve a major role in providing a bridge to the future. As in any transition period, the problems presented to its inhabitants often appear intractable, involve painful adjustments and undesired costs, and test the ingenuity of human beings to break from past patterns and antiquated solutions and to create novel solutions to fit new and unfamiliar circumstances.

The opening chapter to Part V of this bibliography contains references to works on oil and natural gas – references which can be categorized into four distinguishable fuel sources: petroleum, oil shale, and natural gas in its gaseous and liquid forms. Besides offering supplemental histories of petroleum development (Fanning, 1947, 1948; Forber, 1975a and b; Stoff, 1980), those works devoted to crude oil clarify a number of key issues and themes which profile oil's role and problems in the transition from the petroleum to the post-petroleum era. Many of these writings address the difficult problem of estimating the extent of the nation's and the world's oil reserves still available for exploitation in a technologically and economically feasible manner. This concern has produced a wide range of estimates of the remaining global oil reserves (Dam, 1976; Eckbo, 1976; Moody and Geiger, 1975; and Odell, 1973), strategies for exploring for new sources (Erickson, 1970; Menard, 1981), and investigations of the finite limits to oil exploration (Clark, 1973; and Cook, 1976).

Available domestic supplies have been the focus of the research of a number of analysts and writers, including Berg, et al. (1974), Frank and Schanz (1973), Gillette (1974, 1975), and Renshaw and Renshaw (1980). More delimited studies and journal reports on Alaskan oil (Cichetti, 1976; Cooper, 1973; Hwa and Mancke, 1979; Jensen and Ellis, 1967; Rice, 1973; and Tussing, 1971) and off-shore oil sources (Devanney, 1974; Emery, 1975, 1976; Kash, et al., 1974; Kerr, 1979; Mitchell, 1976) are also incorporated into this section. As oil prices continue to rise, investigations into the methods and economic feasibility of secondary and tertiary recovery of oil from old oil fields, such as the one by DeNevers (1965), will undoubtedly become more numerous. The continued dependence of the United States upon foreign oil for supplying a significant portion of its petroleum needs has spurred policy analysts and energy economists to review its oil import policy and to seek ways of reducing the level of such imports (Kuenne, et al., 1974; Newlon and Breckner, 1975; Russell, 1978; Tussing, 1974; and Yergin, 1980).

The post-embargo record, too, has been a subject for scholarly research and investigation. This interest in post-embargo developments has been expressed in general studies of the oil crisis and means of coping with it (Menderhausen, 1976; and Vernon, 1976), analyses of the consequences of higher prices for the world economy (Fried, 1975; and Vallenilla, 1975), and critical investigations of evidence of exploitation by multinational oil corporations of the conditions following the Arab embargo (Allvine, 1974; Aronowitz, 1974; and Fixler and Farrar, 1975). In the political realm, diverse issues of government oil policy – including regulation, depletion allowances, decontrol, and allocation – are probed (Berner and Scoggins, 1973; McDonald, 1976; Renshaw, 1980; and Willrich, 1977).

The potential for extracting oil from oil shale deposits in the United States at competitive market prices has been treated generally by several students and observers of energy developments (Bell, 1948; DeNevers, 1966; Dineen and Cook, 1974; Maugh, 1977; Metz, 1974; Thorne, et al., 1964; and Yen, 1976). In addition, important social and political facets of this technology are given due regard by energy analysts who highlight federal oil shale policy (Moffett, 1976; and U.S., Congress, O.T.A., 1980), explore its economic promise (Stevens and Kalter, 1975), and provide technology assessments of its broad ranging impact (Rees, 1974; and U.S., Congress, O.T.A., 1980). As for natural gas, various writers and analysts take up a number of issues relating to its low risk, high quality nature (Oppenheimer, 1981; and Wilson, 1973), its economics (MacAvoy and Pindyck, 1975), its politics (Nivola, 1980; U.S., Congress, Senate, 1972; and Willrich, 1977), and its industrial development (Peebles, 1980; and Stephenson, 1974). Other authors and researchers (DeNevers, 1967; Drake and Reid, 1977; and Wolf, 1978) illuminate the peculiar benefits and the potentially catastrophic risks of liquid natural gas.

Coal – the focus of the next chapter – remains nature's most ambiguous energy resource. On the one hand, it is available in plentiful supplies in a number of countries, including and especially in the United States, where reserves lasting up to 200 or more years are estimated to exist. Its plentitude makes it a possible bridge fuel to an energy future beyond fossil fuels, while new and existing technological processes for converting coal into liquid and gaseous fuels offer the possibility of coal becoming an alternative energy source. For the United States, its abundant reserves hold out the hope of making America the OPEC of the eighties and nineties.

On the other hand, strip mining operations in America's West scar the natural landscape, while human efforts to reclaim the land oftentimes involve sizeable costs and produce merely cosmetic results. Deep mining procedures in the East cause risks to miners' safety and dangers to their long-term health. Then, too, the combustion of coal for industrial production and electric generation spews an assortment of toxic chemicals and even radioactive substances into the atmosphere, endangering public health and causing environmental problems over wide areas both within and across national boundaries. Nor can one view with equanimity the problems of community life associated with absentee ownership, corporate power, and turbulent business-labor relations. Small wonder that federal policy towards coal vacillates in its commitment to the "rock that burns" as the swing fuel in the energy transition.

The chapter on coal provides many references to works covering the full panoply of issues concerning the promises and perils of this energy source. In the introductory section, various facets of the technology of surface and underground mining are explored by several engineers and social scientists (Berkowitz, 1979; Doyle, 1976; Hawley, 1977; Meyers, 1977; Patterson and Griffin, 1978; and Tetra Tech, 1976). More germane to the centrality of coal production and use for national and even international energy policy is the profusion of writings on the appropriateness of coal as a bridge fuel in the interim period between the age of fossil fuels and the age of alternative energy sources (Business Communications Company Staff, 1976; Comptroller General, 1977; Coombs, 1980; Energy Modeling Forum, 1978; Ezra, 1978; Gordon, 1976b; Greene and Gallagher, 1980; Horwitch, 1979; Meadows and Stanley-Miller, 1975; National Research Council, 1977; Nephew, 1975; Simeons, 1978; Squires, 1974; Wilson, 1980; Yanarella and Yanarella, 1979; and Young, 1975). Whether or not coal can fulfill such a bridging role will depend in part upon the character and tractability of existing constraints to accelerating coal production and utilization, which are probed by other energy policy analysts in this section (Bhutani, et al., 1975; Gordon, 1978b; Hibbard, 1979; Horwitch, 1979; and Monsour, 1976). Similarly important to any reasoned public decision on the role of coal in the energy picture to the year 2000 and beyond is a concern of other technical experts and social scientists – namely, the projected social and environmental impacts of increased coal development (Hitman Associates, 1976; Hultman and Davis, 1977; Krutilla, et al., 1978; Rall, 1977; and U.S., Congress, O.T.A., 1978).

The political context of coal development will no doubt have to undergo substantial change if a public commitment to an equitable, non-exploitative, and environmentally-benign coal policy for the last two decades of the twentieth century is to be realized. Valuable references to studies of the coal industry have already been offered and reviewed in an earlier chapter (Part III, Chapter 2), and are supplemented here by the Noyes Data Corporation's study (1978) of ownership patterns in the industry. In this section, important investigations of coal miners and the United Mine Workers of America are delineated (Coleman, 1969; Densmore, 1977; Dubofsky and VanTine, 1977; Finley, 1972; and Seltzer, 1978). So, too, are examinations of industry-labor relations in the coalfields and at the negotiation table, as well as reports on emergent areas of agreement regarding future coal development (Alexander, 1978; Lane 1977; and Murray, 1978). Any reconsideration of the role of coal in the future will have to come to grips with the issue of the function of state and federal policies in this area, including their responsibilities in the realm of regulation (Binder, 1973; Greenbaum and Harley, 1974; Menzel and Edgman, 1980; Nehring and Zycher, 1976; Rosenbaum, 1978; Schlottman, 1977; and Yanarella and Yanarella, 1979b).

The human costs to miners in the past and those anticipated in the future should not be slighted in any genuine policy debate regarding the future of coal. The next section on coal mine health and safety brings together citations to works examining many subjects in this area. A glimpse at the historical record on coal mine health and safety is offered by McAteer (1973) and by Graebner (1976), and also by an older study by the U.S. Department of the Interior's Coal Mines Administration (1947), though these should be supplemented with the Office of Technology Assessment's study of the direct use of coal (1978) cited in the preceding section. Two recent studies (Gehrs, et al., 1980; and Rall, 1977) contribute estimates of the likely health and safety implications of increased coal utilization. Current federal regulations, moreover, are explored in several works (Kramer and Clague, 1973; Murray, 1973; and U.S., Department of the Interior, Bureau of Mines, 1977) and proposed reforms are suggested in one other (Kent, 1973). Meanwhile, two economists have individually scrutinized the economic impact of health damage to coal miners (Cameron, 1976) and the economic consequences of government regulation upon productivity in the industry (Henderson, 1980).

Despite flagging political enthusiasm for synthetic fuels development in the last year or two, this prospective alternative energy source derived from coal remains a part of national energy plans for meeting America's energy needs in future decades. The history of its development, from its role in Nazi Germany's war economy to its ambiguous status in America's energy policies today, is detailed in a number of works listed in this section (Borkin, 1978; Horwitch, 1978; Swabb, 1978; U.S., A.E.C., 1974; and Vietor, 1978, 1980). The technology of synfuels production involves two separate processes and two distinctive fuels: coal gasification (Cochran, 1976; Hammond, 1976a; Hammond and Ogden, 1975; Hederman, 1978; Lamb, 1977; Massey, 1974; Mills, 1971; Perry, 1974; Schora, 1976; and Squires, 1974) and coal liquefaction (Cochran, 1976; Ellington, 1977; Hammond, 1976a; Nowacki, 1979; and Swabb, 1978). The economics of these complex and exceedingly costly technological processes have been the focus of several studies (Hammond and Ogden, 1975; and Pelofsky, 1977), while detailed assessments of their promise and potential impacts have been offered by others (A.M.E. Technology, 1976; Comptroller General, 1976; Dow, 1981; Hederman, 1978; Marshall, 1980; National Research Council, 1977a and b; Rosenbaum, 1980; Synfuels Interagency Task Force, 1975; U.S., D.O.E., 1979; U.S., E.R.D.A., 1975; and U.S., Library of Congress, 1979). Finally, programs for synfuels development at varying levels of capital investment and public commitment have been advanced (Committee for Economic Development, 1979; Comptroller General, 1977; Exxon, 1980; U.S., E.R.D.A., 1976; and U.S., Library of Congress, 1980).

There was a time when civilian nuclear power was touted as an unalloyed blessing promising us "electricity too cheap to meter" and an environment free of the pollutants and contaminants associated with other energy sources. Presidents spoke stirring words to rally American citizens behind the Nuclear Dream; scientists and engineers designed bold and imaginative plans for realizing the dream of cheap nuclear power; congressmen promoted these projects and funded them with astronomical budgets; and citizens awaited the opening up of the nuclear cornucopia. Yet, by the end of the decade of the seventies, the Nuclear Dream faded as virtually every phase of the nuclear fuel cycle was challenged by significant criticism and political protest. And without a new injection of public enthusiasm and a new dose of federal support, the condition of the nuclear industry appeared likely to go from critical to terminal.

The bibliographical references in the succeeding chapter on nuclear power shed considerable light on the rise and apparent demise of civilian atomic power as an energy source. A sizeable body of literature in this chapter deals with the history of civilian nuclear power development in the United States and, to a lesser extent, in other countries (Allen, 1977; Bupp and Derian, 1978; Burn, 1978; Dahl and Brown, 1951;

Dale, 1979; Dawson, 1976; DeLeon, 1980; Del Sesto, 1979; Fairchild and Landman, 1961; Gandara, 1977; Hewlett and Duncan, 1974; Hodgetts, 1964; Mullenbach, 1963; Pendergrass, 1979; Perry, et al., 1977; Rolph, 1977; Temples, 1980; Thomas, 1956; Wilson, 1979; and Yanarella, 1981). At least since the beginning of the 1970s, a substantial part of the history of civilian nuclear power has been caught up in a rising tide of opposition to a nuclear future, and the scope and character of the nuclear power debate have been discussed in several penetrating surveys and interpretations of this public controversy (American Assembly, 1976; Fenn, 1981; Foley and Van Buren, 1978; Foreman, 1970; Inglis, 1973; Kadiroglu, et al., 1978; Kaiser, 1978; Kleitman, et al., 1979; Lapp, 1977; Murphy, 1976; Myers, 1977; Nelkin, 1971; Pringle, 1979; Schmidt and Bodansky, 1976; Weaver, 1979; Weinberg, 1970, 1977, 1979a and b). In this context of growing uncertainty about the merits of the nuclear option, supporters have felt compelled to restate and reassert the case for fission power development (Beckman, 1976; Bethe, 1976, 1977; Cohen, 1974a and b; Eklund, 1977; Hafele and Manne, 1975; Hoyle, 1977; Ikle, 1976; Lapp, 1977; Seaborg, 1971; and Srouji, 1977). Controversy over the dominant design for nuclear power production -- the light water reactor -- has prompted some nuclear power advocates and agnostics to reflect on the wisdom of the choice of this reactor design (Bupp and Derian, 1978; Lilienthal, 1980; and Perry, et al., 1977) and led others to look to alternative fission reactor designs (Agnew, 1981; Hafemeister, 1978/79; McIntyre, 1975; Manne and Richels, 1980; Nye, 1977; Perlmutter and Kadiroglu, 1978; Rogers, 1971; and Rush, et al., 1977).

The changing economic picture for atomic power development is surveyed in a number of works spanning the period from 1950 to the virtual present (Barnett, 1979; Bupp, et al., 1975; Burn, 1967; Comptroller General, 1979a; Miller, 1976; Mullenbach, 1973; Rossin and Rieck, 1978; Schurr and Marshack, 1950; Sherfield, 1972; and Shurcliff, 1977). Policy analysts will find wide-ranging reviews of the various public policy dimensions of civilian nuclear power -- including nuclear fuels policy (Ahmed, 1979; and Atlantic Council, 1976), environmental considerations (Eicholz, 1976; Karam, et al., 1977; Keating, 1979; and Sagan, 1974), nuclear regulation (Muntzing, 1980; Phillips, 1979; and Rolph, 1977), licensing and siting policies and issues (Green, 1974; Green and Rosenthal, 1963; Klema and West, 1977; Muntzing, 1976; Nelkin, 1974; and Spangler, 1974), public subsidies (Comptroller General, 1979a), decommissioning (Sefcik, 1979), and law and civil liberties (Bloustein, 1964; and Solomon, 1980). In a context of policy stalemate and industry stagnation, several works consider the future of civilian nuclear power in the United States (Bacher, 1977; Comptroller General, 1979b; and Lanouette, 1979).

The gravamen of civilian nuclear power is probably the issue of nuclear reactor safety. The single event which played the greatest role in mobilizing growing fears and evoking new anxieties among the public over reactor safety was undoubtedly the accident at Three Mile Island, near Harrisburg, Pennsylvania. As a powerful symbol, TMI has come to serve as a benchmark for dating phases and aspects of the reactor safety debate. Because of the confused nature and uncertain scope of the accident, the Three Mile Island incident has spawned considerable analysis and interpretation (Kleitman, et al., 1980; Martin, 1980; The Need for Change, 1979; Nuclear Reactor Analysis Center, 1979; Public's Right to Information Staff Report, 1979; Reid, 1979; Rogovin and Frampton, 1979; and Yanarella and Yanarella, 1980), issued in a host of lessons and prescriptions (Fischer, 1981; Kemeny, 1980; McCracken, 1979, 1981; and Peterson, 1980), and prompted varying governmental reforms and industry responses (Kasperson, et al., 1979; and Sylves, 1980).

The history of dispute over nuclear reactor safety goes back to the challenges by scientists within the regulatory process and outside of the bureaucracy (Del Sesto, 1980; Gillette, 1971, 1972a-e; Golay, 1980; Primack, 1975; Primack and von Hippel, 1974; and Sobel, 1978), and a number of analysts, observers, and participants have spotlighted the background and issues of reactor problems, debates, and accidents prior to TMI (Ames, 1978; Fuller, 1975; Stever, 1980; and Yanarella and Yanarella, 1980). Once the political controversy was triggered, a steady stream of government-sponsored studies (Behring, 1973; Gillette, 1977a, 1974; Rasmussen, 1975; Shapley, 1977; U.S., A.E.C., 1974; and von Hippel, 1977), institutional studies (I.A.E.A., 1978, 1979a and b), and academic studies (Chicken, 1981; Graham, 1971; Hendrie, 1976; Holdren, 1974; Lewis, 1980; Rolph, 1979; Russell, 1979; and Sorenson, 1979) was forthcoming.

Closely associated with the issue of the safety of atomic reactors is the concern for the effects of atomic radiation upon human health. No dispute exists within the scientific community regarding the harmful biological effects of enormous releases of atomic radiation. Studies of the short- and long-term consequences of radiation damage from the Hiroshima and Nagasaki bombings in World War II and the impact from nuclear testing by the Superpowers in the fifties and sixties (Barnaby, 1977; Beirly and Klement, 1965; Boffey, 1970; and Committee for the Compilation, 1981) are compelling and chilling in their findings. Rather, the debate among biologists, epidemiologists, health physicists, and other specialists turns on the issue of the biological impact of low-level ionizing radiation associated with the normal operation of atomic reactors or with the periodic venting of radioactive gases from nuclear power plants or with other contexts where human beings are exposed to low-level doses of atomic radiation (Caldicott, 1977; Cohen, 1976; Griffith and Ballentine, 1972; Holden, 1979; Lapp, 1979; Marx, 1979; Morgan, 1978; Roblat, 1978; Sternglass, 1972, 1981; Tamplin and Gofman, 1970; and Thompson and Bibb, 1970).

General reviews on radiation (U.N., 1977; and Woolard and Young, 1979) and on radiation dangers (Ebert and Howard, 1969; Edsall, 1976; and Lindop and Roblat, 1971) do not provide unambiguous means for determinig appropriate radiation standards (see Boffey, 1971; Gillette, 1974c; and Hamilton, 1972) to minimize risks to the general public, workers, or even research scientists. For one thing, a certain level of radioactive exposure to humans is unavoidable, since a small amount of background radiation occurs naturally in the environment (National Council, 1975). The task has been further complicated by theoretical problems, such as whether scientists should embrace a threshold theory (i.e., assume biological damage only occurs above a threshold level of radiation) or a linear theory (i.e., assume biological risk from radiation can be extrapolated downward from findings of high radiation contact to lower and lower radiation exposure). The matter has been made even more difficult by the suppression of pertinent information in the past by government organizations like the Atomic Energy Commission. Thusfar, even the most noteworthy scientific studies (Ad Hoc Population Dose Assessment Group, 1979; Anderson, 1978; Mancuso, et al., 1978; and Najarian, 1978), including the respected BEIR Report ("The Beir Report," 1973; Marshall, 1979; National Research Council, 1972, 1974; and Tamplin, 1973) have failed to bring consensus among scientists on significant issues concerning this serious subject.

The problem of nuclear waste disposal – the topic of the next section -- remains the Achilles heel of civilian nuclear power. Although the nuclear industry is nearly thirty years old, no consensus over a safe and effective method of disposing of high-level radioactive waste from atomic nuclear power plants exists. If none is established, the future of nuclear power is clearly in jeopardy. Within this section, a series of over-views on this difficult matter are presented (Angino, 1977; I.A.E.A., 1978; Lapp, 1977; Lipschutz, 1980; and U.S., E.P.A., 1979), discussions of alternative methods being considered are reviewed (Carter, 1978; DeMarsily, et al., 1977; Hanrahan, 1979; Kerr, 1979a and b; Kubo and Rose, 1973; Lewis, 1971; and Zeller, et al., 1973), and official government policy studies and proposals are enumerated (Hewlett, 1979; Lee, 1980; Maugh, 1979c; U.S., D.O.E., 1979; and U.S., Executive Branch, 1978). Beyond this, some investigators explore the case of the West Valley, New York, storage facili-ties which proved so faulty that they had to be closed (Carter, 1977; Lester and Rose, 1977; and Rochlin, et al., 1978). Several social dimensions to the waste disposal prob-lem (LaPorte, 1979; and Rochline, 1977), including public attitudes towards waste disposal management (Carter, 1977d; Jakimo and Bupp, 1978; and Zinberg, 1979), are illuminated. Reports on the growing inventories of high-level radioactive wastes from civilian and military nuclear programs should be both chastening to anti-nuclear activists who would like to dismiss the problem and alarming to nuclear power propo-nents who might like to think it is easily manageable (Krugman and von Hippel, 1977; and Smith, 1978).

Two short sections on nuclear safeguards and nuclear fuel enrichment and re-processing follow; and their shared concern is with the dangers of nuclear weapons proliferation. The term, nuclear safeguards, has come to signify the quest for means either to inhibit or prevent the proliferation of nuclear weapons as an outgrowth of national civilian atomic power programs or to block the theft of weapons-grade nuclear materials and thus to forestall nuclear terrorism. In this section, proposals for develop-ing nuclear safeguards (Brenner, 1980; Gilinsky, 1977; I.A.E.A., 1978; Lovins and Lovins, 1980; U.S., Congress, O.T.A., 1977; Weiss, 1978; Wilson, 1977; and Wohlstetter, et al., 1979b) address both the issue of nuclear proliferation (Atlantic Council, 1978; Baker, 1975; Coffey and Lambert, 1977; Greenwood, et al., 1977. Guhin, 1976; Johnson, 1977; Lovins, et al., 1980; Pierre and Moyne, 1976; and Wohlstetter, et al., 1979a) and the issue of nuclear theft and terrorism (Cohen, 1976; Leachman and Althoff, 1972; Schlesinger, 1974; and Willrich, 1975). The section on fuel reprocessing and uranium enrichment opens up a number of topics – including the benefits and perils of enrichment as a means of averting a uranium fuel shortage in the near future (Hammond, 1976; and Walsh, 1974), the technology of laser enrichment (Metz, 1977b), and the dangers of nuclear proliferation posed by laser enrichment processes (Casper, 1977; Glackin, 1976; and Krass, 1977).

The concluding section of the chapter on nuclear power provides a panoramic view of the literature on the rise of anti-nuclear protest in the United States. The anti-nuclear opposition is by no means a homogeneous movement with a unified and co-herent political strategy for closing off the nuclear option. Populated by ecologists and conservationists, cultural radicals and eco-technologists, reform liberals and hu-manist conservatives, gentile populists and countercultural romantics, it is held to-gether by the least common denominator – opposition, not so much to the hard tech-nology path and its attendant authoritarian social structure, but to civilian nuclear power with its manifest health and safety hazards. It remains, moreover, riddled with all of the contradictions of a single-issue movement which wants to be broader and more critical and is stymied both by its lack of a theory of political change and by its polyglot membership.

Several of the works contained in this section review the historical development of the anti-nuclear opposition from its origins in the environmental and public interest science movements to its gradual interpenetration with the solar and soft energy paths movement (Del Sesto, 1979; Kasperson, et al., 1979; Nelkin and Fallows, 1978; Novick, 1976; Primack and von Hippel, 1974; and Wasserman, 1979). Others give passionate expression to the weighty case against nuclear power (Berger, 1976; Croal, 1979; Dudderstadt and Kikuchi, 1979; Environmental Action Foundation, 1979; Faulkner, 1977; Gofman, 1979; Gofman and Tamplin, 1973, 1979; Graeub, 1974; Gyorgy, et al., 1979; Hayes, 1976; Junoit, 1979; Lewis, 1972; Nader and Abbotts, 1979; Novick, 1969; Olson, 1976; Reader, 1980; and Stephenson and Zachar, 1979). Two recent sociological analyses offer a profile of members of the no-nuke movement (Davidon, 1979; and Katz and List, 1981). In an effort to repair the politically diffuse and theoretically scattered nature of its strategic outlook, a number of social activists recommend refinements and revisions in no-nuke strategy (Barkan, 1979; Bayer, 1979; Darnovsky, 1979; Jezer, 1977; and Pector, 1979). Finally, a persistent critic of the movement offers a sustained critique of its arguments and its animating purposes and vision (McCracken, 1977, 1981).

1. Fossil Fuels — Oil and Gas

Adelman, Morris A. The World Petroleum Market. Baltimore, Maryland: Published for Resources for the Future by the Johns Hopkins University Press, 1972.

Allvine, Fred C. and Patterson, James M. Highway Robbery: An Analysis of the Gasoline Crisis. Bloomington, Indiana: Indiana University Press, 1974.

Aronowitz, Stanley. Food, Shelter, and the American Dream. New York: Seabury Press, 1974.

Bell, Harold S. Oil Shales and Shale Oils. New York: Van Nostrand Company, 1948.

Berg, R. R., et al. "Prognosis for Expanded U.S. Production of Crude Oil." Science, 184 (April 19, 1974), 331-336.

Berner, Arthur S. and Scoggins, Sue. "Oil and Gas Drilling Programs — Structure and Regulation." George Washington Law Review, 41 (March 1973), 471-504.

Blair, John M. The Control of Oil. Westminster, Maryland: Pantheon Books, 1976; New York: Vintage Books, 1978.

Cichetti, Charles J. Alaskan Oil: Alternative Routes and Markets. Washington, D.C.: Resources for the Future, 1972.

Clark, Colin W. "The Economics of Overexploitation." Science, 181 (August 17, 1973), 630-634.

Cook, Earl. "Limits to Exploitation of Nonrenewable Resources." Science, 191 (February 20, 1976), 677-682.

Cooper, Bryan. Alaska: The Last Frontier. New York: William Morrow & Company, 1973.

Dam, K. W. Oil Resources: Who Gets What How? Chicago, Illinois: University of Chicago Press, 1976.

DeNevers, Noel. "Liquid Natural Gas." Scientific American, 217 (October 1967), 30-37.

_____. "The Secondary Recovery of Petroleum." Scientific American, 213 (July 1965), 34-49.

_____. "Tar Sands and Oil Shales." Scientific American, 214 (February 1966), 21-29.

Devanney, John W., III. "Key Issues in Offshore Oil." Technology Review, 76 (January 1974), 20-25.

Dineen, Gerald U. and Cook, Glenn L. "Oil Shale and the Energy Crisis." Technology Review, 76 (January 1974), 26-33.

Doran, Charles F. Myth, Oil, and Politics: Introduction to the Political Economy of Petroleum. Riverside, New Jersey: Free Press, 1977.

Drake, Elisabeth and Reid, Robert C. "The Importation of Liquefied Natural Gas." Scientific American, 236 (April 1977), 22-29.

Eckbo, P. L. The Future of World Oil. Cambridge, Massachusetts: Ballinger Publishing Company, 1976.

Emery, Kenneth O. "New Opportunities for Offshore Petroleum Exploration." Technology Review, 77 (March/April 1975), 30-33.

_____. "Offshore Oil: Technology ... and Emotion." Technology Review, 78 (February 1976), 30-37.

Engler, Robert. The Brotherhood of Oil: Energy Policy and the Public Interest. Chicago, Illinois: University of Chicago Press, New York: New American Library, 1977.

_____. The Politics of Oil: A Study of Private Power and Democratic Directions. New York: Macmillan, 1961.

Erickson, Edward W. "Crude Oil Prices, Drilling Incentives, and the Supply of New Discoveries." Natural Resources Journal, 10 (January 1970), 27-52.

Fanning, Leonard M. American Oil Operations Abroad. New York: McGraw-Hill, 1947.

_____. The Rise of American Oil. New York: Harper and Brothers, 1948.

Fixler, L. Donald and Ferrar, Robert L. "Oil Pricing Policy Options from the Producer's Point of View - A Theoretical Approach." Energy Policy, 3 (June 1975), 136-143.

127

Forber, R. J. Studies in Early Petroleum History. Westport, Connecticut: Hyperion Press, 1975a.

_____. More Studies in Early Petroleum History. Westport, Connecticut: Hyperion Press, 1975b.

Frank, Helmut J. and Schanz, John J., Jr. "The Future of American Oil and Gas." Annals of the American Academy of Political and Social Science, 410 (November 1973), 24-34.

Fried, Edward R., ed. Higher Oil Prices and the World Economy: The Adjustment Problem. Washington, D.C.: The Brookings Institution, 1975.

Gillette, Robert. "Oil and Gas Resources: Academy Calls USGS Math Misleading." Science, 187 (February 28, 1975), 723-727.

_____. "Oil and Gas Resources: Did USGS Gush Too High?" Science, 185 (July 12, 1974), 127-130.

Hedberg, Hollis D. "Ocean Boundaries and Petroleum Resources." Science, 191 (March 12, 1976), 1009-1018.

Hobbie, Barbara and Mancke, Richard B. "Oil Monopoly Divestiture: A Clash of Media vs. Expert Perceptions. Energy Policy, 5 (September 1977), 232-244.

Hwa, David; Klema, Ernest; and Mancke, Richard. "Markets for Alaskan Oil." Energy Policy, 7 (March 1979), 23-28.

Jensen, E. J. and Ellis, H. S. "Pipelines." Scientific American, 216 (January 1967), 62-77.

Kash, Don E.; White, Irvin L.; et al. Energy Under the Oceans - A Technology Assessment of Outer Continental Shelf Oil and Gas Operations. Folkestone, Kent, England: Bailey Brothers & Swinfen, 1974.

Kerr, Richard A. "Petroleum Exploration: Discouragement About the Atlantic Outer Continental Shelf Deepens." Science, 204 (June 8, 1979), 1069-1072.

Kuenne, Robert E., et al. Intermediate-Term Energy Programs to Protect Against Crude-Petroleum Import Interruptions: Feasible Alternatives, Program Costs, and Operational Methods of Funding. Arlington, Virginia: Institute for Defense Analysis, 1974.

Leach, Gerald. "The Impact of the Motor Car on Oil Reserves." Energy Policy, 1 (December 1973), 195-207.

McDonald, Stephen L. "U.S. Depletion Policy: Likely Effects of Changes." Energy Policy, 4 (March 1976), 56-62.

MacAvoy, Paul W. and Pindyck, Robert S. The Economics of the Natural Gas Shortage (1960-1980). New York: Elsevier-North Holland Publishing Company, 1975.

Manoharan, S. The Oil Crisis: The End of an Era. New York: International Publications Service, 1975.

Maugh, Thomas H., II. "Oil Shale: Prospects on the Upswing ... Again." Science, 198 (December 9, 1977), 1023-1027.

_____. "Tar Sands: A New Fuel Industry Takes Shape." Science, 199 (February 17, 1978), 756-760.

Medvin, N.; Lav, I. J.;and Ruttenburg, S. H. The Energy Cartel: Big Oil vs. the Public Interest. Washington, D.C.: Ruttenburg, Friedman, Kilgallon, Gutchess & Associates., n.d.

Menard, H. William. "Toward a Rational Strategy for Oil Exploration." Scientific American, 244 (January 1981), 55-65.

Mendershausen, Horst. Coping with the Oil Crisis. Baltimore, Maryland: The Johns Hopkins University Press, 1976.

Metz, William D. "Oil Shale: Resource of Low-Grade Fuel." Science, 184 (June 21, 1974), 1271-1275.

Miller, Edward. "Some Implications of Land Ownership Patterns for Petroleum Policy." Land Economics, 49 (November 1973), 414-423.

Mitchell, Edward J., ed. Dialogue on World Oil. Washington, D.C.: American Enterprise Institute for Public Policy Research, 1974.

_____. The Question of Offshore Oil. Washington, D.C.: American Enterprise Institute for Public Policy Research, 1976.

Moffett, J. D. "Federal Oil Shale Policy: An Analysis of Development Alternatives." Houston Law Review, 13 (May 1976), 701-732.

Moody, John D. and Geiger, Robert E. "Petroleum Resources: How Much Oil and Where?" Technology Review, 77 (March/April 1975), 38-45.

Newlon, Daniel H. and Breckner, Norman V. The Oil Security System: An Import Strategy for Achieving Oil Security and Reducing Oil Prices. Lexington, Massachusetts: Lexington Books, 1975.

Nivola, Pietro S. "Energy Policy and the Congress: The Politics of the Natural Gas Policy Act of 1978." Public Policy, 28 (Fall 1980), 491-543.

Odell, Peter R. "The Future of World Oil: A Rejoinder." Geographical Journal, 139 (October 1973), 436-454.

Oppenheimer, Bruce Ian. Oil and the Congressional Process. Lexington, Massachusetts: Lexington Books, 1974.

Oppenheimer, Ernest J. Natural Gas: The New Energy Leader. New York: Pen & Podium Publications, 1981.

Peebles, Malcolm W. H. Evolution of the Gas Industry. New York: New York University Press, 1980.

Penrose, Edith. "The Oil 'Crisis'." Round Table, 254 (April 1974), 135-148.

Ramseier, Rene O. "Oil on Ice." Environment, 16 (April 1974), 6-13.

Rees, T. M. "A Rational Way to Develop Shale Oil." American Journal of Economics and Sociology, 33 (October 1974), 205-267.

Renshaw, Edward F. "The Decontrol of U.S. Oil Production." Energy Policy, 8 (March 1980), 38-49.

_____ and Renshaw, Perry F. "U.S. Oil Discovery and Production: The Projections of M. King Hubbert." Futures, 12 (February 1980), 58-66.

Rice, Richard A. "How to Reach that North Slope Oil." Technology Review, 75 (June 1973), 8-18.

Rushton, William Faulkner. "Big Oil on the Bayou." Environmental Action, 10 (January 1979), 16-23.

Russell, Milton. Limiting Oil Imports. Baltimore, Maryland: The Johns Hopkins University Press, 1978.

Rybczynski, T. M., ed. The Economics of the Oil Crisis. London, England: Published for the Trade Policy Research Centre by Macmillan, 1976.

Sampson, Anthony. The Seven Sisters: The Great Oil Companies and the World They Made. New York: Viking Press, 1975.

Sharbaugh, H. Robert. "Petroleum and Energy." Annals of the American Academy of Political and Social Science, 420 (July 1975), 86-97.

"Sharing the Off-Shore Oil Bonanza." Technology Review, 78 (February 1976), 38-45.

Stephenson, Lee. "Gas Industry 'Aided and Abetted' by FPC, Study Shows." Environmental Action, 6 (September 28, 1974), 12-13.

Stevens, T. H. and Kalter, R. J. "The Economics of Oil Shale Development Policy." Land Economics, 51 (November 1975), 355-364.

Stewart, R. J. "Oil Spills and Offshore Petroleum." Technology Review, 78 (February 1976), 60-67.

Stoff, Michael B. Oil, War, and American Security: The Search for a National Policy on Foreign Oil, 1941-1947. New Haven, Connecticut: Yale University Press, 1980.

Thorne, H. M.; Stanfield, K. E.; Dineen, G. U.;and Murphy, W. I. R. Oil Shale Technology. [U.S. Bureau of Mines Circular 8216] Washington, D.C.: U.S. Government Printing Office, 1964.

Tussing, Arlon R. Alaska Pipeline Report. Juneau, Alaska: Institute of Social, Economic, and Government Research, 1971.

_____ . "Toward a Rational Policy for Oil and Gas Imports." In Harold Wolozin, ed. Energy and the Environment. Morristown, New Jersey: General Learning Press, 1974, pp. 59-88.

United States. Congress. Office of Technology Assessment. An Assessment of Oil Shale Technologies. [OTA-M-18] Washington, D.C.: Office of Technology Assessment, U.S. Congress, June 1980.

United States. Congress. Senate. Committee on Interior and Insular Affairs. Natural Gas Policy Issues. Washington, D.C.: U.S. Government Printing Office, 1972.

Vallenilla, Luis. Oil: The Making of a New Economic Order. New York: McGraw-Hill, 1975.

Vernon, Raymond, ed. The Oil Crisis. New York: W. W. Norton and Company, 1976.

Willrich, Mason. Administration of Energy Shortages: Natural Gas and Petroleum. Cambridge, Massachusetts: Ballinger Publishing Company, 1977.

Wilson, Richard. "Natural Gas is a Beautiful Thing?" Bulletin of the Atomic Scientists, 29 (September 1973), 35-40.

Windsor, Philip. Oil: A Guide Through the Energy Jungle. Boston, Massachusetts: Gambit, 1976.

Wolf, Sidney M. "Liquefied Natural Gas." Bulletin of the Atomic Scientists, 34 (December 1978), 20-25.

Yen, T. F., ed. Science and Technology of Oil Shale. Ann Arbor, Michigan: Ann Arbor Science Publishers, 1976.

Yergin, Daniel, ed. The Dependence Dilemma: U.S. Gasoline Consumption and America's Security. Cambridge, Massachusetts: Harvard Center for International Affairs, 1980.

2. Coal

GENERAL

Alexander, Tom. "A Promising Try at Environmental Detente for Coal." Fortune, (February 13, 1978), 94-102.

Althouse, Ronald. Work, Safety, and Lifestyle Among Southern Appalachian Coal Miners: A Survey of the Men of Standard Mines. Morgantown, West Virginia: Office of Research and Development, Division of Social and Economic Development, Appalachian Center, West Virginia University, 1974.

Arble, Mead. The Long Tunnel: A Coal Miner's Journal. New York: Atheneum Publishers, 1976.

Atwood, Genevieve. "The Strip Mining of Western Coal." Scientific American, 233 (December 1975), 23-29.

Austin, Richard Cartwright. Spoil: A Moral Study of Strip Mining for Coal. New York: National Division, Board of Global Ministries, Methodist Church, 1973.

_____ and Borreli, Peter. The Strip Mining of America: An Analysis of Surface Coal Mining and the Environment. New York: Sierra Club, 1971.

Ayres, Ronald F. Coal: New Markets/New Prices - Ramifications of the Federal Coal Conversion Program. New York: McGraw-Hill, 1977.

Balliet, Lee. 'A Pleasing Tho' Dreadful Sight': Social and Economic Impacts of Coal Production in the Eastern Coalfields. [A Report to the Office of Technology Assessment, Congress of the United States] Washington, D.C.: Office of Technology Assessment, 1978.

Berkowitz, N. An Introduction to Coal Technology. New York: Academic Press, 1979.

Berry, Wendell. "Strip Mining Morality: The Landscaping of Hell." Nation, 202 (January 24, 1966), 96-100.

Bethell, Thomas N. Conspiracy in Coal. Huntington, West Virginia: Appalachian Movement Press, 1972.

Bhutani, J., et al. An Analysis of Constraints on Increased Coal Production. [MTR-6830] McLean, Virginia: The MITRE Corporation, 1975.

Binder, Denis. "A Novel Approach to Reasonable Regulation of Strip Mining." University of Pittsburgh Law Review, 34 (Spring 1973), 339-374.

Binder, Frederick M. Coal Age Empire: Penny Coal and Its Utilization to 1860. Harrisburg, Pennsylvania: Pennsylvania Historical and Museum Commission, 1974.

Box, Thadis, et al. Rehabilitation Potential of Western Coal Lands. Cambridge, Massachusetts: Ballinger Publishing Company, 1974.

Brooks, David B. "Strip Mine Reclamation and Economic Analysis." Natural Resources Journal, 6 (1966), 13-44.

_____. "Strip Mining, Reclamation, and the Public Interest." American Forests, 72 (1966), 51-57.

Brown, Malcolm and Webb, John N. Seven Stranded Coal Towns: A Study of an American Depressed Area. Washington, D.C.: U.S. Government Printing Office, 1941; New York: DaCapo Press, 1971.

Bryant, F. Carleen. The Social Impact of Surface Mining in a Rural Appalachian Community. Knoxville, Tennessee: Appalachian Resources Project, University of Tennessee, 1976.

Bulmer, M. I. A. "Sociological Models of the Mining Community." The Sociological Review, 23 (February 1975), 61-92.

Burness, H. Stuart. The Effect of Uncertainty on the Supply of Coal. [Prepared for the Institute for Mining and Minerals Research] Lexington, Kentucky: Office of Research and Engineering Services, College of Engineering, University of Kentucky, 1976.

Business Communications Company Staff. Future for Coal as Fuel and Chemical. Rev. ed. [E-004] Stamford, Connecticut: Business Communications Company, 1976.

Camplin, Paul, ed. Strip Mining in Kentucky. Frankfort, Kentucky: Kentucky Division of Strip Mining and Reclamation, 1965.

Cannon, James S. Mine Control: Western Coal Leasing and Development. New York: Council on Economic Priorities, 1978.

Caudill, Harry M. My Land Is Dying. New York: E. P. Dutton & Company, 1971.

_____. Night Comes to the Cumberlands. Boston, Massachusetts: Atlantic Monthly Press/Little, Brown, and Company, 1962.

Chaffin, Lillie D. Coal, Energy, and Crisis. New York: Harvey House, 1974.

Christenson, Carroll L. Economic Redevelopment in Bituminous Coal: The Special Case of Technological Advance in United States Coal Mines, 1930-1960. Cambridge, Massachusetts: Harvard University Press, 1962.

Clean Fuels From Coal: Symposium Papers. [Papers presented at the Symposium on Clean Fuels From Coal, ITT Research Institute, Chicago, 1973] Chicago, Illinois: Institute of Gas Technology, 1973.

Clean Fuels From Coal Symposium II. [Papers presented at the Second Symposium on Clean Fuels From Coal, ITT Research Institute, Chicago, 1975] Chicago, Illinois: Institute of Gas Technology, 1975.

Coal Conversion Business. Stamford, Connecticut: Business Communications Company, 1978.

Coal Conversion: Practical and Legal Implications. [Corporate Law and Practice Handbook Series, Vol. 246] New York: Practicing Law Institute, 1977.

Coleman, McAlister. Men and Coal. New York: Farrar & Rinehart, 1943; New York: Arno & The New York Times, 1969.

Coleman, Ron. We Dig Coal: The Story of Coal Mining in Buchanan County, Virginia. Radford, Virginia: Commonwealth Press, 1975.

Comptroller General. U.S. Coal Development - Promises, Uncertainties. [EMD 77-43] Washington, D.C.: General Accounting Office, September 22, 1977.

Conaway, James. "The Last of the West: Hell, Strip It!" Atlantic Monthly, 232 (September 1973), 91-103.

Coombs, Charles. Coal in the Energy Crisis. New York: William Morrow Company, 1980.

Copeland Otis L. and Packer, Paul E. "Land Use Aspects of the Energy Crisis and Western Mining." Journal of Forestry, 70 (November 1972), 671-675.

Cushing, George H. The Human Story of Coal. New York: W. E. Rudge, 1924.

"The Demand for Energy and Appalachia's Coal." [Special Issue] Appalachia, February/March 1972.

Densmore, Raymond E. The Coal Miner of Appalachia. Parsons, West Virginia: McClain Printing Company, 1977.

Doyle, William S. Deep Coal Mining: Waste Disposal Technology. Park Ridge, New Jersey: Noyes Data Corporation, 1976.

_____. Strip Mining of Coal - Environmental Solutions. Park Ridge, New Jersey: Noyes Data Corporation, 1976.

Dubofsky, Melvyn and Van Tine, Warren. John L. Lewis: A Biography. New York: Quadrangle Press, 1977.

Energy Modeling Forum. Coal in Transition: 1980-2000. [EMF Report 2, Volume 1] Stanford, California: Stanford University, Energy Modeling Forum, July 1978.

Erickson, Kai T. Everything In Its Path: Destruction of a Community in the Buffalo Creek Flood. New York: Simon and Schuster, 1976.

Ezra, Derek. Coal and Energy: The Need to Exploit the World's Most Abundant Fossil Fuel. New York: Halsted Press, 1978.

Finley, Joseph E. The Corrupt Kingdom: The Rise and Fall of the United Mine Workers. New York: Simon and Schuster, 1972.

Gaines, Linda, II and Berry, R. Stephen. TOSCA: The Total Social Cost of Coal and Nuclear Power. Cambridge, Massachusetts: Ballinger Publishing Company, 1979.

Gordon, Richard L. Coal and Canada - U.S. Energy Relations. Washington, D. C.: Canadian - American Committee, 1976a.

_____. Coal in the U.S. Energy Market. Lexington, Massachusetts: Lexington Books, 1978a.

_____. "Coal - Swing Fuel." In R. J. Kalter and W. A. Vogely, eds. Energy Supply and Government Policy. Ithaca, New York: Cornell University Press, 1976b, pp. 193-215.

_____. "The Hobbling of Coal: Policy and Regulatory Uncertainties." Science, 200 (April 14, 1978b), 153-158.

_____. U.S. Coal and the Electric Power Industry. Baltimore, Maryland: The Johns Hopkins University Press, 1975.

Greenbaum, Margaret Elaine and Harvey, Curtis E. Surface Mining, Land Reclamation, and Acceptable Standards. Lexington, Kentucky: Institute for Mining and Minerals Research, College of Engineering, University of Kentucky, 1974.

Greenburg, William. "Chewing It Up at 200 Tons a Bite: Strip Mining." Technology Review, 75 (February 1973), 46-55.

135

Greene, Robert P. and Gallagher, J. Michael, eds. Future Coal Prospects: Country and Regional Assessments. [World Coal Study] Cambridge, Massachusetts: Ballinger Publishing Company, 1980.

Hardesty, C. Howard, Jr. "Coal and the Energy Crisis." West Virginia Law Review, 76 (April 1974), 257-266.

Hardt, Jerry. Harlan County Flood Report. Corbin, Kentucky: Appalachia-Science in the Public Interest, 1978.

Harter, Walter. Coal: The Rock That Burns. Nashville, Tennessee: Nelson, Thomas, 1978.

Hawley, Mones, E., ed. Coal: Scientific and Technical Aspects. New York: Academic Press, 1977.

_____. Coal: Social, Economic, and Environmental Aspects. New York: Academic Press, 1976.

Henry, John P., Jr. and Schmidt, Richard A. "Coal: Still Old Reliable?" Annals of the American Academy of Political and Social Science, 410 (November 1973), 35-51.

Hibbard, W. R., Jr. "Policies and Constraints for Major Expansion of U.S. Coal Production and Utilization." In Jack M. Hollander, et al., eds. Annual Review of Energy. Volume 4. Palo Alto, California: Annual Reviews, Inc., 1979, 147-174.

Hittman Associates, Inc. Underground Coal Mining: An Assessment of Technology. Palo Alto, California: Electric Power Research Institute, 1976.

Horwitch, Mel. "Coal: Constrained Abundance." In Robert Stobaugh and Daniel Yergin, eds. Energy Future: Report of the Energy Project at the Harvard Business School. New York: Random House, 1979, pp. 79-107.

Hultman, Charles W. and Davis, Bernard. Implications of a Doubling of Kentucky Coal Production for the State Economy. Lexington, Kentucky: Published for the Kentucky Center for Energy Research by the Office of Research, College of Business and Economics, University of Kentucky, 1977.

Jacobsen, Sally. "The Great Montana Coal Rush." Bulletin of the Atomic Scientists, 29 (April 1973), 37-42.

Jovick, Robert L. Critique of Prospective Coal Development in Eastern Montana. Sidney, Montana: The Economic Development Association of Eastern Montana, 1972.

Kolbash, Ronald Lee. A Study of Appalachia's Coal Mining Communities and Associated Environmental Problems. Ann Arbor, Michigan: University Microfilms International, 1978.

Krohe, James, Jr. Midnight at Noon: A History of Coal Mining in Sangamon County. Springfield, Illinois: Sangamon County Historical Society, 1975.

Krutilla, John V., et al., eds. Economic and Fiscal Impacts of Coal Development: Northern Great Plains. Baltimore, Maryland: The Johns Hopkins University Press, 1978.

Landy, Marc Karnis. The Politics of Environmental Reform: Controlling Kentucky Strip Mining. Washington, D.C.: Resources for the Future; Baltimore, Maryland: Distributed by The Johns Hopkins University Press, 1976.

Lane, Winthrop D. Civil War in West Virginia: The Story of the Industrial Conflict in the Coal Mines. New York: B. W. Huebsch, 1921; Oriole Editions, 1977.

Lantz, Herman R. People of Coal Town. Carbondale, Illinois: Southern Illinois University Press, 1958.

Lewis, Helen M.; Johnson, Linda; and Askins, Donald, eds. Colonialism in Modern America: The Appalachian Case. Boone, North Carolina: Appalachian Consortium Press, 1978.

Lindbergh, Kristina R. and Provorse, Barry L. Coal: A Contemporary Energy Story. Seattle, Washington: Scribe Publishing Corporation, 1978.

Longman, R. C. Appalachian Kentucky: An Exploited Region. Toronto, Ontario: McGraw-Hill Ryerson, 1971.

Meadows, D. and Stanley-Miller, J. "The Transition to Coal." Technology Review, 78 (October/November 1975), 19-29.

Menzel, Donald C. and Edgman, Terry D. "The Struggle to Implement a National Surface Mining Policy." Publius, 10 (Winter 1980), 81-92.

Meyers, Robert A. Coal Desulfurization. New York: Marcel Dekker, 1977.

Miernyk, William H. "Coal and the Future of the Appalachian Economy." Appalachia, October/November 1975, 29-35.

Miller, Saunders. The Economics of Nuclear and Coal Power. New York: Praeger Publishers, 1976.

Monsour, Lucy G. Kentucky's Coal Production Constraints. Lexington, Kentucky: Kentucky Center for Energy Research, April 1976.

Moore, John R., et al. Economics of the Private and Social Costs of Appalachian Coal Production: A Progress Report. Knoxville, Tennessee: Appalachian Resources Project, University of Tennessee, 1973.

137

Mountain Community Union. Land Use and Environmental Rights Committee. You Can't Put It Back: A West Virginia Guide to Strip Mine Opposition. Fairmont, West Virginia: Mountain Community Union, 1976.

Murray, Francis X. Where We Agree: Report of the National Coal Policy Project. 2 Volumes. Boulder, Colorado: Westview Press, 1978.

Muschett, F. Douglas. Coal Development in Montana: Economic and Environmental Impacts. Ann Arbor, Michigan: Department of Geography, University of Michigan, 1977.

National Petroleum Council. Committee on U.S. Energy Outlook. Other Energy Resources Subcommittee. Coal Task Group. U.S. Energy Outlook: Coal Availability: A Report. Washington, D.C.: National Petroleum Council, 1973.

National Research Council. Coal as an Energy Resource: Conflict and Consensus. [Academy Forum] Washington, D.C.: National Academy of Sciences, 1977.

_____. Environmental Studies Board. Underground Disposal of Coal Mine Wastes. Washington, D.C.: National Academy of Sciences, 1975.

Nehring, Richard and Zycher, Benjamin. Coal Development and Government Regulation in the Northern Great Plains: A Preliminary Report. Santa Monica, California: RAND Corporation, 1976.

Nephew, Edmund A. "The Challenge and Promise of Coal." Technology Review, 76 (December 1973), 20-29.

Newman, Monroe O. The Political Economy of Appalachia: A Case Study in Regional Integration. Lexington, Massachusetts: Lexington Books, 1972.

Nicholson, Arthur F. Marketing of Kentucky Coal. Lexington, Kentucky: Kentucky Center for Energy Research, April 1976.

Noyes, R., ed. Coal Resources: Characteristics and Ownership in the U.S.A. Park Ridge, New Jersey: Noyes Data Corporation, 1978.

Osborn, Elburt F. "Coal and the Present Energy Situation." Science, 183 (February 8, 1974), 477-481.

Patterson, Walter C. and Griffin, Richard. Fluidized-Bed Technology: Coming to a Boil. New York: Inform, Inc., 1978.

Peterson, Bill. Coaltown Revisited: An Appalachian Notebook. Chicago, Illinois: Henry Regnery Company, 1972.

Pfleider, Eugene P., ed. Surface Mining. New York: American Institute of Mining, Metallurgical, and Petroleum Engineers, 1972.

Plunkett, H. D. and Bowman, M. J. Elites and Change in the Kentucky Mountains. Lexington, Kentucky: University Press of Kentucky, 1973.

Rall, David P., Director. Report of the Committee on Health and Environmental Effects of Increased Coal Utilization. [Report submitted to the Secretary of the Department of Health, Education, and Welfare] Washington, D.C.: U.S. Department of Health, Education, and Welfare, December 27, 1977.

Randall, Alan, et al. Estimating Environmental Damages from Surface Mining of Coal in Appalachia: A Case Study. Cincinnati, Ohio: Office of Research and Development, U.S. Environmental Protection Agency, January 1978.

Reitze, Arnold W., Jr. "Old King Coal and the Merry Rapists of Appalachia." Case Western Reserve Law Review, 22 (1971), 650-737.

Riddel, Frank S., ed. Appalachia: Its People, Heritage, and Problems. Dubuque, Iowa: Kendall/Hunt Publishing Company, 1974.

Rosenbaum, Walter A. Coal and Crisis: The Political Dilemmas of Energy Management. New York: Praeger Publishers, 1978;

Rosenblum, Stuart M. "The Future of the Coal Substitution Option." Duquesne Law Review, 3 (1975), 581-622.

Rowe, James E., ed. Coal Surface Mining: The Impacts of Reclamation. Boulder, Colorado: Westview Presss, 1979.

Schlottman, Alan M. Environmental Regulation and the Allocation of Coal: A Regional Analysis. New York: Praeger Publishers, 1977.

_____ and Spore, Robert L. Economic Impacts of Surface Mine Reclamation. Knoxville, Tennessee: Appalachian Resources Project, University of Tennessee, 1976.

Schmidt-Bleek, F. and Carlsmith, R. S., eds. Symposium on Coal and Public Policies. Knoxville, Tennessee: Center for Business and Economic Research, College of Business Administration, University of Tennessee, 1972.

Seltzer, Curtis Ian. The United Mine Workers of America and the Coal Operators: The Political Economy of Coal in Appalachia. [Thesis - Columbia University] Ann Arbor, Michigan: University Microfilms International, 1978.

Sheppard, Muriel. Cloud By Day: The Story of Coal and Coke and People. Chapel Hill, North Carolina: University of North Carolina Press, 1947.

Simeons, C. Coal: Its Role in Tomorrow's Technology. Elmsford, New York: Pergamon Press, 1978.

Spaid, Ora. "Forecast: Doubled Coal Production in Appalachia." Appalachia, June/ July 1975, 1-10.

Squires, Arthur M. "Coal: A Past and Future King." Ambio, 3 (1974), 1-14.

Stacks, John F. Stripping. New York: Sierra Club Books, 1972.

The Strip Mine Handbook. Washington, D.C.: Environmental Policy Institute, and Center for Law and Social Policy, 1978.

Tetra Tech, Inc. Energy From Coal: A State-of-the-Art Review. [Report prepared for the Energy Research and Development Administration] Washington, D.C.: U.S. Government Printing Office, 1976.

Toole, K. Ross. The Rape of the Great Plains: Northwest America, Cattle, and Coal. Boston, Massachusetts: Little, Brown, and Company, 1976.

Tyner, Wallace E. and Kalter, Robert J. Western Coal: Promise or Problem? Lexington, Massachusetts: D.C. Heath and Company, 1978.

United States. Congress. Office of Technology Assessment. The Direct Use of Coal: Prospects and Problems of Production and Consumption. [OTA-E-86] Washington, D.C.: U.S. Government Printing Office, April 1979.

_____. A Technology Assessment of Coal Slurry Pipelines. Washington, D.C.: U.S. Government Printing Office, March 1978.

United States. Congress. Senate. Committee on Interior and Insular Affairs. Coal Surface Mining and Reclamation. Washington, D.C.: U.S. Government Printing Office, 1973a.

_____. Factors Affecting the Use of Coal in Present and Future Energy Markets. Washington, D.C.: U.S. Government Printing Office, 1973b.

_____. The Issues Related to Surface Mining: A Summary Review with Selected Readings. Washington, D.C.: U.S. Government Printing Office, 1971.

United States. Council on Environmental Quality. Coal Surface Mining and Reclamation: An Environmental and Economic Assessment of Alternatives. Washington, D.C.: U.S. Government Printing Office, 1973.

United States. Department of the Interior. Surface Mining and Our Environment: A Special Report to the Nation. Washington, D.C.: U.S. Government Printing Office, 1967.

United States. Department of the Interior. Office of Coal Research. Clean Energy from Coal Technology. Washington, D.C.: U.S. Government Printing Office, 1974.

United States. Federal Energy Administration. Project Independence: Final Task Force Report on Coal. Washington, D.C.: U.S. Government Printing Office, November 1974.

Vecsey, George. One Sunset a Week: The Story of a Coal Miner. New York: Saturday Review Press, 1974.

Walton, Daniel R. and Kauffman, Peter W. Preliminary Analysis of the Probable Causes of Decreased Coal Mining Productivity (1969-1976). Reston, Virginia: Management Engineers, Inc., November 7, 1977.

When Coal Was King. Lebanon, Pennsylvania: Applied Arts Publishers, 1970.

Wilson, Carroll. Coal – Bridge to the Future. [World Coal Study] Cambridge, Massachusetts: Ballinger Publishing Company, 1980.

Yanarella, Ernest J. and Yanarella, Ann-Marie, eds. Coal Mine Regulation and Professional Ethics: The Case of Mine Safety. [Proceedings of a Symposium, University of Kentucky, September 27, 1979] Lexington, Kentucky: Social Science/Technology Development Group, University of Kentucky, October 1979d.

_____ . Energy Development in Kentucky: Its Impact Upon Community Life and Higher Education. [Proceedings of a Symposium, University of Kentucky, March 28, 1979] Lexington, Kentucky: Social Science/Technology Development Group, University of Kentucky, April 1979b.

_____ . The Role of Coal in the Energy Picture to the Year 2000: Economic and Political Perspectives. [Proceedings of a Symposium, University of Kentucky, February 9, 1979] Lexington, Kentucky: Social Science/Technology Development Group, University of Kentucky, March 1979a.

_____ . Surface Mining and the Limits of Reclamation: What Is Being Reclaimed? What Is Not? [Proceedings of a Symposium, University of Kentucky, May 3, 1979] Lexington, Kentucky: Social Science/Technology Development Group, University of Kentucky, June 1979c.

Young Gordon. "Will Coal Be Tomorrow's 'Black Gold'?" National Geographic, 148 (August 1975), 234-259.

Cameron, Robert B. An Estimation of the Tangible Costs of Black Lung Disease Related Disability to the Bituminous Coal Mine Operations of Appalachia. [Appalachian Resources Project No. 47] Knoxville, Tennessee: University of Tennessee, 1976.

Gehrs, C. W., et al. "Environmental Health and Safety Implications of Increased Coal Utilization." In M. A. Elliott, ed. Chemistry of Coal Utilization. Suppl. Volume 2. New York: Wiley Interscience, 1980.

Graebner, William. Coal-Mining Safety in the Progressive Period: The Political Economy of Reform. Lexington, Kentucky: Published for the Organization of American Historians by the University Press of Kentucky, 1976.

Henderson, David Richard. The Economics of Safety Legislation in Underground Coal Mining. Ann Arbor, Michigan: University Microfilms International, 1978.

Kent, William H. An Analysis of Appalachian State Coal Mine Health and Safety and Workmen's Compensation Programs: Recommendations for Improvement. Washington, D.C.: Appalachian Regional Commission, 1973.

Kramer, Leo and Clague, Ewan, eds. The Health-Impaired Miner Under Black Lung Legislation. New York: Irvington Publishers, 1973.

McAteer, J. Davitt. Coal Mine Health and Safety: The Case of West Virginia. New York: Praeger Publishers, 1973.

Murray, Robert Kirk. The Act, PL 91-173: Living With the Federal Coal Mine Health and Safety Act of 1969. Denver, Colorado: Rocky Mountain Coal Association, 1973.

National Research Council. Commission on Natural Resources. Committee on Mineral Resources and the Environment. Mineral Resources and the Environment, Supplementary Report: Coal Workers' Pneumoconiosis. Washington, D.C.: National Academy of Sciences, 1975.

Rall, David P., Director. Report of the Committee on Health and Environmental Effects of Increased Coal Utilization. Washington, D.C.: U.S. Department of Health, Education, and Welfare, December 27, 1977.

United Mine Workers of America. Department of Occupational Health. Black Lung. Washington, D.C.: United Mine Workers of America, 1970.

United States. Department of the Interior. Bureau of Mines. The Impact of the Bureau of Mines Coal Mine Health and Safety Research Program, 1970-1977. Washington, D.C.: U.S. Government Printing Office, 1977.

United States. Department of the Interior. Coal Mines Administration. A Medical
   Survey of the Bituminous Coal Industry. Washington, D.C.: U.S. Government
   Printing Office, 1947.

Yanarella, Ernest J. and Yanarella, Ann-Marie, eds. Coal Mine Regulation and Professional Ethics: The Case of Mine Safety. [Proceedings of a Symposium, University of Kentucky, September 27, 1979] Lexington, Kentucky: Social Science/
   Technology Development Group, University of Kentucky, October 1979.

Zahorski, Witold W., ed. Coal Workers' Pneumoconiosis: A Critical Review. Hanover,
   New Hampshire: University Press of New England, 1974.

SYNFUELS

A.M.E. Technology. Social, Economic, and Environmental Impacts of Coal Gasification
   and Liquefaction Plants: Final Report. [Prepared by R. G. Edwards, A. B.
   Broderson, and W. P. Hauser] Lexington, Kentucky: Office of Research and
   Engineering Services, College of Engineering, University of Kentucky, 1976.

Borkin, Joseph. The Crime and Punishment of I. G. Farben. New York: Free Press,
   1978.

Bruder, George F. and Gentile, Carmen L. "State Synfuel Agencies' Role in the Federal
   Program." State Government, 54 (1981), 14-20.

Cochran, Neal P. "Oil and Gas from Coal." Scientific American, 234 (May 1976),
   24-29.

Committee for Economic Development. Helping Insure Our Energy Future: A Program
   for Developing Synthetic Fuel Plants Now. New York: Committee for Economic
   Development, July 29, 1979.

Commoner, Barry. "Once and Future Fuel." New Yorker, October 29, 1979, 106-206.

Comptroller General. First Federal Attempt to Demonstrate a Synthetic Fossil Energy
   Technology - A Failure. [EMD-77-59] Washington, D.C.: U.S. General Accounting Office, August 17, 1977.

_____. Status and Obstacles to Commercialization of Coal Liquefaction
   and Gasification. [RED-76-81] Washington, D.C.: U.S. General Accounting
   Office, 1976.

Dow, Alan. Synthetic Fuels from Coal and Oil Shale. Lexington, Kentucky: Appalachia-
   Science in the Public Interest, 1981.

Ellington, Rex T., ed. Liquid Fuels from Coal. New York: Academic Press, 1977.

Exxon, U.S.A. The Role of Synthetic Fuels in the United States' Energy Future.
   Houston, Texas: Exxon U.S.A., 1980.

Goldman, Gordon Kenneth. Liquid Fuels from Coal, 1972. Park Ridge, New Jersey: Noyes Data Corporation, 1972.

Hammond, Allen L. "Coal Research II: Gasification Faces an Uncertain Future." Science, 193 (August 27, 1976a), 750-753.

_____ . "Coal Research III: Liquefaction Has Far To Go." Science, 193 (September 3, 1976b), 873-875.

Hammond, Ogden and Zimmerman, Martin B. "The Economics of Coal-Based Synthetic Gas." Technology Reveiw, 77 (July/August 1975), 42-51.

Hederman, W. F., Jr. Prospects for the Commercialization of High-BTU Coal Gasification. [Prepared for the Department of Energy] Santa Monica, California: RAND Corporation, 1978.

Horwitch, Mel. "Uncontrolled Growth and Unfocused Growth: Unsuccessful Life Cycles of Large-Scale, Public-Private, Technological Enterprises, With Special Reference to the United States SST Program and the United States Attempt to Develop Synthetic Fuels from Coal." [Paper presented at the Symposium on the Management of Science and Technology, Rio de Janeiro, June 22, 1978; to be published in Interdisciplinary Science Review]

Lamb, G. H. Underground Coal Gasification. Park Ridge, New Jersey: Noyes Data Corporation, 1977.

Marshall, E. "DOE Leads Synfuels Crusade without a Map." Science, 209 (September 12, 1980), 1208-1210.

Massey, Lester G., ed. Coal Gasification. Washington, D.C.: American Chemical Society, 1974.

Mills, Alex G. "Gas from Coal: Fuel of the Future." Environmental Science and Technology, December 1971, 1178-1183.

National Research Council. Assessment of Low and Intermediate BTU Gasification of Coal. Washington, D.C.: National Academy of Sciences, 1977a.

_____ . Assessment of Technology for the Liquefaction of Coal. Washington, D.C.: National Academy of Sciences, 1977b.

"Nazi Coal Conversion Methods Reviewed." Science, 196 (April 29, 1977), 508-509.

Nowacki, Perry. Coal Liquefaction Processes. Park Ridge, New Jersey: Noyes Data Corporation, 1979.

Pelofsky, Arnold H., ed. Synthetic Fuels Processing - Comparative Economics. New York: Marcel Dekker, 1977.

Perry, Harry. "The Gasification of Coal." Scientific American, 230 (March 1974), 19-25.

Richardson, Francis W. Oil from Coal. Park Ridge, New Jersey: Noyes Data Corporation, 1975.

Rosenbaum, Walter A. "Notes from No Man's Land." In Regina Axelrod, ed. Environment, Energy, and Public Policy. Lexington, Massachusetts: Lexington Books, 1980, pp. 61-79.

Schora, Frank C., Jr., et al. Fuel Gases from Coal. New York: MSS Information Corporation; Edison, New Jersey: Distributed by Arno Press, 1976.

Squires, Arthur M. "Clean Fuels from Coal Gasification." Science, 184 (April 19, 1974), 340-346.

Steinberg, Meyer. "Nuclear Power for the Production of Synthetic Fuels and Feedstocks." Energy Policy, 5 (March 1977), 12-24.

Swabb, L. E., Jr. "Liquid Fuels from Coal: From R&D to an Industry." Science, 199 (February 10, 1978), 619-622.

Synfuels Interagency Task Force. Synthetic Fuels Commercialization Program: Draft Environmental Statement. [Report prepared for the President's Energy Resource Council] Washington, D.C.: U.S. Government Printing Office, 1975.

United States. Atomic Energy Commission. Technical Information Center. Coal Processing: Gasification, Liquefaction, Desulfurization: A Bibliography, 1930-1974. Oak Ridge, Tennessee: U.S. Atomic Energy Commission, Office of Information Services, Technical Information Center, 1974.

United States. Department of Energy. Environmental Analysis of Synthetic Liquid Fuels. [DOE/EV-0044] Washington, D.C.: U.S. Department of Energy, July 12, 1979.

United States. Energy Research and Development Administration. Issues Relative to the Development and Commercialization of a Coal-Derived Synthetic Liquids Industry. [Report prepared by the Hudson Institute, Inc.] Croton-on-Hudson, New York: Hudson Institute, 1975.

_____. Synthetic Fuels Commercial Demonstration Program Fact Book. Washington, D.C.: Energy Research and Development Administration, 1976.

United States. Library of Congress. Congressional Research Service. The Pros and Cons of a Crash Program to Commercialize Synfuels. [Report prepared for the House Subcommittee on Energy Development and Applications of the House Committee on Science and Technology] Washington, D.C.: U.S. Government Printing Office, February 1980.

_____ . Synthetic Fuels from Coal: Status and Outlook of Coal Gasification and Liquefaction. [Printed at the request of the Committee on Energy and Natural Resources, United States Senate] [No. 96-17] Washington, D.C.: U.S. Government Printing Office, 1979.

Vietor, Richard H. K. The Synthetic Liquid Fuels Program: Energy Politics in the Truman Era. [Harvard Business School Working Paper No. 78-54] Cambridge, Massachusetts: Graduate School of Business Administration, Harvard University, 1978.

_____ . "The Synthetic Liquid Fuels Program: Energy Politics in the Truman Era." Business History Review, 54 (Spring 1980), 1-34.

3. Nuclear Power

GENERAL

Agnew, Harold M. "Gas-Cooled Nuclear Power Reactors." Scientific American, 244 (June 1981), 55-63.

Ahmed, S. Basheer. Nuclear Fuel and Energy Policy. Lexington, Massachusetts: Lexington Books, 1979.

Allen, Wendy. Nuclear Reactors for Generating Electricity: U.S. Development from 1946-1963. Santa Monica, California: RAND Corporation, June 1977.

The American Assembly, Columbia University. The Nuclear Power Controversy. Arthur W. Murphy, ed. Englewood Cliffs, New Jersey: Prentice-Hall, 1976.

Atlantic Council of the United States. Nuclear Fuels Policy Working Group. Nuclear Fuels Policy. Lexington, Massachusetts: Lexington Books, 1976.

Bacher, Robert. "Nuclear Energy and Our Future." Bulletin of the Atomic Scientists, 33 (March 1977), 63-65.

Barnett, Harold J. Atomic Energy in the United States Economy: A Consideration of Certain Industrial, Regional, and Economic Development Aspects. Stuart Bruchey, ed. New York: Arno Press, 1979.

Basheer, S. Nuclear Fuel and Energy Policy. Lexington, Massachusetts: Lexington Books, 1979.

Beckmann, Peter. The Health Hazards of Not Going Nuclear. Boulder, Colorado: Golem Press, 1976.

Benedict, Manson. "Electrical Power from Nuclear Fission." Bulletin of the Atomic Scientists, 27 (September 1971), 8-16.

Bethe, H. A. "The Necessity of Fission Power." Scientific American, 234 (January 1976), 21-31.

_____ . "The Need for Nuclear Power." Bulletin of the Atomic Scientists, 33 (March 1977), 59-63.

Bloustein, Edward J., ed. Nuclear Energy, Public Policy, and the Law. Dobbs Ferry, New York: Oceana Publications, 1964.

Bupp, Irvin C., Jr. "The Status of Nuclear Power: A Perspective from the United States." Ambio, 5 (1976), 119-123.

_____ and Derian, Jean-Claude. Light Water: How the Nuclear Dream Dissolved. New York: Basic Books, 1978. [Published in paperback as: The Failed Future of Nuclear Power. New York: Harper & Row, 1980]

_____ ; Donsimoni, Marie-Paule; and Treitel, Robert. "The Economics of Nuclear Power." Technology Review, 77 (February 1975), 14-25.

Burn, Duncan. Nuclear Power and the Energy Crisis: Politics and the Atomic Industry. New York: New York University Press, 1978.

_____ . The Political Economy of Nuclear Energy: An Economic Study of Contrasting Organizations in the U.K. and U.S.A., With Evaluation of their Effectiveness. [Research Monograph No. 9] London, England: Institute of Economic Affairs, 1967.

Burwell, C. C.; Ohanian, M. J.; and Weinberg, A. M. "A Siting Policy for an Acceptable Nuclear Future." Science, 204 (June 8, 1979), 1043-1051.

Chapman, Peter F. "Energy Analysis of Nuclear Power Stations." Energy Policy, 3 (December 1975), 285-290.

Cohen, Bernard L. Nuclear Science and Society. Garden City, New York: Anchor Press/Doubleday, 1974a.

_____ . "Perspectives on the Nuclear Debate." Bulletin of the Atomic Scientists, 30 (October 1974b), 35-39.

Comey, David Dinsmore. "Will Idle Capacity Kill Nuclear Power?" Bulletin of the Atomic Scientists, 30 (November 1974), 23-28.

Committee for Economic Development. Nuclear Energy and National Security. New York: Committee for Economic Development, 1976.

Comptroller General. Nuclear Power Costs and Subsidies. [EMD-79-52] Washington, D.C.: U.S. General Accounting Office, July 12, 1979a.

Comptroller General. Questions on the Future of Nuclear Power: Implications and Trade-Offs. [EMD-79-56] Washington, D.C.: U.S. General Accounting Office, 1979b.

Dahl, Robert A. and Brown, Ralph S., Jr. Domestic Control of Atomic Energy. Millwood, New York: Kraus Reprint Company, 1951.

Dale, Alfred. Nuclear Power Development in the U.S. to Nineteen Sixty: A New Pattern in Innovation and Technological Change. Stuart Bruchey, ed. New York: Arno Press, 1979.

Dawson, Frank G. Nuclear Power Development and Management of a Technology. Seattle, Washington: University of Washington Press, 1976.

Deffeyes, K. S. and MacGregor, I. D. "World Uranium Sources." Scientific American, 242 (January 1980), 66-87.

DeLeon, Peter. "Comparative Technology and Public Policy: The Development of the Nuclear Power Reactor in Six Nations." Policy Sciences, 11 (February 1980), 285-308.

Del Sesto, Steven L. Science, Politics, and Controversy: Civilian Nuclear Power in the United States, 1946-1974. Boulder, Colorado: Westview Press, 1979.

Duderstadt, James J. Nuclear Power. New York: Marcel Dekker, 1979.

Eicholz, Geoffrey G. Environmental Aspects of Nuclear Power. Ann Arbor, Michigan: Ann Arbor Science Publishers, 1976.

Eklund, Sigvard. "'We Must Move Forward With All Deliberate Speed'." Bulletin of the Atomic Scientists, 33 (October 1977), 42-47.

Fairchild, Johnson E. and Landman, David, eds. America Faces the Nuclear Age. White Plains, New York: Sheridan House, 1961.

Feiveson, Harold A., et al. "Fission Power: An Evolutionary Strategy." Science, 203 (January 26, 1979), 330-337.

Fenn, Scott. The Nuclear Power Debate: Issues and Choices. New York: Praeger Publishers, 1981.

Flowers, Brian. "Nuclear Power: A Perspective of the Risks, Benefits, and Options." Bulletin of the Atomic Scientists, 34 (March 1978), 21-26, 54-57.

Foley, Gerald and van Buren, Ariane, eds. Nuclear or Not? Choices for Our Energy Future. London, England: Heinemann Educational Books, 1978.

Ford, Fred T., et al. The Nuclear Fuel Cycle. Cambridge, Massachusetts: Union of Concerned Scientists; San Francisco, California: Friends of the Earth, 1974.

Foreman, Harry, ed. Nuclear Power and the Public. Minneapolis, Minnesota: University of Minnesota Press, 1970.

Francis, John and Albrecht, Paul, eds. Facing Up To Nuclear Power. Edinburgh, Scotland: Saint Andrew Press, 1976.

Gaines, Linda, II and Berry, R. Stephen. TOSCA: The Total Social Cost of Coal and Nuclear Power. Cambridge, Massachusetts: Ballinger Publishing Company, 1979.

Gandara, Arturo. Utility Decision-Making and the Nuclear Option. [R-2148-NSF] Santa Monica, California: The RAND Corporation, 1977.

Garvey, Gerald. Nuclear Power and Public Policy. Lexington, Massachusetts: Lexington Books, 1977a.

_____. Nuclear Power and Social Planning: The City of the Second Sun. Lexington, Massachusetts: Lexington Books, 1977b.

Geesaman, Donald P. and Abrahamson, Dean E. "The Dilemma of Fission Power." Bulletin of the Atomic Scientists, 30 (November 1974), 37-41.

Green, Harold P. "Public Participation in Nuclear Plant Licensing: The Great Decision." William and Mary Law Review, 15 (Spring 1974), 503-525.

_____ and Rosenthal, Alan. Government of the Atom: The Integration of Powers. New York: Atherton Press, 1963.

Hafele, Wolf and Manne, Alan S. "Strategies for a Transition from Fossil to Nuclear Fuels." Energy Policy, 3 (March 1975), 3-23.

Hafemeister, David W. "Nonproliferation and Alternative Nuclear Technologies." Technology Review, 81 (December 1978/January 1979), 58-63.

Hahn, Robert W. "An Assessment of the Determination of Energy Needs: The Case of Nuclear Power." Policy Sciences, 13 (February 1981), 9-24.

Hewlett, Richard G. and Duncan, Francis. Nuclear Navy, 1946-1962. Chicago, Illinois: University of Chicago Press, 1974.

Hilbery, Norman. "Nuclear Power in the U.S." Nuclear Industry, 11 (August 1964), 5-12.

Hodgetts, J. E. Administering the Atom for Peace. New York: Atherton Press, 1964.

Hogerton, John F. "The Arrival of Nuclear Power." Scientific American, 218 (February 1968), 21-31.

Hohenemser, Christopher. "The Distrust of Nuclear Power." Science, 196 (April 1, 1977), 25-34.

Hoyle, Fred. Energy or Extinction? The Case for Nuclear Energy. Exeter, New Hampshire: Heinemann Educational Books, 1977.

Huizenga, John R. "Nuclear Fission Revisited." Science, 168 (June 19, 1970), 1405-1413.

Hunt, S. E. Fission, Fusion, and the Energy Crisis. Elmsford, New York: Pergamon Press, 1974, 1979.

Ikle, Fred C. "Illusions and Realities About Nuclear Energy." Bulletin of the Atomic Scientists, 32 (October 1976), 14-17.

Inglis, David Rittenhouse. "Nuclear Energy and the Malthusian Dilemma." Bulletin of the Atomic Scientists, 27 (February 1971), 14-18.

_____. Nuclear Energy: Its Physics and Its Social Challenge. Reading, Massachusetts: Addison-Wesley Publishing Company, 1973.

Institute of Energy Analysis. Economic and Environmental Implications of the U.S. Nuclear Moratorium, 1985-2010. Cambridge, Massachusetts: M.I.T. Press, 1979.

International Atomic Energy Agency. Environmental Effects of Cooling Systems at Nuclear Power Plants. Vienna, Austria: International Atomic Energy Agency, 1975.

Johnson, Gerald W. "Plowshare at the Crossroads." Bulletin of the Atomic Scientists, 26 (June 1970), 83-91.

Kadiroglu, O. K.; Perlmutter, A.; and Scott, L., eds. Nuclear Energy and Alternatives. Cambridge, Massachusetts: Ballinger Publishing Company, 1978.

Kaiser, Karl. "The Great Nuclear Debate." Foreign Policy, 30 (Spring 1978), 83-110.

Karam, R. A., et al., eds. Environmental Impact of Nuclear Power Plants. Elmsford, New York: Pergamon Press, 1977.

Keating, William. Politics, Technology, and the Environment: Technology Assessment and Nuclear Energy. Stuart Bruchey, ed. New York: Arno Press, 1979.

Kleitman, Daniel J.; Rasmussen, Norman C.; Stewart, Richard B.; and Yellin, Joel. "Nuclear Power: Can We Live With It?" Technology Review, 81 (June/July 1979), 32-47.

Klema, Ernest D. and West, Robert L. Public Regulation of Site Selection for Nuclear Power Plants: Present Procedures and Reform Proposals: An Annotated Bibliography. Baltimore, Maryland: Published for Resources for the Future by the Johns Hopkins University Presss, 1977.

Klineberg, Otto. Social Implications of the Peaceful Uses of Nuclear Energy. New York: Unipub, 1964.

Kostuik, J. "Future Uranium Supply and Demand: Industrial and Commercial Considerations." Energy Policy, 5 (June 1977), 122-129.

Kuljian, Harry A. and Kramer, Andrew W. Energy Through Nuclear Reactors. Philadelphia, Pennsylvania: St. Joseph's College Press, 1972.

Lanouette, William J. "Nuclear Power - An Uncertain Future Grows Dimmer Still." National Journal, 11 (April 28, 1979), 676-686.

Lapp, Ralph. The Nuclear Controversy. Greenwich, Connecticut: Reddy Communications, 1975, 1977.

Libby, Leona M. The Uranium People. New York: Charles Scribner's Sons, 1979.

Lieberman, M. A. "United States Uranium Resources - An Analysis of Historical Data." Science, 192 (April 30, 1976), 431-436.

Lilienthal, David. Atomic Energy: A New Start. New York: Harper & Row, Publishers, 1980.

Long, F. A. "Peaceful Nuclear Explosions." Bulletin of the Atomic Scientists, 32 (October 1976), 19-28.

Lucas, N. J. D. "Hydrogen and Nuclear Power." Energy Policy, 4 (March 1976), 25-36.

McIntyre, Hugh C. "Natural-Uranium Heavy-Water Reactors." Scientific American, 233 (October 1975), 17-27.

Mann, Martin. Peaceful Uses of Atomic Energy. New York: Thomas Y. Crowell Company, 1975.

Manne, Alan S. and Richels, Richard G. "Evaluating Nuclear Fuel Cycles." Energy Policy, 8 (March 1980), 3-16.

Meyer, Leo A. Nuclear Power in Industry. 2nd ed. [Original title: Atomic Energy in Industry] Chicago, Illinois: American Technical Society, 1974.

Miller, Marvin. "The Nuclear Dilemma: Power, Proliferation, and Development." Technology Review, 81 (May 1979), 18-29.

Miller, Saunders. The Economics of Nuclear and Coal Power. New York: Praeger Publishers, 1976.

Mullenbach, Philip. Civilian Nuclear Power: Economic Issues and Policy Formulation. New York: The Twentieth Century Fund, 1963.

Muntzing, L. Manning, ed. Nuclear Power and Its Regulation in the United States. Elmsford, New York: Pergamon Press, 1980.

_____. "Siting and Environment: Essentials in an Effective Nuclear Siting Policy." Energy Policy, 4 (March 1976), 3-11.

Murphy, Arthur W., ed. The Nuclear Power Controversy. Englewood Cliffs, New Jersey: Prentice-Hall, 1976.

Myers, Desaix. The Nuclear Power Debate: Moral, Economic, Technical, and Political Issues. New York: Praeger Publishers, 1977.

Nadis, Steve. "Time for a Reassessment." Bulletin of the Atomic Scientists, 36 (February 1980), 37-44.

Nelkin, Dorothy. Nuclear Power and Its Critics: The Cayuga Lake Controversy. Ithaca, New York: Cornell University Press, 1971.

_____. "The Role of Experts in a Nuclear Siting Controversy." Bulletin of the Atomic Scientists, 30 (November 1974), 29-36.

"Nuclear Power in 1980: Its Problems and Prospects. Part I." [Special Issue] Bulletin of the Atomic Scientists, 36 (January 1980).

Nuclear Power Issues and Choices: Report of the Nuclear Energy Policy Study Group. S. N. Keeney, Chairman. Cambridge, Massachusetts: Ballinger Publishing Company for the MITRE Corporation and Ford Foundation, 1977.

Nye, Joseph S., Jr. "Time to Plan for the Next Generation of Nuclear Technology." Bulletin of the Atomic Scientists, 33 (October 1977), 38-41.

Parsegian, V. L. "New Goals for Atomic Energy." Bulletin of the Atomic Scientists, 27 (October 1971), 2-7.

Pendergrass, Bonnie B. Public Power, Politics, and Technology in the Eisenhower and Kennedy Years: The Hanford Dual-Purpose Reactor Controversy, 1956-1962. Stuart Bruchey, ed. New York: Arno Press, 1979.

Perlmutter, Arnold and Kadiroglu, Osman K., eds. Nuclear Energy and Alternatives. Cambridge, Massachusetts: Ballinger Publishing Company, 1978.

Perry, Robert L., et al. Development and Commercialization of the Light Water Reactor, 1946-1976. [R-2180-NSF] Santa Monica, California: RAND Corporation, August 1977.

Phillips, David G. Federal-State Relations and the Control of Atomic Energy. Stuart Bruchey, ed. New York: Arno Press, 1979.

Pringle, Lawrence. Nuclear Power: From Physics to Politics. New York: Macmillan, 1979.

Richardson, Robert A. "The Selling of the Atom." Bulletin of the Atomic Scientists, 30 (October 1974), 28-35.

Rickard, Corwin L. and Dahlberg, Richard C. "Nuclear Power: A Balanced Approach." Science, 202 (November 10, 1978), 581-584.

Rogers, Franklyn C. "Underground Nuclear Power Plants." Bulletin of the Atomic Scientists, 27 (October 1971), 38-41, 51.

Rolph, Elizabeth. Regulation of Nuclear Power: The Case of the Light Water Reactor. [R-2104-NSF] Santa Monica, California: RAND Corporation, June 1977.

Rose, David J. "Nuclear Electric Power." Science, 184 (April 19, 1974), 351-359.

Rossin, A. D. and Rieck, T. A. "Economics of Nuclear Power." Science, 201 (August 18, 1978), 582-589.

Rush, Howard J.; MacKerron, Gordon; and Surrey, John. "The Advanced Gas-Cooled Reactor: A Case Study in Reactor Choice." Energy Policy, 5 (June 1977), 95-105.

Sagan, Leonard A., ed. Human and Ecologic Effects of Nuclear Power Plants. Springfield, Illinois: Charles C. Thomas, Publisher, 1974.

_____ . "Human Costs of Nuclear Power." Science, 177 (August 11, 1972), 487-493.

Schmidt, Fred H. and Bodansky, David. The Fight Over Nuclear Power. San Francisco, California: Albion, 1976.

Schurr, S. H. and Marschak, J. Economic Aspects of Atomic Power. Princeton, New Jersey: Princeton University Press, 1950.

Seaborg, Glenn T. "On Misunderstanding the Atom." Bulletin of the Atomic Scientists, 27 (September 1971), 46-53.

_____ . "Our Nuclear Future - 1995." Bulletin of the Atomic Scientists, 26 (June 1970), 7-14.

Sefcik, Joseph A. "Decommissioning Commercial Nuclear Reactors." <u>Technology Review</u>, 81 (June/July 1979), 56-71.

Shapley, Deborah. "Nuclear Navy: Rickover Thwarted Research on Light Water Reactors." <u>Science</u>, 192 (June 18, 1976), 1210-1213.

_____. "Nuclear Power Plants: Why Do Some Work Better Than Others?" <u>Science</u>, 195 (March 25, 1977), 1311-1313.

Sherfield, Lord, ed. <u>Economic and Social Consequences of Nuclear Energy</u>. London, England: Oxford University Press, 1972.

Shurcliff, Alice W. "The Local Economic Impact of Nuclear Power." <u>Technology Review</u>, 79 (January 1977), 40-47.

Solomon, Norman. "Nuclear Big Brother." <u>The Progressive</u>, 44 (January 1980), 14-21.

Spangler, Miller B. "Environmental and Social Issues of Site Choice for Nuclear Power Plants." <u>Energy Policy</u>, 2 (March 1974), 18-32.

Speth, Gus. "The Nuclear Recession." <u>Bulletin of the Atomic Scientists</u>, 34 (April 1978), 24-27.

Srouji, Jacque. <u>Critical Mass: Nuclear Power, the Alternative to Energy Famine</u>. Nashville, Tennessee: Aurora Publishers, 1977.

Surrey, John. "The Future Growth of Nuclear Power. Part 1: Demand and Supply." <u>Energy Policy</u>, 1 (September 1973), 107-129.

_____. "The Future Growth of Nuclear Power. Part 2: Choices and Obstacles." <u>Energy Policy</u>, 1 (December 1973), 208-224.

_____ and Thomas, Steve. "Worldwide Nuclear Plant Performance." <u>Futures</u>, 12 (February 1980), 3-17.

Temples, James R. "The Politics of Nuclear Power: A Subgovernment in Transition." <u>Political Science Quarterly</u>, 95 (Summer 1980), 239-260.

Thomas, Morgan. <u>Atomic Energy and Congress</u>. Ann Arbor, Michigan: University of Michigan Press, 1956.

United States. Congress. Senate. "Amending the Atomic Energy Act of 1946 as Amended and for Other Purposes." In <u>Legislative History of the Atomic Energy Act of 1954</u>. [PL-83-703; Senate Report No. 83-1699] Washington, D.C.: U.S. Government Printing Office, 1951.

Von Hippel, Frank and Williams, Robert H. "Energy Waste and Nuclear Power Growth." <u>Bulletin of the Atomic Scientists</u>, 32 (December 1976), 18-21, 48-56.

Weaver, Kenneth. "The Promise and Peril of Nuclear Power." National Geographic Magazine, 155 (April 1979), 459-493.

Weinberg, Alvin M., ed. Economic and Environmental Implications of a U.S. Nuclear Moratorium. Cambridge, Massachusetts: M.I.T. Press, 1979.

_____. "Is Nuclear Energy Acceptable?" Bulletin of the Atomic Scientists, 33 (April 1977), 54-60.

_____. "Nuclear Energy and the Environment." Bulletin of the Atomic Scientists, 26 (June 1970), 69-74.

_____. "Nuclear Energy: Salvaging the Atomic Age." Wilson Quarterly, 3 (Summer 1979), 88-112.

_____. "Reflections on the Energy Wars." American Scientist, 66 (March-April 1978), 153-158.

_____. "Social Institutions and Nuclear Energy." Science, 177 (July 7, 1972), 27-34.

Wilkes, Owen and Mann, Robert. "The Story of Nukey Poo." Bulletin of the Atomic Scientists, 34 (October 1978), 32-36.

Wilson, Carroll L. "Nuclear Energy: What Went Wrong?" Bulletin of the Atomic Scientists, 35 (June 1979), 13-17.

Yanarella, Ernest J. "The Politics of the 'Peaceful Atom' from the Manhattan Project to Three Mile Island." Peace and Change, 6 (Spring 1981), 45-58.

REACTOR SAFETY

Ames, Mary. "The Case of the North Anna Nuclear Power Plant: Public Risk and Public Relations." In Outcome Uncertain: Science and the Public Process. Washington, D.C.: Communications Press, 1978, pp. 83-112.

Behring, Charles. "WASH-1250: Is It a Nuclear Whitewash?" Environmental Action, 5 (October 13, 1973), 5-7.

Chicken, J. C. Nuclear Power Hazard Control Policy. Elmsford, New York: Pergamon Press, 1981.

Del Sesto, Steven L. "Conflicting Ideologies of Nuclear Power : Congressional Testimony on Nuclear Reactor Safety." Public Policy, 28 (Winter 1980), 39-70.

Finlayson, Fred C. "A View from the Outside." Bulletin of the Atomic Scientists, 31 (September 1975), 20-25.

Fischer, David W. "Planning for Large-Scale Accidents: Learning from the Three Mile Island Accident." Energy - The International Journal, 6 (January 1981), 93-108.

Fuller, John G. We Almost Lost Detroit. New York: Readers's Digest Press, 1975; New York: Ballantine Books, 1976.

Gillette, Robert. "Nuclear Reactor Safety: A New Dilemma for the AEC." Science, 173 (July 9, 1971), 126-130.

_____. "Nuclear Reactor Safety: At the AEC the Way of the Dissenter is Hard." Science, 176 (May 5, 1972), 492-498.

_____. "Nuclear Safety (I): The Roots of Dissent." Science, 177 (September 1, 1972), 771-776.

_____. "Nuclear Safety (II): The Years of Delay." Science, 177 (September 8, 1972), 867-871.

_____. "Nuclear Safety (III): Critics Charge Conflicts of Interest." Science, 177 (September 15, 1972), 970-975.

_____. "Nuclear Safety (IV): Barriers to Communication." Science, 177 (September 22, 1972), 1080-1082.

_____. "Nuclear Safety: AEC Report Makes the Best of It." Science, 179 (January 26, 1973), 360-363.

_____. "Nuclear Safety: Calculating the Odds of Disaster." Science, 185 (September 6, 1974), 838-839.

_____. "Radiation Spill at Hanford: The Anatomy of an Accident." Science, 181 (August 24, 1973), 728-730.

_____. "Reactor Safety: AEC Conceded Some Points to Its Critics." Science, 178 (November 3, 1972), 482-484.

Golay, Michael W. "How Prometheus Came to be Bound: Nuclear Regulation in America." Technology Review, 82 (June/July 1980), 28-41.

Graham, John. Fast Reactor Safety. New York: Academic Press, 1971.

Hendrie, J. M. "Safety of Nuclear Power." In Jack M. Hollander and Melvin K. Simmons, eds. Annual Review of Energy. Volume 1. Palo Alto, California: Annual Reviews, Inc., 1976, pp. 663-683.

Holdren, John P. "Hazards of the Nuclear Fuel Cycle." Bulletin of the Atomic Scientists, 30 (October 1974), 14-23.

Inglis, David Rittenhouse. "The Hazardous Industrial Atom." Bulletin of the Atomic Scientists, 26 (February 1970), 50-54.

International Atomic Energy Agency. Design for Safety of Nuclear Power Plants. New York: Unipub, 1979.

_____. Nuclear Power and Its Fuel Cycle: Nuclear Safety. Volume 5. New York: Unipub, 1978.

_____. Safety in Nuclear Power Plant Operation, Including Commissioning and Decommissioning. New York: Unipub, 1979.

Is Nuclear Power Safe? Washington, D.C.: American Enterprise Institute for Public Policy Research, 1975.

Kasperson, Roger, et al. "Institutional Responses to Three Mile Island." Bulletin of the Atomic Scientists, 35 (December 1979), 20-24.

Kemeny, John G. "Saving American Democracy: The Lessons of Three Mile Island." Technology Review, 83 (June/July 1980), 64-75.

Kleitman, Daniel J.; Rasmussen, Norman C.; Stewart, Richard B.; and Yellin, Joel. "Nuclear Power: Can We Live With It?" Technology Review, 81 (June/July 1979), 32-47.

Kouts, Herbert J. C. "The Future of Reactor Safety Research." Bulletin of the Atomic Scientists, 31 (September 1975), 32-37.

Lewis, E. E. Nuclear Power Reactor Safety. New York: Wiley Interscience, 1977.

Lewis, Harold. "The Safety of Fission Reactors." Scientific American, 242 (March 1980), 53-65.

McCracken, Samuel. "The Harrisburg Syndrome." Commentary, 67 (June 1979), 27-37.

_____. The War Against the Atom. New York: Basic Books, 1981.

Martin, Daniel W. Three Mile Island: Prologue or Epilogue? Cambridge, Massachusetts: Ballinger Publishing Company, 1980.

The Need for Change: The Legacy of TMI -- Report of the President's Commission on the Accident at Three Mile Island. John Kemeny, Chairman. Washington, D.C.: Executive Office, October 31, 1979.

Nuclear Safety Analysis Center. Analysis of Three Mile Island – Unit 2 Accident. Palo Alto, California: Electric Power Research Institute, July 1979.

Peterson, Russell W. "Lessons of Three Mile Island." Technology in Society, 2 (1980), 295-302.

Primack, Joel. "Nuclear Reactor Safety: An Introduction to the Issues." Bulletin of the Atomic Scientists, 31 (September 1975), 15-19.

_____ and von Hippel, Frank. "Nuclear Reactor Safety. Bulletin of the Atomic Scientists, 30 (October 1974), 5-12.

Public's Right to Information Task Force. Staff Report to the President's Commission on the Accident at Three Mile Island. Washington, D.C.: U.S. Government Printing Office, 1979.

Rasmussen, Norman C. "The Safety Study and Its Feedback." Bulletin of the Atomic Scientists, 31 (September 1975), 25-28.

Reid, Robert G. "The View from Middletown." SIPI Scope (March-June 1979), 11-13.

Rogovin, M. and Frampton, G. T. Three Mile Island: A Report to the Commissioner and to the Public. [NUREG/CR-1250] 2 Volumes. Washington, D.C.: U.S. Nuclear Regulatory Commission, April 6, 1979.

Rolph, Elizabeth S. Nuclear Power and the Public Safety: A Study in Regulation. Lexington, Massachusetts: Lexington Books, 1979. [Originally published as: Regulation of Nuclear Power: The Case of the Light Water Reactor. [R-2104-NSF] Santa Monica, California: RAND Corporation, June 1977.]

Russell, C. R. Reactor Safeguards. Elmsford, New York: Pergamon Press, 1979.

Shapley, Deborah. "Reactor Safety: Independence of Rasmussen Study Doubted." Science, 197 (July 1, 1977), 29-31.

Sheridan, Thomas B. "Human Error in Nuclear Power Plants." Technology Review, 82 (February 1980), 22-33.

Sills, David L.; Wolf, C. P.; and Shelanski, Vivien B., eds. Accident at Three Mile Island: The Human Dimensions. Boulder, Colorado: Westview Press, 1981.

Sobel, Lester, A., ed. Atomic Energy and the Safety Controversy. New York: Facts on File, 1978.

Sorenson, Bent. "Nuclear Power: The Answer That Became a Question: An Assessment of Accident Risks." Ambio, 8 (1979), 10-17.

Stever, Donald W., Jr. Seabrook and the Nuclear Regulatory Commission: The Licensing of a Nuclear Power Plant. Hanover, New Hampshire: University Press of New England, 1980.

Sylves, Richard T. "Carter's Nuclear Licensing Reform vs. Three Mile Island." Publius, 10 (Winter 1980), 69-80.

Thomas, Steve and Surrey, John. "What Makes Nuclear Power Plants Break Down?" Techology Review, 83 (May/June 1981), 56-67.

"Three Mile Island and the Future of Nuclear Power." IEEE Spectrum [Special Issue] 16 (November 1979).

United States. Atomic Energy Commisssion. Reactor Safety Study: An Assessment of Accident Risks in U.S. Commercial Nuclear Power Plants. [WASH-1400] Norman Rasmussen, Chairman. Washington, D.C.: Atomic Energy Commission, August 1974 [Draft]; Nuclear Regulatory Commission, October 1975 [Final].

von Hippel, Frank. "Looking Back on the Rasmussen Report." Bulletin of the Atomic Scientists, 33 (February 1977), 42-47.

Webb, Richard E. The Accident Hazards of Nuclear Power Plants. Amherst, Massachusetts: University of Massachusetts Press, 1976.

Weingast, Barry. "Congress, Regulation, and the Decline of Nuclear Power." Public Policy, 28 (Spring 1980), 231-255.

Welch, B. L. "Deception on Nuclear Power Risks - A Call for Action. Bulletin of the Atomic Scientists, 36 (September 1980), 50-54.

Yanarella, Ann-Marie and Yanarella, Ernest J., eds. Three Mile Island and Marble Hill: Nuclear Power and Political Control. [Proceedings of a Symposium, University of Kentucky, October 18, 1979] Lexington, Kentucky: University of Kentucky, Office of the Dean of Undergraduate Studies, 1980.

RADIATION EFFECTS

Ad Hoc Population Dose Assessment Group. Population Dose and Health Impact of the Accident at Three Mile Island Nuclear Station. [A preliminary assessment for the period March 28 - April 7, 1979] Washington, D.C.: U.S. Government Printing Office, 1979.

Anderson, T. C. "Radiation Exposures of Hanford Workers: Critique of the Mancuso, Stewart and Kneale Report." Health Physics, 35 (1978), 743-750.

Barnaby, Frank. "The Continuing Body Count at Hiroshima and Nagasaki." Bulletin of the Atomic Scientists, 33 (December 1977), 48-53.

"The BEIR Report: Effects on Populations of Exposure to Low Levels of Ionizing Radiation." Bulletin of the Atomic Scientists, 29 (March 1973), 47-49.

Beirly, Eugene W. and Klement, Alfred W., Jr. "Radioactive Fallout from Nuclear Weapons Tests." Science, 147 (February 26, 1965), 1057-1060.

Boffey, Philip M. "Hiroshima/Nagasaki - Atomic Bomb Casualty Commission Perseveres in Sensitive Studies." Science, 168 (May 8, 1970), 679-683.

_____. "Radiation Standards: Are the Right People Making Decisions?" Science, 171 (February 26, 1971) , 780-783.

Caldicott, Helen. Nuclear Madness: What You Can Do. Ed. by Nahum Stickin. Brookline Massachusetts: Autumn Press, 1979.

Carter, Luther J. "Uranium Mill Tailings." Science, 202 (October 13, 1978), 191-195.

Cohen, Bernard L. "Environmental Impacts of Nuclear Power Due to Radon Emissions." Bulletin of the Atomic Scientists, 32 (February 1976), 61-63.

_____. "What Is the Misunderstanding All About?" Bulletin of the Atomic Scientists, 35 (February 1979), 53-56.

Comey, David Dinsmore. "The Legacy of Uranium Tailings." Bulletin of the Atomic Scientists, 31 (September 1975), 43-45.

Committee for the Compilation of Materials on Damage Caused by the Atomic Bombs in Hiroshima and Nagasaki. Hiroshima and Nagasaki: The Physical, Medical, and Social Effects of the Atomic Bombings. New York: Basic Books, 1981.

Comptroller General. The Environmental Protection Agency Needs Congressional Guidance and Support to Guard the Public in a Period of Radiation Proliferation. [CED-78-27] Washington, D.C.: U.S. General Accounting Office, January 20, 1978.

Ebert, Michael and Howard, Alma, eds. Current Topics in Radiation Research. Amsterdam, The Netherlands: North-Holland; New York: Wiley Interscience, 1969.

Edsall, John T. "Toxicity of Plutonium and Some Other Actinides." Bulletin of the Atomic Scientists, 32 (September 1976), 27-37.

Environmental Policy Institute. Radiation Standards and Public Health. Washington, D.C.: Environmental Policy Institute, 1978.

Fetter, Steven A. and Tsipis, Kosta. "Catastrophic Releases of Radioactivity." Scientific American, 244 (April 1981), 41-47.

Gillette, Robert. "Plutonium (I): Questions of Health in a New Industry." Science, 185 (September 20, 1974), 1027-1032.

_____. "Plutonium (II): Watching and Waiting for Adverse Effects." Science, 185 (September 27, 1974), 1141-1143.

161

Gillette, Robert. "'Transient' Nuclear Workers: A Special Case for Standards." Science, 186 (October 11, 1974), 125-129.

Griffiths, Joel and Ballantine, Richard. Silent Slaughter. Chicago, Illinois: Henry Regnery Company, 1972.

Hamilton, L. D. "On Radiation Standards." Bulletin of the Atomic Scientists, 28 (March 1972), 30-33.

Holden, Constance. "Low-Level Radiation: A High-Level Concern." Science, 204 (April 13, 1979), 155-158.

International Atomic Energy Agency. Nuclear Power and Its Fuel Cycle: Radioactivity Management. Volume 4. New York: Unipub, 1978.

Lapp, Ralph E. The Radiation Controversy. Greenwich, Connecticut: Reddy Communications, 1979.

Lindop, Patricia J. and Rotblat, J. "Radiation Pollution of the Environment." Bulletin of the Atomic Scientists, 27 (September 1971), 17-24.

Mancuso, T. F.; Stewart, A.; and Kneale, G. "Radiation Exposures of Hanford Workers Dying from Cancer and Other Causes." Health Physics, 33 (1977), 369-385.

Marshall, Eliot. "NAS Study on Radiation Takes the Middle Road." Science, 204 (May 18, 1979), 711-714.

Marx, Jean L. "Low-Level Radiation: Just How Bad Is It?" Science, 204 (April 13, 1979), 160-164.

Morgan, Karl Z. "Cancer and Low-Level Ionizing Radiation." Bulletin of the Atomic Scientists, 34 (September 1978), 30-41.

Najarian, Thomas. "The Controversy Over the Health Effects of Radiation." Technology Review, 81 (November 1978), 74-82.

National Council on Radiation Protection and Measurement. Natural Background Radiation in the United States. [NRCP Report No. 45] Washington, D.C.: National Council on Radiation Protection and Measurement, 1975.

National Research Council. Division of Medical Sciences. Committee on the Biological Effects of Ionizing Radiation. The Effects on Populations of Exposure to Low Levels of Ionizing Radiation: Report of the Advisory Committee on the Biological Effects of Ionizing Radiation. [The BEIR Report] Washington, D.C.: National Academy of Science, 1972, 1974.

Rotblat, J. "The Risks for Radiation Workers." Bulletin of the Atomic Scientists, 34 (September 1978), 41-46.

Sternglass, Ernest J. Low-Level Radiation. New York: Ballantine Books, 1972.

_____. Secret Fallout: Low-Level Radiation from Hiroshima to Three Mile Island. New York: McGraw-Hill, 1981.

Tamplin, Arthur. "The BEIR Report: A Focus on Issues." Bulletin of the Atomic Scientists, 29 (May 1973), 19-20.

_____ and Gofman, John W. "The Radiation Effects Controversy." Bulletin of the Atomic Scientists, 26 (September 1970), 2, 5-8.

Thompson, Theos J. and Bibb, William R. "The AEC Position: Response to Gofman and Tamplin." Bulletin of the Atomic Scientists, 26 (September 1970), 9-12, 48.

United Nations. Scientific Committee on the Effects of Atomic Radiation. Sources and Effects of Ionizing Radiation, 1977. New York: United Nations, 1977.

Woolard, Robert F. and Young, Eric R. Health Dangers of the Nuclear Fuel Chain and Low-Level Ionizing Radiation: A Bibliograpy/Literature Review. Cambridge, Massachusetts: Physicians for Social Responsibility, 1979.

NUCLEAR WASTE DISPOSAL

Angino, Ernest E. "High-Level and Long-Lived Radioactive Waste Disposal." Science, 198 (December 2, 1977), 885-890.

Carter, Luther J. "Congressional Committees Ponder Whether to Give States a Right of Veto Over Radioactive Waste Repositories." Science, 200 (June 9, 1978), 1136-1137.

_____. "Nuclear Wastes: Popular Antipathy Narrows Search for Disposal Sites." Science, 197 (September 23, 1977), 1265-1266.

_____. "Nuclear Wastes: The Science of Geologic Disposal Seen as Weak." Science, 200 (June 9, 1978) 1135-1137.

_____. "Radioactive Wastes: Some Urgent Unfinished Business." Science, 195 (February 18, 1977), 661-666, 704.

_____. "West Valley: The Question Is Where Does Buck Stop On Nuclear Wastes." Science, 195 (March 1977), 1306-1308.

Cohen, Bernard L. "The Disposal of Radioactive Wastes from Fission Reactors." Scientific American, 236 (June 1977), 21-31.

Deese, David A. Nuclear Power and Radioactive Waste: A Sub-Seabed Disposal Option? Lexington, Massachusetts: Lexington Books, 1978.

DeMarsily, G., et al. "Nuclear Waste Disposal: Can the Geologist Guarantee Isolation? Science, 197 (August 5, 1977), 519-527.

Hambleton, William W. "The Unsolved Problem of Nuclear Wastes." Technology Review, 74 (March/April 1972), 15-19.

Hanrahan, David. "Hazardous Wastes: Current Problems and Near-Term Solutions." Technology Review, 82 (November 1979), 20-31.

Hebel, Charles L., et al. Report to the American Physical Society [by the Study Group on Nuclear Fuel Cycles and Waste Management]. Washington, D.C.: American Physical Society, 1977.

Hewlett, Richard G. Federal Policy for Disposal of Radioactive Waste from Commercial Nuclear Power Plants. Washington, D.C.: U.S. Department of Energy, March 1979.

International Atomic Energy Agency. Radioactive Wastes. Vienna, Austria: International Atomic Energy Agency, 1978.

Jakimo, Alan and Bupp, Irvin C. "Nuclear Waste Disposal: Not in My Backyard." Technology Review, 80 (March/April 1978), 64-72.

Kerr, Richard A. "Geologic Disposal of Nuclear Wastes: Salt's Lead is Challenged." Science, 204 (May 11, 1979), 603-606.

_____. "Nuclear Waste Disposal: Alternatives to Solidification in Glass Proposed." Science, 204 (April 20, 1979), 289-291.

Krugman, Hartmut and von Hippel, Frank. "Radioactive Wastes: A Comparison of U.S. Military and Civilian Inventories." Science, 197 (August 26, 1977), 883-884.

Kubo, Arthur S. and Rose, David J. "Disposal of Nuclear Wastes." Science, 182 (December 21, 1973), 1205-1211.

LaPorte, Todd. "Nuclear Wastes: Increasing Scale and Sociopolitical Impacts." In Charles T. Unseld, et al., eds. Sociopolitical Effects of Energy Use and Policy. Washington, D.C.: National Academy of Sciences, 1979, pp. 355-373.

Lapp, R. E. Radioactive Waste: Society's Problem Child. Greenwich, Connecticut: Reddy Communications, 1977.

Lee, K. N. "A Federalist Strategy for Nuclear Waste Management." Science, 208 (May 16, 1980), 679-684.

Lester, Richard K. and Rose, David J. "The Nuclear Wastes at West Valley, New York." Technology Review, 79 (May 1977), 20-29.

Lewis, Richard S. "The Radioactive Salt Mine." Bulletin of the Atomic Scientists, 27 (June 1971), 27-30.

Lipschutz, Ronnie. Radioactive Waste: Politics, Technology, and Risk. Cambridge, Massachusetts: Ballinger Publishing Company, 1980.

Management of Plutonium-Containing Solid Wastes. Paris, France: Organization for Economic Cooperation and Development, Nuclear Energy Agency, 1975.

Maugh, Thomas H., II. "Burial is Last Resort for Hazardous Wastes." Science, 204 (June 22, 1979), 1295-1298.

_____. "Hazardous Waste Technology is Available." Science, 204 (June 1, 1979), 930-933.

_____. "Toxic Waste Disposal A Growing Problem." Science, 204 (May 25, 1979), 819-923.

Rochlin, Gene I. "Nuclear Waste Disposal: Two Social Criteria." Science, 195 (January 7, 1977), 23-31.

_____; Held, Margery; Kaplan, Barbara G.; and Kruger, Lewis. "West Valley: Remnant of the AEC." Bulletin of the Atomic Scientists, 34 (January 1978), 17-26.

Smith, Kirk R. "Military Uses of Uranium: Keeping the U.S. Energy Accounts." Science, 201 (August 18, 1978), 609-611.

Sweet, William "Unresolved: The Front End of Nuclear Waste Disposal." Bulletin of the Atomic Scientists, 35 (May 1979), 44-48.

United States. Department of Energy. Report to the President by the Interagency Review Group on Nuclear Waste Management. John M. Deutch, Chairman. [TID 28817; Draft] Springfield, Virginia: National Technical Information Service, October 1978.

United States. Department of Energy. Directorate of Energy Research. Report of the Task Force for Review of Nuclear Waste Management. [DOE/ER-0004/D] Washington, D.C.: U.S. Department of Energy, February 1978.

United States. Environmental Protection Agency. Everybody's Problem: Hazardous Waste. Washington, D.C.: U.S. Government Printing Office, 1979.

United States. Executive Branch. Office of Science and Technology Policy. Alternative Technology Strategies for the Isolation of Nuclear Wastes: Report of Subgroup One. Washington, D.C.: U.S. Government Printing Office. September 1978.

Zeller, E. J.; Saunders, D. F.; and Angino, E. E. "Putting Radioactive Wastes on Ice: A Proposal for an International Radionuclide Depository in Antarctica." Bulletin of the Atomic Scientists, 29 (January 1973), 4-9, 50-52.

Zinberg, Dorothy. "The Public and Nuclear Waste Management." Bulletin of the Atomic Scientists, 35 (January 1979), 34-39.

NUCLEAR SAFEGUARDS

Atlantic Council Working Group on Nuclear Fuels Policy. Nuclear Power and Nuclear Weapons Proliferation. 2 Volumes. Ed. by John E. Gray and Joseph W. Harned. Boulder, Colorado: Westview Press, 1978.

Baker, Steven J. Commercial Nuclear Power and Nuclear Proliferation. [An Occasional Paper of the Cornell University Peace Studies Program] Ithaca, New York: Cornell University Peace Studies Program, 1975.

Brenner, Michael. Splicing the Atom: The Remaking of U.S. Non-Proliferation Policy. New York: Cambridge University Press, 1980.

Coffey, Joseph I. and Lambert, Richard D., eds. Nuclear Proliferation: Prospects, Problems, and Proposals. Philadelphia, Pennsylvania: American Academy of Political and Social Science, 1977.

Cohen, Bernard L. "The Potentialities of Terrorism." Bulletin of the Atomic Scientists, 32 (June 1976), 34-35.

Comey, David Dinsmore. "The Perfect Trojan Horse." Bulletin of the Atomic Scientists, 32 (June 1976), 33-34.

Dunn, L. A. "Nuclear 'Gray Marketing'." International Security, 1 (Winter 1977), 107-118.

Edelhertz, Herbert and Walsh, Marilyn. The White-Collar Challenge to Nuclear Safeguards. Lexington, Massachusetts: Lexington Books, 1978.

Feiveson, Harold A. and Taylor, Theodore B. "Security Implications of Alternative Fission Futures." Bulletin of the Atomic Scientists, 32 (December 1976), 14-18, 46-48.

Flood, Michael. "Nuclear Sabotage." Bulletin of the Atomic Scientists, 32 (October 1976), 29-36.

Franko, Lawrence G. "U.S. Regulation of the Spread of Nuclear Technologies Through Supplier Power." Law and International Business, 10 (1978), 1180-1204.

Gilinsky, Victor. "Plutonium, Proliferation, and Policy." Technology Review, 79 (February 1977), 58-65.

Greenwood, Ted, et al. Nuclear Proliferation: Motivation, Capabilities, and Strategies for Control. New York: McGraw-Hill, 1977.

Guhin, Michael A. Nuclear Paradox: Security Risks of the Peaceful Atom. Washington, D.C.: American Enterprise Institute for Public Policy Research, 1976.

International Atomic Energy Agency. Nuclear Power and Its Fuel Cycle: Nuclear Power and Public Opinion and Safeguards. New York: Unipub, 1978.

Johnson, Brian. "Nuclear Power Proliferation: Problems of International Control." Energy Policy, 5 (September 1977), 179-194.

Krieger, David. "Terrorists and Nuclear Technology." Bulletin of the Atomic Scientists, 31 (June 1975), 28-34.

Leachman, Robert B. and Althoff, Philip,eds. Preventing Nuclear Theft: Guidelines for Industry and Government. New York: Irvington Publishers, 1972.

Lovins, Amory B. and Lovins, L. Hunter. Energy/War: Breaking the Nuclear Link. New York: Harper & Row, Publishers, 1980.

_____ and Ross, Leonard. "Nuclear Power and Nuclear Bombs." Foreign Affairs, 58 (Summer 1980), 1137-1177.

Nye, Joseph S., Jr. "Balancing Nonproliferation and Energy Security." Technology Review, 81 (December 1978/January 1979), 48-57.

Pierre, Andrew J. and Moyne, Claudia W. Nuclear Proliferation: A Strategy for Control. New York: Foreign Policy Association, 1976.

Rose, David J. and Lester, Richard K. "Nuclear Power, Nuclear Weapons, and International Stability." Scientific American, 238 (April 1978), 45-58.

Schleimer, Joseph D. "The Day They Blew Up San Onofre." Bulletin of the Atomic Scientists, 30 (October 1974), 24-27.

Shapley, Deborah. "Plutonium: Reactor Proliferation Threatens a Nuclear Black Market." Science, 172 (April 9, 1971), 143-146.

United States. Congress. Office of Technology Assessment. Nuclear Proliferation and Safeguards. New York: Praeger Publishers, 1977.

Weiss, Leonard. "Nuclear Safeguards: A Congressional Perspective." Bulletin of the Atomic Scientists, 34 (March 1978), 27-33.

Williams, Frederick C. and Deese, David A., eds. Nuclear Nonproliferation: The Spent Fuel Problem. Elmsford, New York: Pergamon Press, 1979.

Willrich, Mason. "Terrorists Keep Out." Bulletin of the Atomic Scientists, 31 (May 1975), 12-16.

Wilson, Richard. "How to Have Nuclear Power Without Weapons Proliferation." Bulletin of the Atomic Scientists, 33 (November 1977), 39-44.

Wohlstetter, Albert, et al. Swords from Plowshares: The Military Potential of Civilian Nuclear Energy. Chicago, Illinois: University of Chicago Press, 1979.

_____; Gilinsky, Victor; Gillette, Robert; and Wohlstetter, Roberta. Nuclear Policies: Fuel Without the Bomb. Cambridge, Massachusetts: Ballinger Publishing Company, 1979.

REPROCESSING AND ENRICHMENT

Bebbington, William P. "The Reprocessing of Nuclear Fuels." Scientific American, 235 (December 1976), 30-41.

Casper, Barry M. "Laser Enrichment: A New Path to Proliferation." Bulletin of the Atomic Scientists, 33 (January 1977), 28-41.

Fialka, John. "'Strange Alchemy' in Barnwell, S. C." Environmental Action, 7 (August 16, 1975), 4-8.

Geesaman, Donald P. "Plutonium and the Energy Decision." Bulletin of the Atomic Scientists, 27 (September 1971), 33-36.

Glackin, James J. "The Dangerous Drift in Uranium Enrichment." Bulletin of the Atomic Scientists, 32 (February 1976), 22-29.

Hammond, Allen L. "Uranium: Will There Be a Shortage or an Embarrassment of Enrichment?" Science, 192 (May 28, 1976), 866-867.

Krass, Allan S. "Laser Enrichment of Uranium: The Proliferation Connection." Science, 196 (May 13, 1977), 721-731.

Metz, William D. "Laser Enrichment: Time Clarifies the Difficulty." Science, 191 (March 19, 1976), 1162-1163, 1193.

_____. "Reprocessing Alternatives: The Options Multiply." Science, 196 (April 15, 1977), 284-287.

_____. "Reprocessing: How Necessary Is It for the Near Term?" Science, 196 (April 1, 1977), 43-45.

Speth, Gus; Tamplin, Arthur and Cochran, Thomas. "Plutonium Recycle or Civil Liberties? We Can't Have Both." Environmental Action, 6 (December 7, 1974), 10-13.

Walsh, John. "Fuel Reprocessing Still the Focus of U.S. Nonproliferation Policy." Science, 201 (August 25, 1978), 692-697.

_____. "Uranium Enrichment: Both the Americans and Europeans Must Decide Where to Get the Nuclear Fuel of the 1980's." Science, 184 (June 14, 1974), 1160-1161.

ANTI-NUCLEAR PROTEST

Barkan, Steven E. "Strategic, Tactical, and Organizational Dilemmas of Protest Movements Against Nuclear Power." Social Problems, 27 (October 1979), 19-37.

Bayer, Michael. "Nationalization and the Anti-Nuclear Movement: Response to Pector." Socialist Review, No. 45 (May-June 1979), 129-130.

Berger, John. Nuclear Power - The Unviable Option: A Critical Look at Our Energy Alternatives. Palo Alto, California: Ramparts Press, 1976.

Clark, Wilson. "Pulling the Plug on the 'Energy Crisis'." Environmental Action, 4 (January 20, 1973), 11-14.

Croall, Stephen. The Anti-Nuclear Handbook. New York: Pantheon Books, 1979.

Darnovsky, Marcy. "A Strategy for the Anti-Nuclear Movement: Response to Pector." Socialist Review, No. 45 (May-June 1979), 119-127.

Davidon, A. M. "United States Anti-Nuclear Movement." Bulletin of the Atomic Scientists, 35 (December 1979), 45-49.

Del Sesto, Steven L. "The Commercialization of Civilian Nuclear Power and the Evolution of Opposition: The American Experience, 1960-1974." Technology in Society, 1 (1979), 301-328.

Duderstadt, James and Kikuchi, Chihiro. Nuclear Power: Technology on Trial. Ann Arbor, Michigan: University of Michigan Press, 1979.

Ebbin, Steven and Kasper, Raphael. Citizen Groups and the Nuclear Power Controversy: Uses of Scientific and Technical Information. Cambridge, Massachusetts: M.I.T. Press, 1974.

Environmental Action Foundation. Shut 'em Down: The Case for a Nuclear Moratorium. New York: Harper & Row, Publishers, 1979.

Faulkner, Peter T., ed. Silent Bomb: A Guide to the Nuclear Energy Controversy. New York: Random House, 1977.

Feldman, Dede. "The Yellowcake Connection." Environmental Action, 11 (June 1979), 5-9.

Gillette, Robert. "Nuclear Power: Hard Times and a Questioning Congress." Science, 187 (March 21, 1975), 1058-1062.

Gofman, John W. "Irrevy": An Irreverent, Illustrated View of Nuclear Power: From Blunderland to Seabrook IV. San Francisco, California: Committee for Nuclear Responsibility, 1979.

_____. "Nuclear Power and Ecocide: An Adversary View of New Technology." Bulletin of the Atomic Scientists, 27 (September 1971), 28-32.

_____. "Reacting to Reactors - Part 1: The 'Peaceful Atom': Time for a Moratorium." Environmental Action, 4 (November 25, 1972), 11-15.

_____ and Tamplin, Arthur R. Poisoned Power: The Case Against Nuclear Power Plants. London, England: Chatto & Windus, 1973.

Goodin, Robert E. "No Moral Nukes." Ethics, 90 (April 1980), 417-448.

Graeub, Ralph. The Gentle Killers: Nuclear Power Stations. Levittown, New York: Transatlantic Arts, 1974.

Gravel, Senator Mike "Reacting to Reactors - Part II: Moratorium Politics: Finding the Critical Mass." Environmental Action, 4 (December 9, 1972), 9-13.

Gyorgy, Anna and Friends. No Nukes: Everyone's Guide to Nuclear Power. Boston, Massachusetts: South End Press, 1979.

Hayes, Denis. Nuclear Power: The Fifth Horseman. Washington, D.C.: Worldwatch Institute, 1976.

Jezer, Marty. "The Socialist Potential of the No-Nuke Movement." Radical American, 11 (September-October 1977), 63-69.

Jungk, Robert. The New Tyranny: How Nuclear Power Enslaves Us. New York: Grosset & Dunlap, 1979.

Kasperson, Roger, et al. "Public Opposition to Nuclear Energy: Retrospect and Prospect." In Charles T. Unseld, et al., eds. Sociopolitical Effects of Energy Use and Policy. Washington, D.C.: National Academy of Sciences, 1979, 259-292.

Katz, Neil H. and List, David C. "Seabrook: A Profile of Anti-Nuclear Activists, June 1978." Peace and Change, 7 (Spring 1981), 59-69.

Lewis, Richard S. The Nuclear Power Rebellion: Citizens vs. the Atomic Industrial Establishment. New York: Viking Press, 1972.

Lifton, Robert Jay. "Nuclear Energy and the Wisdom of the Body." Bulletin of the Atomic Scientists, 32 (September 1976), 16-20.

170

McCracken, Samuel. "The War Against the Atom." Commentary, 64 (September 1977), 33-47.

_____. The War Against the Atom. New York: Basic Books, 1981.

Nader, Ralph and Abbotts, John. The Menace of Atomic Energy. New York: W. W. Norton & Company, 1979.

Nelkin, Dorothy. "Anti-Nuclear Connections: Power and Weapons." Bulletin of the Atomic Scientists, 37 (April 1981), 36-40.

_____ and Fallows, Susan. "The Evolution of the Nuclear Debate: The Role of Public Participation." In Jack M. Hollander, Melvin K.Simmons and David O. Wood, eds. Annual Review of Energy. Volume 3. Palo Alto, California: Annual Reviews, Inc., 1978, pp. 275-312.

Novick, Sheldon. The Careless Atom. Boston, Massachusetts: Houghton Mifflin, 1969.

_____. The Electric War: The Fight Over Nuclear Power. San Francisco, California: Sierra Club Books, 1976.

Olson, McKinley C. Unacceptable Risk: The Nuclear Power Controversy. New York: Bantam Books, 1976.

Pector, Jeff. "The Nuclear Power Industry and the Anti-Nuclear Movement." Socialist Review, No. 42 (November-December 1978), 9-35.

Primack, Joel and von Hippel, Frank. Advice and Dissent: Scientists in the Political Arena. New York: New American Library, 1974.

Reader, Mark, comp. and ed. Atom's Eve: Ending the Nuclear Age. New York: McGraw-Hill, 1980.

Robinson, Gail. "Bursting the Nuclear Bubble." Environmental Action, 10 (May 1979), 23-25.

Stephenson, Lee and Zachar, George, eds. Accidents Will Happen: The Case Against Nuclear Energy. New York: Perennial Library, 1979.

Walsh, John. "Opposition to Nuclear Power: Raising the Question at the Polls." Science, 190 (December 5, 1975), 964-966.

Wasserman, Harvey. Energy War: Reports From the Front. Westport, Connecticut: Lawrence Hill & Company, Publishers, 1979.

Part VI
Alternative Energy Sources

Whether the post-petroleum era is just around the corner or whether it can be postponed for another half century or so, the prospect of the eventual exhaustion of the globe's conventional fossil fuels and even of its uranium reserves is a real one. Bridge or swing fuels may allow for a smooth transition, just as international conflict and/or policy stalemate may make for a rough passage. In any case, humankind must begin now to look toward alternative energy sources to supply the world's energy needs in the future. The sixth part of this bibliographical guide is organized around five possible sources of alternative energy supply currently being explored at various stages of research and development and, perhaps, commercialization. The eventual choice of one or a combination of these alternative energy sources will be determined by a complex process involving political, economic, and social considerations, as much as by technical matters.

The general references collected in the chapter on <u>conservation</u> give the reader the distinct impression that the potential for grappling with the energy crisis through conservation measures and energy-efficiency technologies is so vast that no rational political system could dismiss it as the foremost alternative energy source available (Abelson, 1974; Alliance to Save Energy, 1980; Dryden, 1975; Dumas, 1976; El-Mallakh and El-Mallakh, 1978; Hayes, 1976; Healy and Hertzfeld, 1976; Kaufman and Daly, 1977; Marshall, 1976; Mauss and Ullman, 1979; Nordhaus, 1979; Ross and Williams, 1980, 1981; Sant, 1979, 1981; Schipper, 1975; Shinskey, 1978; Socolow, 1977; Taylor, 1979; Veziroglu, 1977; Watson, 1978; and Williams, 1975). Energy conservation appears to be the "ultimate resource," providing the "easy energy path" to a "conserver society." In truth, no such national or global political consensus exists, in part because broader cultural and ideological barriers have tended to promote a facile identification of conservation with pain, sacrifice, and discomfort (such as raising thermostats in the summer and lowering them in the winter), in part because free-market capitalist ideology and its state-socialist counterpart have remained attached in an almost compulsive fashion to the production ethic and to supply-side solutions to energy problems. By contrast, the preponderant weight of opinion and interpretation in these books and articles suggests that conservation is better conceived as energy efficiency manifested in the more effective use of appropriate energy sources, through better architectural design, improved automobile engines, better electrical motors, and generally the proper matching of energy sources to appropriate end-uses.

The more technological aspects of energy conservation are detailed in Berg

(1974), Kovach (1974), Over and Sjoerdsma (1974), Schoen, et al. (1975), and Veziroglu (1978). The possibilities and limitations of these and other applications for specific realms of energy use are explored by many engineers and policy analysts, including research and interpretation found in such areas as: transportation and fuel conservation (Ayres and McKenna, 1972; Berry and Fels, 1973; Cohn, 1975; Grey, et al., 1978; Hirst, 1973, 1976; Jerome, 1972; McGillivray and Kemp, 1975; Post and Post, 1973; Ross and Williams, 1977; Tien, et al., 1975; Wildhorn, et al., 1975; and Yergin, 1980), residential conservation (Haimes, 1980; Hirst and Carney, 1977, 1978; Hirst and Hannon, 1979; Socolow, 1979; Williams and Ross, 1979), industrial conservation and energy efficiency (Berg, 1974, 1976, 1978; Beray and Makino, 1974; Committee for Consumer Policy, 1976; Dean, 1980; Giftopoulos, et al., 1974; Myers and Nakamura, 1978), including co-generation (Governor's Commission on Cogeneration, 1978; Harleman, 1971; Karkheck, 1977; and Williams, 1977, 1978), land use and agriculture (Friedrich, 1978; Harwood, 1977; Johnson, 1977; and Keyes, 1976), and municipal waste (Abert, 1978; Detweiler, 1978; and Kramer, 1973). The importance of architectural design to energy efficiency and to the foundation of a conserver society is clarified in a number of stimulating works (American Institute of Architects, 1974; Burberry, 1978; Caudill, et al., 1974; Flavin, 1980; Knowles, 1974; Rafalik, 1974; Snell, 1976; and Steadman, 1975).

Neither the social and political dimensions of energy conservation nor the role of the social sciences in fostering and evaluating the conservation potential are neglected. Shippee (1981) and Socolow (1978) consider what the social sciences can contribute to energy conservation, while the Center for Science in the Public Interest (1977), Craig, et al. (1976), and Crossley (1979) elucidate some of the social aspects of energy conservation. Convergent with these concerns are the analyses of Myers and Nakamura (1978) and Morris (1975), which offer policy evaluations of various conservation efforts. Lastly, the facilitating role of government policy in the area of energy conservation is given due regard and critical attention (Comptroller General, 1978; Council of State Governments, 1976; Council on Environmental Quality, 1975; Dunkerley, 1980; Flavin, 1971; Gibbons and Chandler, 1978; Russell, 1979; Sawhill, 1979; and U.S., D.O.E., 1980).

Virtually all parties in the energy debate concede the importance of the sun as the ultimate source of most of the earth's potential energy. What is disputed is the significance of solar energy as an alternative energy source in the near- to middle-term. Solar energy, high technologists and other hard energy path advocates allege, is an unproven technology still in its infancy; it is an energy source only beginning to spawn an industry and to generate a market. It must await further R & D to bring it to maturity and may require assimilation and guidance by the megacorporations of our political economy in order to realize its full commercial potential. Consequently, its role in satisfying America's energy needs in the year 2000 or even 2025 will be a rather modest one.

In contrast to this skeptical view of the solar prospect, most of the general surveys of the awakening science and technoloy of solar energy present a brighter future for the solar option if certain conditions are met and certain obstacles can be overcome (Behrman, 1976; Bossong, 1980; Boyles, 1978; Brinkworth, 1973; Halacy, 1973; Hayes, 1977a and b; Hoke, 1978; Kendall, et al., 1980; Keyes, 1975; Kreider and Kreider, 1981; Lovins, 1977; Lyons, 1978; McDaniels, 1979; Merrill and Gage, 1978; Messel, 1978; Metz and Hammond, 1978; Rapp, 1981; Reed, 1980; Robertson and Robertson, 1977; Schwartz, 1975; Silverstein, 1977; Stanford Research Institute,

1980; U.S., Council on Environmental Quality, 1979; U.S., D.O.E., 1978, 1979; Williams, 1978, and Wilson and Brown, 1979). This prognosis stems less from a realization that solar energy applications have been around for at least 2500 years than from a recognition of the vast array of solar technologies presently available or soon to be commercially feasible (Ezra, 1975; Fan, 1978; Kendall, et al., 1980; McVeigh, 1977; Meinel and Meinel, 1976; Veziroglu, 1979; and von Hippel and Williams, 1975). Particularly relevant here is the example of solar water and space heating (Bezdek, et al., 1979; Duffie and Beckman, 1976; Hirschberg, 1976; Patton, 1975; and Shurcliff, 1976), which, when viewed in terms of its passive (design) and active (technological) forms, is capable of meeting the energy needs of citizens throughout the United States at competitive prices.

On the other hand, social advocates do not dismiss the sizeable legal and institutional barriers hobbling the emergence of the Solar Age (Blissett, 1978-79; Hayes, 1979; Kellman, 1980; Solar State Legislation, 1978; Solarcal Council, 1979; U.S., Congress, O.T.A., 1978). Problems associated with the phasing of decentralized solar electrical systems into the presently highly centralized electric utilities network have been of special concern to supporters of the solar route to our energy future (see Asburg, 1977; Dickson, et al., 1977; Feldman and Wirtshafter, 1980; and Kellman, 1980; as well as other relevant citations in Part III, Chapter 2). This attentiveness to the formidable impediments to realizing the potentiality of solar energy has prompted some analysts to examine critically government policies and programs for advancing solar energy (Comptroller General, 1978; Ezra, 1975; Hayes, 1979; Maize, 1977; Marshall, 1979; Nelson, 1977; and Reuly, 1976), where the combat between highly centralized/capital-intensive forms of solar technology (Hammond and Metz, 1977; Herendeen, et al., 1979; Marinelli, 1979; Metz, 1977; and U.S., Congress, O.T.A., 1981) and decentralized/labor-intensive forms (Center for Science in the Public Interest, 1976; and Henderson, 1981) is being waged. Such concerns have also led other energy policy analysts to explore the manner in which the political economy of the corporate state (Keyes, 1977; Munson, 1979; and Reece, 1979a and b) and its technologically-oriented cultural horizon (Reid and Ihara, 1978) pose even more fundamental obstacles to bringing the solar prospect and its democratic social vision to fruition.

Within this chapter, the full spectrum of issues in the solar literature is represented. More scientific and technical dimensions of solar energy are touched on generally by those authors who examine solar R&D (Ann Arbor Science, 1977; Daniels and Duffie, 1955; and Pesko, 1975), as well as those who probe the status of and prospects for solar electricity through photovoltaic technology (Backus, 1976; Chalmers, 1976; Costello and Rappaport, 1980; Ehrenreich, 1979; Kelly, 1978; Merrigan, 1975; and Wolf, 1976) and the relationship of solar energy and photosynthesis (Broda, 1976; Calvin, 1974; Marzola and Bartholomew, 1979; and Poole and Williams, 1976). More generally, references to the various other forms of energy customarily subsumed under the solar label are listed – including tidal and OTEC (Duff, 1978; Duffie and Beckman, 1974; Hagen, 1976; Isaacs and Schmidt, 1980; Knight, et al., 1977; Whitmore, 1978; and Zener, 1974, 1976), wind (Boer, 1976; Golding, 1975; Gustavson, 1979; Hickok, 1975; Inglis, 1975, 1978; Merriam, 1977; Reynolds, 1970; Simmons, 1975; Sorenson, 1976; Torrey, 1976; and Wade, 1974), geothermal (Armstead, 1978; Barner, 1972; Bowen and Groh, 1971; Combs, 1975; Ellickson, et al., 1978; Garnish, 1976; and Robson, 1974), biomass (Anderson, 1979; Boer, 1976; Burwell, 1978; Lipinsky, 1978; Marzola and Bartholomew, 1979; and Maugh, 1972), and gasohol (Chambers, et al., 1979; Reed and Lerner, 1973; and Wigg, 1974).

Within the social realm, the economics of solar energy is treated broadly (Bezdek, et al., 1979; Feldman and Wirtschafter, 1980; Hyman, 1978; and Kreith and West, 1979) and in relation to the budding solar industry (Bereny, 1977; Bereny and Howell, 1979; and Pesko, 1975). Two excellent works (Butti and Perlin, 1980; and Flavin, 1980) highlight the practical and aesthetic dimensions of solar architecture from antiquity to the present and near future. Finally, Blissett (1978-79) and Schiefel, et al. (1978) undertake different types of social analysis to explore key social dimensions of the commercialization and sale of solar technologies.

No alternative energy technology has been as controversial as breeder reactors. "To breed or not to breed" has been a consuming political issue and a sensitive matter for presidential decision-making for nearly a decade. The immediate focus of the debate has been the question of whether or not the United States should complete the Clinch River (Tennessee) fast-breeder reactor (Comptroller General, 1979; U.S., Library of Congress, 1978); but, behind this matter lies a host of more basic issues concerning the implications for public health and safety, the advent of a plutonium economy, the dangers of accelerating nuclear proliferation, and the possibilities of nuclear theft and blackmail flowing from a positive decision (Feiveson, et al., 1976; Fuller, 1975; Gillette, 1974; Graham, 1971; Lovins, 1973, 1979; Roblat, 1977; Shapley, 1978; Smith, 1974; Speth, et al., 1974, 1975; and Weinberg, 1977). General perspectives on the breeder reactor as an alternative energy source have been opened up by Cochran (1974) and Hafele, et al. (1977), and surveys of alternative technological designs of breeder reactors have been presented by Bump (1967), Coch, et al. (1965), Seaborg and Bloom (1970), and Spinrad (1978). Other analysts have examined the economics of breeder reactors (Bupp and Derian, 1974; Chow, 1977, 1980; and Pearson, 1979). Given the highly politicized nature of the technological development of this energy alternative, studies on technological planning and government programs related to the development of this capital-intensive/high-technological option are essential to understanding the institutional and other sources of continuing support for this alternately lavishly-praised and roundly-criticized energy technology (Comptroller General, 1979, 1981; Hammond, 1972, 1973; Johnson, et al., 1976; Natchez and Bupp, 1973; Richels and Plummer, 1977; Shapley, 1977; and U.S., Library of Congress, 1978).

An even more distant energy alternative being pursued along the hard energy path is fusion power. General introductions to the principles underlying current fusion research (Artsimovich, 1970; Hagler and Kristiansen, 1977; Hunt, 1980; and U.S., Library of Congress, 1978), as well as the present course and possible alternative designs for obtaining energy from controlled fusion processes (Clarke, 1980; Lidsky, 1972; Metz, 1972-1978) are dominant concerns of many of these references. In addition, the currently favored program -- known as the TOKAMAK -- is amply reviewed by other sources (Coppi and Rem, 1972; Emmett, et al., 1974; Furth, 1979; Murakami and Eubank, 1979; and Steiner and Clarke, 1978). As with breeder reactors, so too with fusion power, a concern over its safety (Chen, 1967) and worries over its connection with weapons developments (Gillette, 1975; Holdren, 1978b) form two key elements of the critique of a complex and esoteric energy technology whose clouded future lies – if at all – somewhere over the rainbow (Gough and Eastlund, 1971; Holdren, 1978a; Kulcinski, 1974; Parkins, 1978; and Post and Ribe, 1974).

Hydrogen, the last energy alternative of the far-off future examined in this part, promises humankind a safe and virtually limitless energy source if and when the secret of unleashing its potential in a technologically and economically feasible manner can be discovered. Overviews of this dark horse in the energy alternatives competition

175

are provided by Hoffman (1981), Klimis (1974), and Veziroglu (1975). Given the enormous technological and economic obstacles it must surmount before its commercial feasibility can be estimated and a rough timetable for its practical application can be sketched, its prospects remain even more murky than other alternative energy sources and even more subject to wide-ranging speculation (Caprioglio, 1974; Maugh, 1972; and Winsche, et al., 1973).

1. Energy Conservation

Abelson, Philip H., ed. Energy: Use, Conservation, and Supply. Washington, D.C.: American Association for the Advancement of Science, 1974.

_____ and Hammond, Allen L. eds. Energy II: Needs, Conservation, and Supply. Washington, D.C.: American Association for the Advancement of Science, 1978.

Abert, James G., et al. "The Economics of Resource Recovery from Municipal Solid Waste." Science, 183 (March 15, 1974), 1052-1058.

Abrahamson, Bernard J., ed. Conservation and the Changing Direction of Economic Growth. Boulder, Colorado: Westview Press, 1978.

Alliance to Save Energy. The Dynamics of Energy Efficiency. Washington, D.C.: Alliance to Save Energy, 1980.

American Institute of Architects Research Corporation. Energy Conservation in Building Design. Washington, D.C.: American Institute of Architects, 1974.

American Society of Civil Engineers, comp. Conservation and Utilization of Water and Energy Resources. New York: American Society of Civil Engineers, 1979.

Appel, John and MacKenzie, James J. "How Much Light Do We Really Need?" Bulletin of the Atomic Scientists, 30 (December 1974), 18-24.

Ayres, Robert U. and McKenna, Richard P. Alternatives to the Internal Combustion Engine - Impacts on Environmental Quality. Baltimore, Maryland: The Johns Hopkins University Press, 1972.

Benoit, Emile. "First Steps to Survival." Bulletin of the Atomic Scientists, 32 (March 1976), 41-48.

Berg, Charles A. "Conservation in Industry." Science, 184 (April 19, 1974), 264-270.

Berg, Charles A. "Energy Conservation Through Effective Utilization." Science, 181 (July 13, 1973), 128-138.

_____. "Potential for Energy Conservation in Industry." In Jack M. Hollander and Melvin K. Simmons, eds. Annual Review of Energy. Volume 1. Palo Alto, California: Annual Reviews, Inc., 1976, pp. 519-534.

_____. "Process Innovation and Changes in Industrial Energy Use." Science, 199 (February 10, 1978), 608-614.

_____. "A Technical Basis for Energy Conservation." Technology Review, 76 (February 1974), 14-23.

Berry, R. Stephen and Fels, Margaret F. "The Energy Cost of Automobiles." Bulletin of theAtomic Scientists, 29 (December 1973), 11-17, 58-60.

_____ and Makino, Hiro. "Energy Thrift in Packaging and Marketing." Technology Review, 76 (February 1974), 32-43.

Boretsky, Michael. "Opportunities and Strategies for Energy Conservation." Technology Review, 79 (July/August 1977), 56-62.

Burberry, Peter. Building for Energy Conservation. New York: Halsted Press, 1978.

Caudill, William W.; Lawyer, Frank D.; and Bullock, Thomas A. A Bucket of Oil: The Humanistic Approach to Building Design for Energy Conservation. Boston, Massachusetts: CBI Publishing Company, 1974.

Center for Science in the Public Interest. Ninety-Five Ways to a Simple Lifestyle. Albert J. Fritsch, et al., eds. Bloomington, Indiana: Indiana University Press, 1977; Garden City, New York: Doubleday & Company, 1977.

Cohn, Charles C. "Improved Fuel Economy for Automobiles." Technology Review, 77 (February 1975), 44-52.

Committee on Consumer Policy. The Energy Label: A Means of Energy Conservation. Paris, France: Organization for Economic Cooperation and Development, 1976.

Comptroller General. U.S. Energy Conservation Could Benefit from Experience of Other Countries. Washington, D.C.: U.S. General Accounting Office, Janaury 10, 1978.

Council of State Governments. Energy Conservation: Policy Considerations for the States. Lexington, Kentucky: Council of State Governments, 1976.

Council on Environmental Quality. The Good News About Energy. Washington, D.C.: U.S. Government Printing Office, 1979.

Craig, Paul P.; Darmstadter, Joel; and Rattien, Stephen. "Social and Institutional Factors in Energy Conservation." In Jack M. Hollander and Melvin K. Simmons, eds. Annual Review of Energy. Volume 1. Palo Alto, California: Annual Reviews, Inc., 1976, 535-551.

Crossley, David J. "The Role of Popularization Campaigns in Energy Conservation." Energy Policy, 7 (March 1979), 57-68.

Cunningham, William H. and Lopreato, Sally C. Energy Use and Conservation Incentives: A Study of the Southwestern United States. New York: Praeger Publishers, 1977.

Darmstadter, Joel. Conserving Energy: Prospects and Opportunities in the New York Region. Baltimore, Maryland: The Johns Hopkins University Press, 1975.

Dean, Norman L. Energy Efficiency in Industry. Cambridge, Massachusetts: Ballinger Publishing Company, 1980.

Detweiler, Raymond F. Wastes Can Produce Cheap Energy and Save Vitally Needed Material Resources. Souderton, Pennsylvania: E. & E. Publishing Company, 1978.

Dryden, I. G., ed. The Efficient Use of Energy. Guildford, England: IPC Science and Technology Press, 1975.

Dumas, Lloyd J. The Conservation Response: Strategies for the Design and Operation of Energy-Using Systems. Lexington, Massachusetts: Lexington Books, 1976.

Dunkerley, Joy. "Energy Use Trends in Industrial Countries: Implications for Conservation." Energy Policy, 8 (June 1980), 105-115.

El-Mallakh, Ragaei and El-Mallakh, Dorothea. Energy Options and Conservation. Boulder, Colorado: International Research Center for Energy and Economic Development, 1978.

Flavin, Christopher. Energy and Architecture: The Solar and Conservation Potential. Washington, D.C.: Worldwatch Institute, 1980.

Ford, K. W., et al., eds. Efficient Use of Energy. [The APS Studies on the Technical Aspects of the More Efficient Use of Energy] New York: American Institute of Physics, 1975.

Freeman, David S. "Toward a Policy of Energy Conservation." Bulletin of the Atomic Scientists, 27 (October 1971), 8-12.

Friedrich, R. A. Energy Conservation for American Agriculture. Cambridge, Massachusetts: Ballinger Publishing Company, 1978.

Gibbons, John H. and Chandler, William U. "A National Energy Conservation Policy." Current History, 75 (July/August 1978), 13-15.

Gillette, Robert. "In Energy Impasse, Conservation Keeps Popping Up." Science, 187 (January 10, 1975), 42-45.

Governor's Commission on Cogeneration. Cogeneration: Its Benefits to New England. Boston, Massachusetts: Commonwealth of Massachusetts, 1978.

Grey, Jerry, et al. "Fuel Conservation and Applied Research." Science, 200 (April 14, 1978), 135-142.

Gyftopoulos, Elias P.; Lazarides, Lazarus J.; and Widener, Thomas P. Potential Fuel Effectiveness in Industry. Cambridge, Massachusetts: Ballinger Publishing Company, 1974.

Haimes, Y. Y. Energy Auditing and Conservation: Methods, Measurements, Management, and Case Studies. Washington, D.C.: Hemisphere Publishing Corporation, 1980.

Hannon, Bruce. "Energy Conservation and the Consumer." Science, 189 (July 11, 1975), 95-102.

_____. "Energy, Labor, and the Conservor Society." Technology Review, 79 (March/April 1977), 47-53.

_____. "Options for Energy Conservation." Technology Review, 76 (February 1974), 24-31.

Harleman, Donald R. F. "Heat - The Ultimate Waste." Technology Review, 74 (December 1971), 44-51.

Harwood, Corbin Crews. Using Land to Save Energy. Cambridge, Massachusetts: Ballinger Publishing Company, 1977.

Hayes, Denis. Energy: The Case for Conservation. [Worldwatch Paper No. 4] Washington, D.C.: Worldwatch Institute, January 1976.

Healy, Robert G. and Hertzfeld, Henry R. Energy Conservation Strategies. Washington, D.C.: Conservation Foundation, 1976.

Hirst, Eric. "Transportation Energy Conservation Policies." Science, 192 (April 2, 1976), 15-20.

_____. "Transportation Energy Use and Conservation Potential." Bulletin of the Atomic Scientists, 29 (November 1973), 36-42.

Hirst, Eric and Carney, Janet. "Effects of Federal Residential Energy Conservation Programs." Science, 199 (February 24, 1978), 845-851.

_____. "Residential Energy Conservation: Analysis of U.S. Federal Programmes." Energy Policy, 5 (September 1977), 211-222.

Hirst, Eric and Hannon, Bruce. "Effects of Energy Conservation in Residential and Commercial Buildings." Science, 205 (August 17, 1979), 656-661.

_____ and Moyers, John C. "Efficiency of Energy Use in the United States." Science, 179 (March 30, 1973), 1299-1304.

Hitch, Charles, ed. Energy Conservation and Economic Growth. Boulder, Colorado: Westview Press, 1978.

Jerome, John. The Death of the Automobile. New York: W. W. Norton & Company, 1972.

Johnson, Warren A. "Energy Conservation in Amish Agriculture." Science, 198 (October 28, 1977), 373-378.

Karkheck, J., et al. "Prospects for District Heating in the United States." Science, 195 (March 11, 1977), 948-955.

Kaufman, Alvin and Daly, Barbara M. Alternative Energy Conservation Strategies: An Appraisal. Washington, D.C.: Congressional Research Service, Library of Congress, April 29, 1977.

Keyes, Dale. "Energy and Land Use: An Instrument of U.S. Conservation Policy?" Energy Policy, 4 (September 1976), 225-236.

Knowles, Ralph L. Energy and Form: An Ecological Approach to Urban Growth. Cambridge, Massachusetts: M.I.T. Press, 1974.

Kovach, Eugene G., ed. Technology of Efficient Energy Utilization. Elmsford, New York: Pergamon Press, 1974.

Kramer, Eugene. "Energy Conservation and Waste Recycling: Taking Advantage of Urban Congestion." Bulletin of the Atomic Scientists, 29 (April 1973), 13-18.

Kreith, Frank and West, Ronald E. Economics of Solar Energy and Conservation Systems. Volumes I-III. West Palm Beach, Florida: CRC Press, 1979.

Large, David B. Hidden Waste. Washington, D.C.: Conservation Foundation, 1976.

Lincoln, G. A. "Energy Conservation." Science, 180 (April 13, 1973), 155-162.

McGillivray, Robert G. and Kemp, Michael A. Alternative Strategies for Reducing Gasoline Consumption by Private Automobiles. Washington, D.C.: Urban Institute, 1975.

Marshall, James. Going, Going, Gone? The Waste of Our Energy Resources. East Rutherford, New Jersey: Coward, McCann, & Geoghegan, 1976.

Mauss, E. A. and Ullmann, J. E., eds. Conservation of Energy Resources. [Annals of the New York Academy of Sciences, Vol. 324] New York: New York Academy of Sciences, 1979.

Morris, Deane N. Evaluation of Measures for Conserving Energy. [P-5477] Santa Monica, California: RAND Corporation, 1975.

Myers, J. G. and Nakamura, L. Saving Energy in Manufacturing: The Post-Embargo Record. Cambridge, Massachusetts: Ballinger Publishing Company, 1978.

Nordhaus, W. D. The Efficient Use of Energy Resources. New Haven, Connecticut: Yale University Press, 1979.

Novick, David A. A World of Scarcities. New York: Halsted Press, 1976.

Over, J. A. and Sjoerdsma, A. C. Energy Conservation: Ways and Means. The Hague, The Netherlands: Future Shape of Technology Foundation, 1974.

Post, Richard F. and Post, Stephen F. "Flywheels." Scientific American, 229 (December 1973), 17-23.

Rafalik, Dianne. "Architecture's Towering Energy Costs." Environmental Action, 5 (April 13, 1974), 11-14.

Ross, Marc and Williams, Robert H. Energy and the Invisible Hand. New York: McGraw-Hill, 1980.

_____. "Energy Efficiency: Our Most Underrated Energy Resource." Bulletin of the Atomic Scientists, 32 (November 1976), 30-38.

_____. Our Energy: Regaining Control. New York: McGraw-Hill, 1981.

_____. "The Potential for Future Fuel Conservation." Technology Review, 79 (February 1977), 48-57.

Rubin, Milton D. "Plugging the Energy Sieve." Bulletin of the Atomic Scientists, 30 (December 1974), 7-17.

Russell, Joe W., Jr. Economic Disincentives for Energy Conservation. Cambridge, Massachusetts: Ballinger Publishing Company, 1979.

Sant, Roger. The Least-Cost Energy Strategy: Minimizing Consumer Costs. Arlington, Virginia: Carnegie-Mellon Institute, 1979.

Sawhill, John, ed. Energy Conservation and Public Policy. New York: Prentice-Hall, 1979.

Schipper, Lee. "Conservation is Here." Bulletin of the Atomic Scientists, 36 (February 1980), 55-58.

_____. Energy Conservation: Its Nature, Hidden Benefits, and Hidden Barriers. Berkeley, California: Lawrence Berkeley Laboratories and Energy and Resources Group, University of California at Berkeley, June 1, 1975.

_____. "Raising the Productivity of Energy Utilization." In Jack M. Hollander and Melvin K. Simmons, eds. Annual Review of Energy. Volume 1. Palo Alto, California: Annual Reviews, Inc., 1976, pp. 455-518.

_____ and Darmstadter, Joel. "The Logic of Energy Conservation." Technology Review, 80 (January 1978), 41-50.

Schoen, Richard; Hirschberg, Allen S.;and Weingart, Jerome. New Energy Technologies for Buildings. James Stein, ed. Cambridge, Massachusetts: Ballinger Publishing Company, 1975.

Shinskey, F. G. Energy Conservation Through Control. New York: Academic Press, 1978.

Shippee, Glenn. "Energy Conservation and the Role of the Social Sciences." Energy Policy, 9 (March 1981), 32-38.

Snell, Jack E., et al. "Energy Conservation in New Housing Design." Science, 192 (June 25, 1976), 1305-1311.

Socolow, Robert H. "The Coming Age of Conservation." In Jack M. Hollander, Melvin K. Simmons and David O. Wood, eds. Annual Review of Energy. Volume 2. Palo Alto, California: Annual Reviews, Inc., 1977, pp. 239-290.

_____, ed. Saving Energy in the Home: Princeton's Experiments at Twin Rivers. Cambridge, Massachusetts: Ballinger Publishing Company, 1978.

Steadman, Philip. Energy, Environment, and Building. Cambridge, England: Cambridge University Press, 1975.

Taylor, Vince. The Easy Path Energy Plan. Cambridge, Massachusetts: Union of Concerned Scientists, 1979.

Tien, John K.; Clark, Roy W.;and Malu, Mahendra K. "Reducing the Energy Investment in Automobiles." Technology Review, 77 (February 1975), 38-43.

United States. Department of Energy. Office of the Assistant Secretary for Conservation and Solar Energy. U.S. Conservation Strategy - An Interim Report. [Draft] Washington, D.C.: Department of Energy, February 1, 1980.

Van Gool, Willem. "Fundamental Aspects of Energy Conservation Policy." Energy - The International Journal, 5 (May 1980), 429-444.

Veziroglu, T. Nejat, ed. Energy Conservation. Elmsford, New York: Pergamon Press, 1977.

_____ . Solar Energy and Conservation: Technology, Commercialization, and Utilization. Elmsford, New York: Pergamon Press, 1978.

Vogt, Frederick, ed. Energy Conservation & Use of Solar and Other Renewable Energy: Proceedings. Elmsford, New York: Pergamon Press, 1981.

Watson, Jane W. Conservation of Energy. New York: Franklin Watts, 1978.

Widmer, Thomas F. and Gyftopoulos, Elias P. "Energy Conservation and a Healthy Economy." Technology Review, 79 (June 1977), 30-40.

Wildhorn, Sorrel; Burright, Burke K.; Enns, John H.;and Kirkwood, Thomas F. How to Save Gasoline: Public Policy Alternatives for the Automobile. Cambridge, Massachusetts: Ballinger Publishing Company, 1975.

Williams, Robert H., ed. The Energy Conservation Papers. Cambridge, Massachusetts: Ballinger Publishing Company, 1975.

_____ . "Industrial Cogeneration." In Jack M. Hollander, Melvin K. Simmons, and David O. Wood, eds. Annual Review of Energy. Volume 3. Palo Alto, California: Annual Reviews, Inc., 1978, pp. 313-356.

_____ . The Potential for Electricity Generation as a Byproduct of Industrial Steam Production in New Jersey. Princeton, New Jersey: Center for Environmental Studies, 1977.

_____ and Ross, Marc. "Drilling for Oil and Gas in Our Homes." Technology Review, 82 (March/April 1979), 24-37.

Yergin, Daniel, ed. The Dependence Dilemma: U.S. Baseline Consumption and America's Security. Cambridge, Massachusetts: Harvard Center for International Affairs, 1980.

2. Solar

Anderson, Russell E. Biological Paths to Self-Reliance: A Guide to Biological Solar Energy Conversion. New York: Van Nostrand Reinhold, 1979.

Ann Arbor Science Special Task Group. 1977 Solar Energy and Research Directory. Ann Arbor, Michigan: Ann Arbor Science Publishers, 1977.

Antal, Michael J., Jr. "Tower Power: Producing Fuels from Solar Energy." Bulletin of the Atomic Scientists, 32 (May 1976), 58-62.

Armstead, H. Christopher. Geothermal Energy: Its Past, Present, and Future Contributions to the Energy Needs of Man. New York: Halsted Press, 1978.

Asburg, Joseph G. and Mueller, Ronald O. "Solar Energy and Electric Utilities: Should They Be Interfaced?" Science, 195 (February 4, 1977), 445-450.

Backus, Charles E. Solar Cells. New York: John Wiley & Sons, 1976.

Barnea, Joseph. "Geothermal Power." Scientific American, 226 (January 1972), 70-77.

Behrman, Daniel. Solar Energy: The Awakening Science. Boston, Massachusetts: Little, Brown and Company, 1976.

Bereny, Justin A. Survey of the Emerging Solar Energy Industry. San Mateo, California: Solar Energy Information Services, 1977.

_____ and Howell, Yvonne. Survey and Directory of the Emerging Solar Energy Industry, 1980-1981. San Mateo, California: Solar Energy Information Services, 1979.

Bezdek, Roger H., et al. "Economic Feasibility of Solar Water and Space Heating." Science, 203 (March 23, 1979), 1214-1220.

Blissett, Marlan.Toward a Solar America: An Institutional Assessment of On-Site Solar Technologies. [Solar Technology Assessment Policy Research Project Report] Austin, Texas: Lyndon B. Johnson School of Public Affairs, University of Texas at Austin, 1978-1979.

Blum, S. L. "Tapping Resources in Municipal Solid Waste." Science, 191 (February 20, 1976), 669-675.

Bockris, John O'M. Energy: The Solar Hydrogen Alternative. New York: John Wiley & Sons, 1975.

Boer, K. W., ed. Agriculture, Biomass, Wind, New Developments. Elmsford, New York: Pergamon Press, 1976.

Bossong, Ken, et al. Solar Compendium. 2 Volumes. Washington, D.C.: Citizens Energy Project, 1980.

_____ . Solar Critique. Washington, D.C.: Citizens Energy Project, 1981.

Bowen, Richard G. and Groh, Edward A. "Geothermal - The Earth's Primordial Energy." Technology Review, 74 (October/November 1971), 42-48.

Boyle, Godfrey. Living on the Sun: Harnessing Renewable Energy for an Equitable Society. Salem, New Hampshire: Marion Boyars, Publishers, 1978.

Brinkworth, B. J. Solar Energy for Man. New York: John Wiley & Sons, 1972; Halsted Press, 1973.

Broda, E. "Solar Power: The Photochemical Alternative." Bulletin of the Atomic Scientists, 32 (March 1976), 49-52.

Burwell, C. C. "Solar Biomass Energy: An Overview of U.S. Potential." Science, 199 (March 10, 1978), 1041-1048.

Butti, Ken and Perlin, John. A Golden Thread: 2500 Years of Solar Architecture and Technology. Palo Alto, California: Cheshire Books, 1980.

Calvin, Melvin. "Solar Energy by Photosynthesis." Science, 184 (April 19, 1974), 375-381.

Capturing the Sun Through Bioconversion. Washington, D.C.: Washington Center for Metropolitan Studies, 1976.

Caputo, Richard S. "Solar Power Plants: Dark Horse in the Energy Stable." Bulletin of the Atomic Scientists, 33 (May 1977), 46-56.

Center for Science in the Public Interest. Solar Energy: One Way to Citizen Control. Albert J. Fritsch, ed. [CSPI Energy Series No. II] Washington, D.C.: Center for Science in the Public Interest, 1976.

Chalmers, Bruce. "The Photovoltaic Generation of Electricity." Scientific American, 235 (October 1976), 34-43.

Chambers, R. S., et al. "Gasohol: Does It or Doesn't It Produce Positive Net Energy?" Science, 206 (November 16, 1979), 789-795.

Cheremisinoff, Paul N. and Regino, Thomas C. Principles and Applications of Solar Energy. Ann Arbor, Michigan: Ann Arbor Science Publishers, 1978.

Combs, James B. "The Geology and Geophysics of Geothermal Energy." Technology Review, 77 (March/April 1975), 46-49.

Comptroller General. The Magnitude of the Federal Solar Energy Program and the Effects of Different Levels of Funding. [EMD-78-27] Washington, D.C.: U.S. General Accounting Office, February 2, 1978.

Costello, Dennis and Rappaport, Paul. "The Technological and Economic Development of Photovoltaics." In Jack M. Hollander, et al., eds. Annual Review of Energy. Volume 5. Palo Alto, California: Annual Reviews, Inc., 1980, pp. 335-350.

Daniels, Farrington. Direct Use of the Sun's Energy. New Haven, Connecticut: Yale University Press, 1964, 1975; New York: Ballantine Books, 1974.

_____ and Duffie, John A., eds. Solar Energy Research. Madison, Wisconsin: University of Wisconsin Press, 1955.

DeWinter, F. Description of the Solar Energy R & D Programs of Many Nations. J. W. deWinter, ed. San Mateo, California: Solar Energy Information Services, 1979.

Dickson, Charles; Eichen, Marc; and Feldman, Stephen. "Solar Energy and U.S. Public Utilities: The Impact on Rate Structure and Utilization." Energy Policy, 5 (September 1977), 195-210.

Duff, George F. D. "Tidal Power in the Bay of Fundy." Technology Review, 81 (November 1978), 34-43.

Duffie, John A. and Beckman, William A. Solar Energy Thermal Processes. New York: Wiley Interscience, 1974.

_____. "Solar Heating and Cooling." Science, 191 (January 16, 1976), 143-149.

Ehrenreich, Henry (Chairman). Principle Conclusions of the American Physical Society Study Group on Solar Photovoltaic Energy Conversion. New York: American Physical Society, 1979.

Ellickson, Phyllis, et al. Balanced Energy and the Environment: The Case for Geothermal Development. [R-2274-DOE] Santa Monica, California: RAND Corporation, 1978.

Ewers, William L. Solar Energy: A Biased Guide. Northbrook, Illinois: Domus Books,
1977.

Ezra, Arthur A. "Technology Utilization: Incentives and Solar Energy." Science, 187
(February 28, 1975), 707-713.

Fan, John C. C. "Plugging Into the Sun." Technology Review, 80 (August/September
1978), 14-37.

Feldman, Stephen L. and Wirtshafter, Robert M. On the Economics of Solar Energy.
Lexington, Massachusetts: Lexington Books, 1980.

Flavin, Christopher. Energy and Architecture: The Solar and Conservation Potential.
Washington, D.C.: Worldwatch Institute, 1980.

Ford, Norman C. and Kane, Joseph W. "Solar Power." Bulletin of the Atomic Scientists, 27 (October 1971), 27-31.

Garnish, J. D. "Geothermal Energy as an 'Alternative' Source." Energy Policy, 4
(June 1976), 130-143.

Golding, E. W. The Generation of Electricity by Windpower. New York: Halsted
Press, 1976.

Goodenough, John B. "The Options for Using the Sun." Technology Review, 79
(October/November 1976), 62-71.

Gustavson, M. R. "Limits to Wind Power Utilization." Science, 204 (April 6, 1979),
13-17.

Hagen, Arthur W. Thermal Energy from the Sea. Park Ridge, New Jersey: Noyes
Data Corporation, 1976.

Halacy, Daniel S., Jr. The Coming Age of Solar Energy. Rev. ed. New York: Harper &
Row, Publishers, 1973.

Hammond, Allen L. and Metz, William D. "Solar Energy Research: Making Solar After
the Nuclear Model?" Science, 197 (July 15, 1977), 241-244.

Harris, Michael. "Reinventing the Water Wheel." Environmental Action, 11 (June
1979), 24-28.

Hayes, Denis. Energy: The Solar Prospect. [Worldwatch Paper 11] Washington, D.C.:
Worldwatch Institute, 1977a.

Hayes, Denis. Rays of Hope: The Transition to a Post-Petroleum World. New York: W. W. Norton & Company, 1977b.

_____. The Solar Energy Timetable. Washington, D.C.: Worldwatch Institute, 1978.

Hayes, Gail B. Solar Access Law: Protecting Access to Sunlight for Solar Energy Systems. Cambridge, Massachusetts: Ballinger Publishing Company, 1979.

Henderson, Hazel. The Politics of the Solar Age: Alternatives to Economics. Garden City, New York: Doubleday & Company, 1981.

Herendeen, R. A.; Kary, T.; and Rebitzer, J. "Energy Analysis of the Solar Power Satellite." Science, 205 (August 3, 1979), 451-454.

Hickok, Floyd. Handbook of Solar and Wind Energy. Boston, Massachusetts: Cahners Books, 1975.

Hildebrandt, Alvin F. and Jant-Hull, Lorin L. "Power With Heliostats." Science, 197 (September 16, 1977), 1139-1146.

Hirshberg, Alan S. "Public Policy for Solar Heating and Cooling." Bulletin of the Atomic Scientists, 32 (October 1976), 37-45.

Hoke, John. Solar Energy. Rev. ed. New York: Franklin Watts, 1978.

Hyman, Mark, Jr. "Solar Economics Comes Home." Technology Review, 80 (February 1978), 28-35.

Inglis, David Rittenhouse. Wind Power and Other Energy Options. Ann Arbor, Michigan: University of Michigan Press, 1978.

_____. "Wind Power Now." Bulletin of the Atomic Scientists, 31 (October 1975), 20-26.

Isaacs, J. D. and Schmitt, W. R. "Ocean Energy: Forms and Prospects." Science, 207 (January 18, 1980), 265-273.

Kellman, B. "De-Utilizing the Energy Industry: Planning the Solar Transition." UCLA Law Review, 28 (October 1980), 1-51.

Kelly, Henry. "Photovoltaic Power Systems: A Tour Through the Alternatives." Science, 199 (February 10, 1978), 634-643.

Kendall, Henry, et al. Energy Strategies: Toward a Solar Future. Cambridge, Massachusetts: Ballinger Publishing Company, 1980.

Keyes, John. Harnessing the Sun. Dobbs Ferry, New York: Morgan and Morgan, 1975.

Keyes, John. The Solar Conspiracy. Dobbs Ferry, New York: Morgan and Morgan, 1977.

Knight, H. Gary, et al, eds. Ocean Thermal Energy Conversion: Legal, Political, and Institutional Aspects. Lexington, Massachusetts: Lexington Books, 1977.

Kreider, Jan F. and Kreith, Frank, eds. Solar Energy Handbook. New York: McGraw-Hill, 1981.

Kreith, Frank and West, R. E. Economics of Solar Energy and Conservation Systems. Volumes I-III. West Palm Beach, Florida: CRC Press, 1979.

Lipinsky, E. S. "Fuels from Biomass: Integration With Food and Materials Systems." Science, 199 (February 10, 1978), 644-651.

Lovins, Amory B. Soft Energy Paths. Cambridge, Massachusetts: Ballinger Publishing Company, 1977.

Lyons, Stephen, ed. Sun: A Handbook for the Solar Decade. San Francisco, California: Friends of the Earth, 1978.

McDaniels, D. K. The Sun: Our Future Energy Source. New York: John Wiley & Sons, 1979.

McVeigh, J. C. Sun Power: An Introduction to the Applications of Solar Energy. New York: Praeger Publishers, 1977.

Maize, Kennedy P. "Government R & D Programs: A Look at the Sunny Side." Environmental Action, 8 (March 12, 1977), 4-8.

Marinelli, Janet. "The Edsel of the Solar Age." Environmental Action, 11 (July/August 1979), 20-24.

Marshall, Eliot. "The Solar Institute: Hobbled by DOE?" Science, 203, (March 23, 1979), 1226-1228.

Marzola, D. L. and Bartholomew, D. P. "Photosynthetic Pathway and Biomass Energy Production." Science, 205 (August 10, 1979), 555-559.

Maugh, Thomas H., II. "Fuel From Wastes: A Minor Energy Source." Science, 178 (November 10, 1972), 599-602.

Meinel, Aden B. and Meinel, Marjorie P. Applied Solar Energy: An Introduction. Reading, Massachusetts: Addison-Wesley Publishing Company, 1976.

_____. "Is It Time for a New Look at Solar Energy?" Bulletin of the Atomic Scientists, 27 (October 1971), 32-37.

Merriam, Marshal F. "Wind Energy for Human Needs." Technology Review, 79 (January 1977), 28-39.

Merrigan, Joseph A. Sunlight to Electricity: Prospects for Solar Energy Conversion by Photovoltaics. Cambridge, Massachusetts: M.I.T. Press, 1975.

Merrill, Richard and Gage, Thomas. Energy Primer: Solar, Water, Wind, and Biofuels. New York: Delta, 1978; Menlo Park, California: Portola Institute, 1978.

Messel, H. and Butler, S. T., eds. Solar Energy. Elmsford, New York: Pergamon Press, 1978.

Metz, William D. "Solar Thermal Electricity: Power Tower Dominates Research." Science, 197 (July 22, 1977), 353-356.

_____ and Hammond, Allen L. Solar Energy in America. Washington, D.C.: American Association for the Advancement of Science, 1978.

Minan, John H. and Lawrence, William H., eds. Legal Aspects of Solar Energy. Lexington, Massachusetts: Lexington Books, 1981.

Morrow, Walter F., Jr. "Solar Energy: Its Time Is Near." Technology Review, 76 (December 1973), 30-43.

Munson, Richard. "Ripping Off the Sun." The Progressive, 43 (September 1979), 12-15.

Nelson, B. E. "The Congress and Solar Energy." Energy Communications, 3 (1977), 601-617.

Parker, Blaine F., ed. Kentucky Solar Energy Handbook. Lexington, Kentucky: Institute for Mining and Minerals Research, 1979.

Patton, Arthur R. Solar Energy for Heating and Cooling of Buildings. Park Ridge, New Jersey: Noyes Data Corporation, 1975.

Pesko, Carolyn, ed. Solar Directory. Ann Arbor, Michigan: Ann Arbor Science Publishers, 1975.

Poole, Alan D. and Williams, Robert H. "Flower Power: Prospects for Photosynthetic Energy." Bulletin of the Atomic Scientists, 32 (May 1976), 48-58.

Rapp, Donald. Solar Energy. Englewood Cliffs, New Jersey: Prentice-Hall, 1981.

Reece, Ray. "Elipsing the Solar Inventor." Environmental Action, 10 (May 1979a), 6-11.

_____ . The Sun Betrayed: A Study of the Corporate Seizure of Solar Energy Development. Boston, Massachusetts: South End Press, 1979b.

Reed, Millard. Solar Energy in Tomorrow's World. New York: Julius Messner, 1980.

Reed, T. B. and Lerner, R. M. "Methanol: A Versatile Fuel for Immediate Use." Science, 182 (December 28, 1973), 1299-1304.

Reid, Herbert G. and Ihara, Randal H. "Technocracy, Democracy, and Solar Energy." In Helios Symposium Proceedings, Volume II. Albany, New York: State University of New York at Albany, 1978, pp. 283-292.

Reuyl, John S., et al. A Preliminary Social and Environmental Assessment of the ERDA Solar Energy Program, 1975-2020. Volumes I and II. Menlo Park, California: The Stanford Research Institute, 1976.

Rex, Robert W. "Geothermal Energy – The Neglected Energy Option." Bulletin of the Atomic Scientists, 27 (October 1971), 52-56.

Reynolds, John. Windmills and Watermills. New York: Praeger Publishers, 1970.

Robertson, Vincent and Robertson, Roin. Alternate Energy - Solar Energy. Albuquerque, New Mexico: Alternate Energy Publishing Company, 1977.

Robson, Geoffrey. "Geothermal Electricity Production." Science, 184 (April 19, 1974), 371-375.

Schiefel, Dennis; Costello, Dennis; Posner, David; and Witholder, Robert. The Market Penetration of Solar Energy: A Model Review Workshop Summary. Golden, Colorado: Solar Energy Research Institute, January 1978.

Schwartz, Marie Sokol, ed. Harvesting the Sun's Energy: A Solar Collection. Fullerton, California: Designs III Printing, 1975.

Shurcliff, William A. "Active-Type Solar Heating Systems for Houses: A Technology in Ferment." Bulletin of the Atomic Scientists, 32 (February 1976), 30-36.

Silverstein, Michael. Once and Future Resource: A History of Solar Energy. Newtonville, Maine: Environmental Design and Research Center, 1977.

Simmons, D. M. Wind Power. Park Ridge, New Jersey: Noyes Data Corporation, 1975.

Solar Energy. New York: Unipub, 1978.

Solar Energy Progress. Stamford, Connecticut: Business Communications Company, 1977.

Solar State Legislation. Rockville, Maryland: National Solar Heating and Cooling Information Center, Janauary 1978.

Solarcal Council. Toward a Solar California: The Solarcal Council Action Program. Sacramento, California: State of California, January 1979.

Sorenson, Bent. "Wind Energy." Bulletin of the Atomic Scientists, 32 (September 1976), 38-45.

Stanford Research Institute. Solar Energy in America's Future. San Mateo, California: Solar Energy Information Services, 1980.

Stickley, R. A. Solar Power Array for the Concentration of Energy (SPACE). Northfield, Minnesota: Sheldahl, Inc., 1974.

Torrey, Volta. Wind Catchers: American Windmills of Yesterday and Tomorrow. Brattleboro, Vermont: Stephen Greene Press, 1976.

United Nations. New Sources of Energy: Solar Two. Seattle, Washington: Cloudburst Press, 1978.

United States. Congress. Office of Technology Assessment. Application of Solar Technology to Today's Energy Needs. [OTA-E-66] Washington, D.C.: Office of Technology Assessment, U.S. Congress, June 1978.

_____. Solar Power Satellites. [OTA-E-144] Washington, D.C.: Office of Technology Assessment, U.S. Congress, August 1981.

United States. Council on Environmental Quality. Solar Energy: Progress and Promise. Washington, D.C.: U.S. Government Printing Office, April 1978; San Mateo, California: Solar Energy Information Services, 1979.

United States. Department of Energy. Solar Energy: A Status Report. San Mateo, California: Solar Energy Information Services, 1979.

_____. Solar Energy Domestic Policy Review. San Mateo, California: Solar Energy Information Services, 1978.

Veziroglu, T. Nejat, ed. Solar Energy and Conservation: Technology, Commercialization, and Utilization. Elmsford, New York: Pergamon Press, 1978.

Von Hippel, Frank and Williams, Robert H. "Solar Technologies." Bulletin of the Atomic Scientists, 31 (November 1975), 25-30.

_____. "Toward a Solar Civilization." Bulletin of the Atomic Scientists, 33 (October 1977), 12-15.

Wade, Nicholas. "Windmills: The Resurrection of an Ancient Energy Technology." Science, 184 (June 7, 1974), 1055-1058.

Weisz, P. B. and Marshall, J. F. "High Grade Fuels from Biomass Farming: Potentials and Constraints." Science, 205 (October 5, 1979), 24-29.

Whitmore, William F. "OTEC: Electricity from the Ocean." Technology Review, 81 (October 1978), 58-63.

Wigg, E. E. "Methanol as a Gasoline Extender: A Critique." Science, 186 (November 29, 1974), 785-790.

Williams, J. Richard. Solar Energy: Technology and Applications. Ann Arbor, Michigan: Ann Arbor Science Publishers, 1974.

Williams, Robert H., ed. Toward a Solar Civilization. Cambridge, Massachusetts: M.I. T. Press, 1978.

Wilson, David A. Creating Energy. Black Mountain, North Carolina: Lorien House, 1979.

Wilson, J. I. and Brown, H. G. Solar Energy. New York: Crane-Russak Company, 1979.

Wolf, Martin. "Photovoltaic Solar Energy Conversion." Bulletin of the Atomic Scientists, 32 (April 1976), 26-33.

Zener, Clarence. "Solar Sea Power." Bulletin of the Atomic Scientists, 32 (January 1976), 17-24.

_____ , et al. Solar Sea Power. Pittsburgh, Pennsylvania: Carnegie-Mellon University, July 1974.

3. Breeder Reactors

Bump, T. R. "A Third Generation of Breeder Reactors." Scientific American, 216 (May 1967), 25-33.

Bupp, Irvin C. and Derian, Jean-Claude. "The Breeder Reactor in the U.S.: A New Economic Analysis." Technology Review, 76 (July/August 1974), 26-37.

Chow, Brian G. "Comparative Economics of the Breeder and Light Water Reactor." Energy Policy, 8 (December 1980), 293-307.

_____ . "The Economic Issues of the Fast Breeder Reactor Program." Science, 195 (February 11, 1977), 551-556.

Cochran, T. B. The Liquid Metal Fast Breeder Reactor. Baltimore, Maryland: The Johns Hopkins University Press, 1974.

Comptroller General. The Clinch River Breeder Reactor: Should Congress Continue to Fund It? [EMD-79-62] Washington, D.C.: U.S. General Accounting Office, May 7, 1979.

_____ . The Department of Energy's Water-Cooled Breeder Program — Should It Continue? [EMD-81-46] Washington, D.C.: U.S. General Accounting Office, March 25, 1981.

Feiveson, Harold A.; Taylor, Theodore B.; von Hippel, Frank; and Williams, Robert H. "The Plutonium Economy: Why We Should Wait and Why We Can Wait." Bulletin of the Atomic Scientists, 32 (December 1976), 10-14.

Fuller, John. We Almost Lost Detroit. New York: Reader's Digest Press, 1975; New York: Ballantine Books, 1976.

Gillette, Robert. "Breeder Reactor Debate: The Sun Also Rises." Science, 184 (May 10, 1974), 650-651.

Graham, John. Fast Reactor Safety. New York: Academic Press, 1971.

Hafele, W.; Holdren, J. P.; Kessler, G.; and Kulcinski, G. L. Fusion and Fast Breeder Reactors. Laxenburg, Austria: International Institute for Applied Systems Analysis, 1977.

Hammond, Allen L. "Breeder Program: Bethe Panel Calls for Reorientation." Science, 182 (December 21, 1973), 1236-1237.

_____. "Breeder Reactors: Power for the Future." Science, 174 (November 19, 1971), 807-810.

_____. "The Breeder's Costs, Or How the AEC Approaches Critical Math." Environmental Action, 4 (May 13, 1972), 11-14.

Johnson, Leland L., et al. Alternative Institutional Arrangements for Developing and Commercializing Breeder Reactor Technology. [R-2069-NSF] Santa Monica, California: RAND Corporation, November 1976.

Koch, L. J.; Kim, F. S.; Wnesch, G. W.; Branyon, C. E.; and Alexander, E. L. "Sodium-Cooled Fast Breeder Reactors." In Proceedings of the Third International Conference on the Peaceful Uses of Atomic Energy. Volume VI: Nuclear Reactors. New York: United Nations, 1965.

Lovins, Amory B. "The Case Against the Fast Breeder Reactor: An Anti-Nuclear Establishment." Bulletin of the Atomic Scientists, 29 (March 1973), 29-35.

_____. "Thorium Cycles and Proliferation." Bulletin of the Atomic Scientists, 35 (February 1979), 16-32.

Natchez, Peter B. and Bupp, Irvin C. "Policy and Priority in the Budgetary Process." American Political Science Review, 67 (September 1973), 951-963.

Pearson, Lynn F. "Breeder Economics and the Political Influence." Energy Policy, 7 (March 1979), 77-78.

Richels, Richard G. and Plummer, James L. "Optimal Timing of the U.S. Breeder." Energy Policy, 5 (June 1977), 106-121.

Rotblat, Joseph, ed. Nuclear Reactors: To Breed or Not to Breed. London, England: Taylor & Francis, 1977.

Seaborg, Glenn T. and Bloom, Justin L. "Fast Breeder Reactors." Scientific American, 223 (November 1970), 13-21.

Shapley, Deborah. "Breeder Reactors: Fast Flux Fuel Rods Subject of Silkwood Charges." Science, 199 (March 3, 1978), 956-958.

_____. "Engineer's Memo Stirs Doubts on Clinch River Breeder." Science, 197 (July 22, 1977), 350-352.

Smith, A. Robert. "The Breeder Reactor: Another SST?" Bulletin of the Atomic Scientists, 30 (October 1974),12-13.

Speth, J. Gustave; Tamplin, Arthur R.; and Cochran, Thomas B. "Bypassing the Breeder." Environmental Action, 6 (April 12, 1975), 10-13.

_____. "Plutonium: An Invitation to Disaster." Environmental Action, 6 (November 23, 1974), 4-8.

_____. "Plutonium Recycle: The Fateful Step." Bulletin of the Atomic Scientists, 30 (November 1974), 15-22.

Spinrad, Bernard I. "Alternative Breeder Reactor Technologies." In Jack M. Hollander, Melvin K. Simmons, and David O. Wood, eds. Annual Review of Energy. Volume 3. Palo Alto, California: Annual Reviews, Inc., 1978, pp. 147-180.

United States. Library of Congress. Congressional Research Service. Breeder Reactors: The Clinch River Project. [Issue Brief No. IB77088] Washington, D.C.: U.S. Government Printing Office, December 1978.

Vandryes, Georges A. "Superphoenix: A Full-Scale Breeder Reactor." Scientific American, 236 (March 1977), 26-35.

Weinberg, Alvin M. "To Breed or Not To Breed." Across the Board: The Conference Board Magazine, 14 (September 1977), 4-23.

4. Fusion

Artsimovich, L. A. "Controlled Nuclear Fusion: Energy for the Distant Future." Bulletin of the Atomic Scientists, 26 (June 1970), 47-55.

Chen, Francis. "The Leakage Problem in Fusion Reactors." Scientific American, 217 (July 1967), 76-91.

Clarke, J. F. "The Next Step in Fusion: What It Is and How It Is Being Taken." Science, 210 (November 28, 1980), 967-972.

Coppi, Bruno and Rem, Jan. "The Tokamak Approach in FusionResearch." Scientific American, 227 (July 1972), 65-75.

Emmett, John L., et al. "Fusion Power by Laser Implosion." Scientific American, 230 (June 1974), 24-37.

Fortescue, Peter. "Comparative Breeding Characteristics of Fusion and Fast Reactors." Science, 196 (June 17, 1977), 1326-1329.

Furth, Harold P. "Progress Toward a Tokamak Fusion Reactor." Scientific American, 241 (August 1979), 40-49.

Gillette, Robert. "Laser Fusion: An Energy Option, But Weapons Simulation Is First." Science, 188 (April 4, 1975), 30-34.

Gough, William C. and Eastlund, Bernard J. "The Prospects of Fusion Power." Scientific American, 224 (February 1971), 50-64.

Hafele, W.; Holdren, J. P.; Kessler, G.; and Kulcinski, G. L. Fusion and Fast Breeder Reactors. Laxenburg, Austria: International Institute for Applied Systems Analysis, 1977.

Hagler, M. O. and Kristiansen, M. An Introduction to Controlled Thermonuclear Fusion. Lexington, Massachusetts: Lexington Books, 1977.

Holdren, John P. "Fusion Energy in Context: Its Fitness for the Long Term." Science, 200 (April 14, 1978a), 168-180.

_____. "Fusion Power and Nuclear Weapons: A Significant Link?" Bulletin of the Atomic Scientists, 34 (March 1978b), 4-5.

Hunt, S. E. Fission, Fusion, and the Energy Crisis. 2nd ed. Elmsford, New York: Pergamon Press, 1980.

Kulcinski, Gerald L. "Fusion Power: An Assessment of Its Potential Impact in the U.S.A." Energy Policy, 2 (June 1974), 104-125.

Lidsky, Lawrence M. "The Quest for Fusion Power." Technology Review, 74 (January 1972), 10-21.

Metz, William D. "Energy Research: Accelerator Builders Eager to Aid Fusion Work." Science, 194 (October 15, 1976), 307-309.

_____. "Fusion Research (I): What Is the Program Buying the Country?" Science, 192 (June 25, 1976), 1320-1323.

_____. "Fusion Research (II): Detailed Reactor Studies Identify More Problems." Science, 193 (July 2, 1976), 38-40, 76.

_____. "Fusion Research (III): New Interest in Fusion-Assisted Breeders." Science, 193 (July 23, 1976), 307-309.

_____. "Laser Fusion: A New Approach to Thermonuclear Power." Science, 177 (September 29, 1972), 1180-1182.

_____. "Laser Fusion: One Milepost Passed - Millions More to Go." Science, 186 (December 27, 1974), 1193-1195.

_____. "Laser Fusion Report Plays Down Power Potential, Plays Up the Need for University and Industrial Research." Science, 188 (April 4, 1975), 32-33.

_____. "Nuclear Fusion: The Next Big Step Will Be a Tokamak." Science, 187 (February 7, 1975), 421-423.

_____. "Report of Fusion Breakthrough Proves to be a Media Event." Science, 201 (September 1, 1978), 792-794.

Mills, G. Fusion Power Plant. Princeton, New Jersey: Princeton University Plasma Physics Lab, August 1974.

Murakami, Masanori and Eubank, Harold P. "Recent Progress in Tokamak Experiments." Physics Today, 32 (May 1979), 25-32.

Parkins, W. E. "Engineering Limitations of Fusion Power Plants." Science, 199 (March 31, 1978), 1403-1408.

Post, R. F. "Fusion Power: The Uncertain Certainty." Bulletin of the Atomic Scientists, 27 (October 1971), 42-48.

_____ and Ribe, F. L. "Fusion Reactors as Future Energy Sources." Science, 186 (November 1, 1974), 397-407.

Rose, David J. "Controlled Nuclear Fusion: Status and Outlook." Science, 172 (May 21, 1971), 797-806.

_____ and Feirtag, Michael. "The Prospect for Fusion." Technology Review, 79 (December 1976), 20-43.

Steiner, Don and Clarke, John F. "The Tokamak: Model T Fusion Reactor." Science, 199 (March 31, 1978), 1395-1403.

United States. Library of Congress. Congressional Research Service. Fusion Power: Potential Energy Source. [Issue Brief No. IB76047] Washington, D.C.: U.S. Government Printing Office, October 1978.

5. Hydrogen

Caprioglio, P. "Prospects for a Hydrogen Economy." Energy Policy, 2 (September 1974), 244-245.

Hoffman, Peter. The Forever Fuel: The Story of Hydrogen. Boulder, Colorado: Westview Press, 1981.

Institute of Gas Technology. Hydrogen for Energy Distribution. Chicago, Illinois: Institute of Gas Technolgy, 1979.

Jones. Lawrence W. "Liquid Hydrogen as a Fuel for the Future." Science, 174 (October 22, 1971), 367-370.

Klimis, Ivan F. "The Hydrogen Alternative: Experience and Strategy." Energy Policy, 2 (September 1974), 245-248.

Lucas, N. J. D. "Hydrogen and Nuclear Power." Energy Policy, 4 (March 1976), 25-36.

Mathis, David A. Hydrogen Technology for Energy. Park Ridge, New Jersey: Noyes Data Corporation, 1977.

Maugh, Thomas H., II. "Hydrogen: Synthetic Fuel of the Future." Science, 178 (November 24, 1972), 849-852.

Veziroglu, T. N., ed. Hydrogen Energy. 2 volumes. New York: Plenum Press, 1975.

Winsche, W. E., et al. "Hydrogen: Its Future Role in the Nation's Energy Economy." Science, 180 (June 29, 1973), 1325-1332.

Part VII
Energy Policies in
International Perspective

If the image of the earth as a global village propounded by the late Marshall McLuhan remains something of a hyperbole, there still is a great deal of truth in this description. An international economic order exists which, to a greater or lesser degree, ties most nations of the First and Third Worlds into a complex, hierarchical system of commercial relations among nation states, regional economic associations, and multinational enterprises. A mass communications system extends over much of the globe, offering possibilities for the emergence of a world culture flowing from the interpenetration of parochial cultures or for the extension of cultural imperialism over native cultures hitherto suffused with values running counter to the rationalistic, instrumentalist, and production-oriented culture of the West. Even the realm of international politics is less a world characterized by incidental, unpredictable outbreaks of conflicts among nation states inhabiting a Hobbesian state of nature than it is an order with a well-known and easily-definable structure of conflicts persisting over time and reproduced in part by unequal political and economic arrangements and in part by antagonistic politico-economic ideologies and cultural visions.

In the light of the above remarks, it should not be surprising that an analysis of energy policies in international perspective should show the same tendencies toward organized coherence, the same structure of hierarchical relations, and the same patterns of systemic control in the international energy domain existing side by side with idiosyncratic, novel, and even semi-autonomous features and tendencies in individual nation states and sub-national groupings around the world. One complicating factor to this overview, of course, is that since 1973 the consolidation of the Organization of Petroleum Exporting Countries (OPEC) has set in motion a series of changes which, if they have not had the consequence of transforming the old international order, have unsettled past patterns within the international political economy and have set loose forces in the world requiring significant adjustment to the political power of these rich but underdeveloped nations and to the international ramifications upon industrialized and developing nations of the dramatic rise of oil prices triggered by that event.

The overriding purpose of the seventh part of this bibliographical guide is to gather together references from the rapidly expanding literature on energy problems, policies, and prospects of other nations and regions of the world in order to assist the researcher in seeing these national and regional energy policies in their uniqueness and generality. Some nations have persisted in energy strategies and policy-making derived from the pre-embargo days of cheap, undisrupted, and apparently unlimited supplies of crude oil (consider Leon Lindberg's apposite term, "energy syndrome"), while other nations have haltingly embarked on alternative energy strategies – oftentimes in conjunction with promoting more democratic forms of policy-making – based upon the realities of the OPEC/Big Oil duopoly and the age of ecological scarcity. Many nations are caught in a transitory phase where both strategies are pursued, sometimes in succession, sometimes simultaneously. The contradictory energy policies which result reflect both the eroding authority of old values and priorities and the continuing political and economic power and influence of ruling elites which will allow the post-petroleum era to emerge only on terms which can be coopted -- i.e., where new requirements and new technologies can be absorbed into the prevailing order without materially altering the existing structure of influence and rewards.

The first chapter of this part offers references to general issues and common problems in the forging of international energy policies and strategies for grappling with the energy crisis in its global framework. Overviews of these considerations are presented by a number of scholars, including Blaney (1973), Mikdashi (1974), Montbrial (1979), and Vernon (1976). Within the economic realm, students of the international energy system have explored important subjects like energy use in comparative perspective (Darmstadter, 1973; Darmstadter, et al., 1977; and Dunkerley, 1978) and world energy supply and demand now and in the future (Allen, et al., 1981; Basile, 1976, 1977; Beaujean, et al., 1977; Darmstadter, 1973; Darmstadter, et al., 1971; Emery, 1977; Grenon, 1977; Gustavson, 1971; Hafele, 1981a and b; Hellman, 1973; Ion, 1978; MITRE Corporation, 1971; Nordhaus and Goldstein, 1978; O.E.C.D., 1977; Parker, 1975; Pindyck, 1979; Sassin, 1980; Siddiqui, 1976; Thomas, 1973; U.N., 1956; Wilson, 1977; and World Energy Conference, 1978). Among those social scientists and policy analysts sensitive to the interweaving of political power and economic clout into a patterned structure at the international level, there has been substantial interest in energy issues in the international political economy (Friedland, et al., 1975; Jenson, 1970; Joyner, 1975; Katz, 1977; Lindberg, 1977c; McKie, 1974; Park, 1976; Parvin, 1976; Ridgeway, 1973; Tanzer, 1974, 1975; and Vernon, 1975), as well as in the role of multinational oil corporations within this politico-economic order (Akins and Friis, 1974; Barnes, 1972; Chandler, 1970; Evan, 1970; Hansen, 1972; Jacoby, 1974; Mitchell, 1974; Odell, 1974, 1975; Parvin, 1976; Rigin, 1974; and Wyant, 1977). Another economic issue of concern to other writers is the problem of energy and international economic development (I.A.E.A., 1970; Lindberg, 1977d; Schurr, et al., 1971; and Sherbing, 1979b).

Politically, the international impact of rising energy prices (Committee for Economic Development, 1975) has prompted a host of social and political analysts to explore comparative policy responses to the energy (or oil) crisis (Erickson and Waverman, 1974; Evans, 1979; Halvorsen and Thornton, 1978; Lawrence and Heisler, 1980; Lindberg, 1977a, b, and e; Mangone, 1977; Steinberg and Yager, 1977), while it has provoked other investigators to examine international energy policies and politics in general (Choucri, 1976; Dunkerley, 1980; Ebinger, 1978; Lawrence and Heisler, 1980; Lovins, 1974, 1975; Murray and LaViolette, 1977; T.B. Taylor Associates, 1978; Willrich, 1976) and comparative resource-specific programs (DeLeon, 1979; DeWinter, 1979; Eklund, 1970; Ferris, 1974; Hodgetts, 1964; and Murray and LaViolette, 1977). The seriousness of the social and economic consequences of increasing oil prices, uncertain supply, and eventual resource exhaustion upon national and regional economies has been a powerful stimulus to the search for cooperative international energy strategies (Akins, 1973; Chayes and Lewis, 1977; Cowen, 1977; DeCarmoy, et al,

1974; Hoffmann and Johnson, 1981; Lantze, 1976; Levy, 1974; Maddox, 1975; Solem, 1976; Thomson and Harbert, 1975; Thring, 1973; and Woodcliffe, 1975) and the quest for international solutions to specific energy problems (American Nuclear Society, 1973; Brubaker, 1975; Dreshhoff, et al., 1974; Harris, 1978; Johnson, 1977; and Scheinman and Curtis, 1977). The new energy circumstances of the seventies have also triggered new sources of international tension and exacerbated old ones (Hurewitz, 1976; Morganthau, 1975; Schurr, et al, 1971; Schwadran, 1974; and Willrich, 1975). Finally, these conditions have evoked a new awareness of the impact of energy production and use upon the global environment (Bolin, 1977; Dickstein, 1974; Handl, 1975; International Institute for Environmental Affairs, 1973; Kelley, 1977; and Lundqvist, 1974) and a heightened concern for developing new energy technologies appropriate to global energy policy (Grenon, 1974; Saltzman, 1977; and U.N., 1972).

The chapter of references on the energy situation in Canada draws a collective portrait of a nation relatively rich in indigenous energy resources moving with some success in fashioning a national energy program for overcoming the impact of OPEC and of ecological scarcity. The relative success of Canada's developing energy policies and the multitude of options it retains for the future are highlighted by many authors (Brooks, 1978; Canada, Energy Policy Sector, 1976; Canada, Federal Government, 1973; Crane, 1973a and b; Crowe, 1976; Erickson and Waverman, 1974; Fischer and Keith, 1977; Hartley, 1979; Helliwell, 1979; Hooker, 1980; McLin, 1967; MacKillop, 1978; Mitzman, 1976; Powell, 1980; Science Council of Canada, 1979; and Ziemba and Schwartz, 1980), whose analyses and interpretations can be assessed by comparing studies of Canada's energy problems (Crane, 1973b; Downs, 1977; Goldstein, 1981; Jackson, 1980; Laxer, 1974; Swanick, 1976; and Sykes, 1973) with those works devoted to its energy needs and resources (Canada, National Energy Board, 1969; Daub and Petersen, 1981; Denny, 1978; and Shrum, 1972).

Besides its large oil and natural gas deposits (Berry, 1974; Crommelin, 1978; Hamilton, 1973a and b; Helliwell, 1977; Rohmer, 1973; Smith, 1969; and U.S., Congress, 1977), Canada has an ongoing nuclear power program founded upon an alternative nuclear reactor design – the CANDU – which does not require enriched uranium fuel (Boyd, 1974; DeLeon, 1979; Eggleston, 1965; Foster and Stewart, 1972; Gray, 1972; Hodgetts, 1964; Lewis, 1959; Morrison and Wonder, 1978; and Robertson, 1978). Solar and other renewable energy resources (Argue, et al., 1978; Clark, 1972; Hollands and Orgill, 1977; and Middleton Associates, 1976) and conservation measures (Canada, Energy Policy Sector, 1977; Jackson, 1980; and Science Council of Canada, 1977) have increasingly been injected into the policy and public debate over Canada's energy future beyond oil and the atom. As a highly-industrialized society, Canada's past and present energy needs have caused environmental problems; the nature of these problems and their management are discussed by some analysts (Biswas, 1974; Lundqvist, 1974; Macdonald, 1979; and Smil, 1974). Lastly, Canada's energy riches have precipitated policy recommendations urging closer cooperation between Canada, the United States, and, in some cases, Mexico in promoting a North American energy policy (Galway, 1972; Gordon, 1976; Grayson, 1981; Greenwood, 1974; Scheinman, 1977; and Symposium, 1972).

The next chapter on the energy context of Western Europe focuses on a region and a community of nations which, as a generalization, are energy-poor in their natural resource base. This condition is reflected in the higher energy prices paid by West Europeans, the greater strides made in energy conservation, and the more unsettling impact of the Arab oil embargo upon their economies. The initial section develops a general perspective on the West European energy situation. There, the reader will find references to a number of broad topics, including Western Europe and the energy crisis (Burgess, 1974; Cicco, 1975; The Community, 1974; Community Energy Policy, 1974; European Community, 1980; Prodi and Clo, 1975; Stringelin, 1975; Western Europe's Energy Crisis, 1973), energy policy and European Community organizations (Bailey, 1976; Brondel and Morton, 1977; Burchard, 1970; DeCarmoy, 1977b; Energy Policies, 1975; European Economic Community, 1974a, b, and c; Galway, 1972; Gordon, 1971; Harris and Davies, 1980; Jensen, 1967; Lovins, 1978; Lucas, 1977; Mauther, 1974; O.E.C.D., 1968, 1969; Simonet, 1975, 1976, 1977; and Spaak, 1973), and West European energy needs (Brodman and Hamilton, 1979; DeCarmoy, 1977a; Energy Research Unit, 1973; Kouris, 1976; Kouris and Robinson, 1977; O.E.C.D., 1975, 1976a; and Ray and Robinson, 1978), energy resources (Alting von Geusau, 1975; Lubell, 1963; and Odell, 1973, 1974, 1976), and future energy options (Hafele, 1974, 1978; and Menderhausen, 1980).

The acuteness of Western Europe's dependency on foreign oil and the economic and security implications of its vulnerability to supply cutoffs have given impetus to some nations in the region either to redouble their efforts to exploit indigenous oil (Odell, 1972, 1973, 1974; and O.E.C.D., 1973b) or coal (Gordon, 1971; U.N., 1978) or to speed the pace of their energy R&D programs for developing alternative energy technologies (I.E.A., 1980; O.E.C.D., 1970, 1975; Simeon, 1978; and U.N., 1974). For many West European nations, conservation programs and energy-efficient technologies provide the quickest and easiest means of alleviating their energy plight (Conservation, 1976; Griffin, 1979; O.E.C.D., 1976; Roberts, 1979; van Victor, 1978; and Yergin, 1975). Solar and other renewables have also found favor in some parts of Western Europe (Eggers-Lura, 1979; Gross, 1974; and Palz and Steemers, 1981). On the other hand, energy technologies of the hard energy path appear more attractive to the political and economic elites of most European countries. This technological bias is evidenced in the many national nuclear power programs populating the continent (Donnelly, 1972; Foratom, 1974; Gueron, 1970; Lucas, 1976; Nau, 1974; O.E.C.D., 1969; Patterson, 1979; Polach, 1964, 1969; Some Implications, 1975; and Zaleski, 1976) and in the energy R&D programs directed to advancing breeder technology (Metz, 1975, 1976a and b). Interestingly, technocratic strategies for solving the energy crisis in Western Europe have been countered in recent years by the rise of citizen action groups seeking to widen participation in energy policy-making (Hawkes, 1977; Nelkin, 1977, 1980; Patterson, 1979; and Sweet, 1977). A major stimulus to this development is the growing recognition of the costs of energy technologies to human health and the environment (O.E.C.D., 1974, 1977; and Perrson, 1976).

As references in the next section demonstrate, the United Kingdom's energy picture mirrors the West European energy scene in some respects but departs from it in others (Bailey, 1973). Unlike most of the countries on continental Europe, the United Kingdom possesses native coal reserves (Bailey, 1974; Battelle Institute, 1976; Benson and Neville, 1976; Berkovitch, 1977; Bullock, 1976; Galloway, 1978; Griffin, 1977; Haynes, 1953; Jackson, 1974; Jenkins, 1971; Jevons, 1906; Leifchild, 1968; Nef, 1966; Robinson, 1975; Sassin and Sadnicki, 1977; U.K., D.O.E., 1974; and Vielvoye, et al., 1976), which it has exploited since the beginning of the industrial revolution. In addition, it possesses substantial oil and natural gas reserves, which it shares with other nations in the North Sea region. Paralleling other West European nations, it has long pursued a civilian nuclear program, although not without dissent

(Brookes, 1978; Carl, 1972; Counter Information Service, 1979; DeLeon, 1978; Development, 1979; Farmer, 1971; Flowers, 1976; Foley and van Buren, 1978; Gowing, 1964, 1967; Hill, 1975; Hodgetts, 1964; Hodgetts, et al., 1975; Jeffery, 1980; Joskow, 1977; Nicholson, 1973; Patterson, 1973, 1977, 1978; Pearce, 1980; Pearce, et al., 1979; Peierls, 1970; Pocock, 1977; Sailor, 1972; Stott and Taylor, 1980; Structure, 1963; U.K., Central Office of Information, 1976; Wayne, 1980; Williams, 1980; and Wonder, 1976). In addition, Great Britain, too, has begun to flirt with R&D programs and demonstration projects for advancing solar and wind technologies (Chapman, 1977; Lalor, 1975; Royal Institute, 1975; and U.K., D.O.E., 1977a and c), just as it has taken a renewed interest in energy conservation (Energy Conservation, 1974, 1979; Patterson, 1979; Pearce, 1980; U.K., D.O.E., 1978a, 1979b; and U.K., National Economic Development Office, 1974) and in co-generation (Combined Heat and Power Group, 1979; and U.K., D.O.E., n.d.). Of particular interest to political analysts is Britain's public ownership of its energy industries (Cook, 1980; Haynes, 1953; and Heald, 1981).

The United Kingdom has fashioned an energy policy which, given its energy resources (discussed above and more generally by Fermie, 1980), problems (Bailey, 1972, 1975; Foley, 1976; and Hines, et al., 1975), and present and anticipated needs (Bussanyi, 1979; Chesshire and Buckley, 1976), may prove adequate for the near term, but is likely to be short-sighted as the era beyond oil, coal, and nuclear power approaches (Bailey, 1972, 1976; Brookes, 1978; Chapman, 1975, 1976; Chesshire, et al., 1977; Cook and Surrey, 1977; Day, et al., 1980; DeCarmoy, 1977; Evans, 1975; Greater London Council, 1978; Harris and Davies, 1979; Heald, 1981; Kolbe, 1976; Leach, et al., 1979; Lewis, 1979; Marshall, 1980; Political and Economic Planning, 1966; Posner, 1973; Robinson, 1975; Stretch, 1961; U.K., D.O.E., 1976a, 1977a, 1979d; and Walsh, 1973). The future direction of British energy policy may be reflected in studies of its current energy R&D programs (Energy Technologies, 1979; Holloman, 1975; Joskow, 1977; Leighton, 1975; Owens, 1980; Ray and Uhlmann, 1979; Surrey and Walker, 1975; and U.K., D.O.E., 1976b, 1979b).

France's poverty of indigenous energy resources (Gamblin, 1968; and Scargill, 1973), in combination with its technocratic tradition and administration (Gilpin, 1968; Hoffman, 1963; and McArthur, 1969), make it particularly prone to capital-intensive, high-technology solutions to the energy crisis (Holloman, 1975; Joskow, 1977; and Pheline, 1975). Perhaps more than any other European nation state, France has been extremely vulnerable to economic instability and social upheaval from oil supply disruptions; thus, it is not surprising that the impact of the Arab oil embargo and sky-rocketing oil prices (Chevalier, 1973; Desprairies, 1972; Menderhausen, 1976; and Rondot, 1976) should have mobilized its political energies and channeled its scientific-technological resources into expanding its civilian atomic power program (Bagarry, 1976; Bupp, 1980; Bupp and Derian, 1978; Chelet, 1975; DeLeon, 1978; Feldon, 1976; Goldschmidt, 1964; Hodgetts, 1964; Joskow, 1977; Nuclear Power, 1979; Scheinman, 1965; Thiriet, 1976; and Wade, 1980) and into feverishly developing the technology of fast breeder reactors (Metz, 1975; Vandryes, 1977; and Zaleski, 1980). Nor is it startling that its solar and wind technology demonstration projects are modeled after central power station/high-technology designs of the hard energy path (Ledoux, 1977; Magnien, 1977; Solar Energy, 1976; Wind Power, 1974).

French energy policies and future options thus reflect its technocratic/administrative state apparatus and its hard energy technology orientation in significant ways (DeCarmoy, 1977, 1979; Gilsbach, 1965; Grenon, 1972; Lucas, 1977, 1978, 1979;

206

Puiseux, 1973; Rapport, 1975; Reflexions, 1975; Roset, 1976; Saumon and Puiseux, 1977; Vilain, 1969; Vizon, 1973; and Yergin, 1978). One of the few countertrends in the country lies in the emergence of public opposition to France's nuclear and breeder reactor programs – opposition which has developed from native political sources and as part of the larger West European and international no-nukes movement (Laponche, 1976; and Nelkin and Pollak, 1980a and b). As in other nations, so too in France, environmental consciousness has compelled more critical attention to the interaction between energy production and use and environmental impact (Energie, 1974; and Rapport, 1974).

A western industrialized society with a robust capitalist economy built upon the pyres of World War II, the Federal Republic of Germany has long struggled to find technological means for reducing its dependence upon foreign energy sources. Its domestic coal resources (Battelle Institute, 1976; Nephew, 1972; and Sassin, et al., 1977) have been used as a fuel for industrial production and served as the basis for Germany's coal liquefaction program for running the Nazi military juggernaut (Borkin, 1978). Spurred by the consequences of the oil embargo and the accompanying oil crisis (Die Energiekrise, 1974; Dolzinski and Ziesing, 1976; and Menderhausen, 1976), West Germany embarked upon a vigorous energy R&D program (Joskow, 1977; and Stocker, 1975) which has sought to foster hard path technologies like nuclear power (DeLeon, 1978; Gerwin, 1964; Joskow, 1977; Nuclear Power, 1972; and Winnacker and Wirtz, 1979) and breeder reactors (Keck, 1980, 1981) and led to small scale programs in solar and wind power (Hoerster, 1976; Hutter, 1973; and Weingart, 1975). Simultaneously, it has worked strenuously to try to solve the waste management problem in the nuclear fuel cycle (Krugman, 1978), while its nuclear export policies have provoked great concern in the international community over their possible implications for nuclear proliferation (Cervenka and Rogers, 1978; Gally, 1976; Louranc, 1976; and Wonder, 1977). Its nuclear and breeder programs have been the object of considerable political debate, and have given rise to a large and sometimes successful anti-nuclear movement (Foster, 1980; and Nelkin and Pollak, 1980a and b). The main features of its resultant energy policies and an outline of its prospective energy options are catalogued in a number of informative works (Arntzen and Schmitt, 1973; Bussel, et al., 1976; Carl, 1977; DeCarmoy, 1977; Denton, 1977; Dullekes, 1976; Fells, 1977; Groner, 1974; Lantzke, 1975; and Wellhofer, 1980).

The energy situation in the nations comprising Scandinavia has elicited a great deal of interest and attention among social scientists and policy analysts, as witnessed by the many references in this section dealing with Scandinavian energy policies and options (Blegaa, et al., 1977; Davidson, 1977; Finland, Council of State, 1981; Finland, Energy Department, 1978; Hambraeus and Stillesjo, 1977; Lehtonen, 1977; Lonnroth, 1977; Lonnroth, et al., 1977, 1980; Lucas, 1978; Nathan, 1981; O.E.C.D., 1969; and Wellhofer, 1980) and its comparatively frugal energy use (Comparison, 1977; Doernberg, 1975; Kristoferson, 1973; and Schipper, 1978b). While the Danes, Swedes, Finns, and Norwegians have exhibited common concerns for energy conservation (Korsgaard, 1978; Sahr, 1979; and Schipper, 1976, 1978a) and energy efficient measures like cogeneration and district heating (Elemenius, 1974; Lind, 1979; and Margen, 1978), and while their energy programs have made protection of their natural environment a high priority (Almer, et al., 1974; Institut for Atomenergi, 1975; and Lundqvist, 1974), the literature reveals a greater variation than might be expected in energy resources and prospective options within this region. The economy of Sweden, for example, was severely affected by the ramifications of the Arab oil embargo of 1973;

as a result, national political commitments were made to accelerate Sweden's developing nuclear power program (Elbek, 1978; Johansson and Steen, 1978a; Miettinen, 1976; and Nuclear R&D, 1976). This in turn precipitated in the mid-seventies a full-scale national debate (Abrahamson, et al., 1980; Anderson, 1980; Nuclear, 1976; Pearson and Nyden, 1980; Rydberg, 1980; Surrey and Huggett, 1976; and Svedin, 1975) and in 1980 led to a country-wide referendum over the future of its atomic power policy (Barnaby, 1980; Daleus, 1975; and Sahr, 1980). Norway, on the other hand, has been largely cushioned from spiraling oil prices and the threat of oil supply disruptions by its ability to exploit oil and natural gas reserves in the North Sea (Aamo, 1976; Ager-Hannsen, 1980; Ausland, 1975; Kjolberg, 1974; Squires, 1975; and Swiss, 1978).

Among the most politically open and socially advanced areas in the West, Scandinavia in its search for future energy sources has directed its energy R&D programs not only to hard energy technologies (Holloman, 1975; Kristoferson, 1978; and Nuclear R&D, 1976), but to soft path applications as well (Engstrom, 1975; Exploring Wind Power, 1974; Hinrichsen and Cawood, 1976; Johansson and Steen, 1978b and c; Kober, 1978; Kvinnsland, 1972; Litell, 1978; Saab-Scania, 1977; Sorenson, 1978; and Tinnin, 1979). It has also generated broad-ranging citizen action groups and movements, as well as significant experiments in public participation in the energy realm – experiments which may serve as models for widening the policy-making arena in other advanced industrial societies (Nelkin, 1977; Pearson and Nyden, 1980; Surrey and Huggett, 1976).

A profile of the energy scene in Italy can be drawn from the short section following the one on Scandinavia. A nation wracked by political instability and economic imbalance between North and South, Italy has confronted its sizeable energy problems (Fogagnalo, 1976; I Problemi, 1971; Jorio, 1976; and Paccione, 1976) with energy policies (Italy's National Energy Plan, 1975; and Mazzanti, 1980) apparently suffering from the same sense of uncertainty and political indecision which characterizes its economic program generally. Among the energy topics examined by analysts in this section are Italy's nuclear power program (Baker, 1973; Hodgetts, 1964; Some Considerations on the Italian Nuclear Development Strategy, 1976), its oil industry (Frankel, 1976), its solar prospects (Robotti, 1975), its breeder reactor development program (Some Considerations of Fast Breeder, 1976), and some aspects of its electric generation system (Angelini, 1974; and Valtorta, 1980).

The two concluding sections to this chapter on Western Europe focus on research and analyses of energy issues bearing on the North Sea and on other West European countries whose references were too few to include in an independent section.

As the most economically advanced of the nations of the Second World, with enormous political power and sizeable energy resources, the Union of Soviet Socialist Republics compels considerable interest for energy researchers. Moreover, because of its long-term adversarial relationship with the United States and the other Western capitalist nations, and because of the possible international implications of the inchoate age of global resource scarcity, studies of the Soviet energy landscape and policy terrain have had considerable influence in the shaping of American foreign and military policies in recent years. Equally significant, its policy of rapid economic transformism, which guided its economic movement from semi-feudal to advanced industrial status in scarcely four decades, has embued its ruling elites with an almost Promethean faith in the potentiality of modern science and technology – a faith which has motivated it to launch vigorous and well-funded high-technology/capital-intensive energy R&D programs.

A number of recognized scholars and specialists of Soviet affairs have investigated fundamental aspects of the Soviet energy situation – including its energy use (Block, 1977), its energy supply and demand balance (Campbell, 1976, 1978; Dienes, 1975; and Hoffmann, 1979), and its prevailing energy problems (Chesshire and Huggett, 1975; Dienes, 1977; and Stern, 1980). The extent of its abundance of conventional energy resources is conveyed in the works on its energy resources generally (Dienes, 1976, 1978; Economic Intelligence Unit, 1973; Elliot, 1975; Gerasimov, 1971; Goldman, 1980; Hardt, 1973; Hodgkins, 1961; Hopkins, et al., 1973; Kirkpatrick, 1979; Olson and Berentsen, 1981; Perokhin, 1975; Shabad, 1976; Slocum, 1974; U.S., C.I.A., 1977, 1978; U.S., Congress, 1978; and Zybenko, 1968), its oil and natural gas supplies (Campbell, 1968, 1976; Goldman, 1975, 1977a and b, 1980), and its coal reserves (Surface Coal Mining, 1976). The Soviet Union's unqualified optimism toward the promise of the "scientific-technological revolution" of the twentieth century has been manifested in a nuclear power program noteworthy for its lack of adequate public health and safety measures and for its accident risks (DeLeon, 1978; Dollezhal and Koryakin, 1980; Duffy, 1978; Emelyanov, 1971; Grunbaum, 1976; Kramish, 1959a and b; Medvedev, 1979; Petrosyants, 1975; Pryde and Pryde, 1974; Roy, 1981; and U.S., A.E.C., 1970). This boundless faith in the miracles of modern science and technology is also evidenced in its development of the breeder reactor (Fastbreeder Progress, 1975), as well as in its work on fusion power (Holcomb, 1969). Not significantly departing from the hard path model of technological innovation in its nuclear and advanced energy R&D programs, the Soviet Union has also begun exploring solar technology and other renewable energy sources (Fishkis, 1976; Grunbaum, 1978; and Hydroelectric Power, 1974).

In the policy realm, scholars have probed several dimensions of Soviet policy-making bearing on energy questions and have sought to clarify the parameters of Soviet energy policies and its future alternatives (Block, 1977; Carlson, 1975; Dienes and Shabad, 1979; Grenon, 1977; Gumpel, 1973, 1979; Kirkpatrick, 1979; Mangone, 1977; and Wright, 1974, 1975). Despite its relative energy wealth, the Soviet Union has caused intrabloc strains by cutbacks in its energy supplies to its East European satellites – cutbacks flowing generally from consequences of the energy crisis and particularly from the enduring Soviet policy of placing highest priority on its own internal economic development (Maddock, 1980). Energy has been a factor of increasing importance in Soviet foreign policy, and recent events -- such as political instability in the Persian Gulf area and the growing dependence of some West European nations and, increasingly, Japan on Soviet oil and natural gas supplies -- will only heighten the prominence of this element in its foreign relations (Goldman, 1977b; Gumpel, 1973; Horelick, 1975; Hunter, 1969; Klinghoffer, 1975, 1977; Landis, 1973; and Russell, 1976). One last issue which Soviet policy makers have reluctantly begun to acknowledge in formulating their energy programs is the nexus of energy development and environmental protection (Bush, 1972; Federenko and Gofman, 1973; and Kramer, 1974).

The references dealing with the Soviet energy terrain are supplemented by citations on energy as a problem focus in Eastern Europe in the succeeding chapter. The most general references examine the energy situation, present and prospective, in Eastern Europe as a region (Dienes, 1976; Guha, 1977; Hoffmann, 1979; Hopkins, et al., 1973; Korda and Moravcik, 1976; Kramer, 1975, 1979; Polach, 1969, 1970;

Russell, 1976; and Wasowski, 1969). Others illuminate the peculiar problems and indigenous circumstances of individual East European nations (Carter, 1976; Dobozi, 1977; Haberstroh, 1978; Houden, 1978; Matusek, 1972; Pearton, 1971; and Suica, 1971). Finally, the availability or development of particular energy resources in Eastern Europe, including coal (Froelich, 1975; and Hoffmann, 1979), oil (Oil Drilling, 1976; and Pearton, 1971), and nuclear power (Developing Nuclear Power, 1974; Mathieson, 1980; New Nuclear Power Plants, 1977; Polach, 1968; and Wilcyznski, 1974), is analyzed.

The next chapter lists references to a growing corpus of writings which explores the general context of the Middle East and the specific background and development of the Arab oil cartel. In the process, it opens up general vistas on oil politics, economic development, and international and regional conflict in an area of the world which is the cradle of Western civilization and may yet become the graveyard of world humanity. The successful cartelization of most of the world's non-Communist oil reserves achieved by the Arab and non-Arab members of the Organization of Petroleum Exporting Countries (OPEC) in the early seventies after years of internal negotiation and preparation has had profound implications for the region, the industrialized West, and the developing nations. It has also placed high on the agenda of individual OPEC members the issue of how to turn their mountains of petrodollars into economic development and external investments so that their economic future in a post-petroleum world will be assured. Their enormous wealth in addition raises key questions about their role and responsibilities in supporting the economic advance of the nations of the Third World.

The historical development of OPEC and its towering politico-economic status and influence are subjects which have been exhaustively researched and debated (Allen, 1979; Anthony, 1975; Arnaoot, 1974; Choucri, 1980; El-Mallakh, 1977, 1978, 1981; Ezzatti, 1978; Field, 1972, 1976; Hansen, 1976; Johany, 1980; Keiser, 1975; Krasner, 1973; Landis, 1980; Mancke, 1975; Mikdashi, 1972, 1974, 1977; Moran, 1976-77, 1978, 1981; Mosley, 1973; OPEC Official Resolutions, 1980; Penrose, 1975; Rouhani, 1971; Rustow, 1974; Rustow and Mungo, 1976; Schmalensee, 1976; Stocking, 1970; Vicker, 1974; and Willett, 1979). Important research has been devoted to analyzing the wide-ranging impact of the rise of OPEC on the world economy and on the economies of individual nations (Brookes, 1975; Choucri, 1980; Farmanfarmaian, et al., 1975; Paust and Blaustein, 1974, 1977; Sherbing, 1976; Shihata, 1976, 1981; Sobhan, 1980; Stork, 1975; Tomeh, 1977; and Williams, 1976), as well as the tightly-knit relationship between multi-national oil corporations and the OPEC cartel (Adelman, 1972-73, 1976; Al-Chalabi, 1980; Al-Otaiba, 1975; RAND, 1975; and Rustow, 1974). In addition, the relationship between OPEC and the West -- both in its more antagonistic phases and in its emerging interdependent one – is the focus of other studies found in this chapter (Bach, 1978; Campbell and Caruso, 1972; El-Mallakh, 1970; Enders, 1975; Gholamnezhad, 1981; Hamilton, 1962; Klebanoff, 1974; Landis, 1980; Lubell, 1963; Mikdashi, 1977; Schurr, 1971; U.S., Congress, 1976; and Willrich and Mossavarrahmani, 1980).

Also covered in this chapter are energy policies of specific Mideast nations (Barger, 1975; Elshafei, 1979; El-Mallakh, 1980; and Mossavar-Rahmani, 1980), and the role and dynamics of oil in regional and international conflicts (Abir, 1974; Ali, 1976; Horelick, 1975; Hurewitz, 1976; Issawi, 1963, 1973; Price, 1976; Remba, 1975; and Shwadran, 1973, 1974, 1977). The possibilities for economic development afforded by the transfer of a significant portion of the world's wealth to the treasuries of the

OPEC nations are investigated by a host of scholars (Alnasraui, 1967; Al-Otaiba, 1977; Anthony, 1975; Gueron, 1978; DeCarmoy, 1974; Edens, 1979; Ghazi, 1977; Hazelton, 1976; Katouzian, 1978; McLachlan and Ghorban, 1979; Mabro and Monroe, 1974; Oweiss, 1977; Seifert, 1973; Sharshar, 1978; and Stone, 1977).

The energy situation in Latin America is characterized by a mosaic of energy-rich and energy-poor countries, with differing capabilities for shaking the legacy of colonial dominion and underdevelopment and achieving economic advancement. The introductory section of the next chapter on Latin America and its most note-worthy energy-endowed countries lists general references which provide an overview of the region's energy resources (Muller, 1978) -- including oil (Balestrini, 1971), coal (Kostlowski, 1979), and nuclear power (Centro Nuclear, n.d.; Eichner, 1979; and Sabato, 1977). Several works consider the impact of the energy crisis and the consolidation of OPEC into a powerful world cartel upon Latin America (Muller, 1978; Rubicek, 1974; and Street, 1978), while others elucidate the contours and possibilities of cooperative energy strategies unfolding among nations in the area (Hammond, 1978; and Increased Latin American Solidarity, 1979).

The prospects for the economic development of Mexico, as well as the ambiguity of its political position in Latin America and among the developing nations, were dramatically altered by the recent discovery of huge oil and natural gas reserves within its territory (Diaz and Serano, 1979; Grayson, 1977; Mancke, 1979a and b; Metz, 1978; Stewart-Gordon, 1979; and U.S., Library of Congress, 1978). Although its nationalized petroleum industry has a fairly long history (Bermudez, 1967; Kane, 1981; Powell, 1972; Sandeman, 1978; and Williams, 1979), these latest discoveries have spurred considerable interest in the potential for inter-American energy cooperation, as noted by Grayson (1981a and b). Several Latin American specialists have investigated the historical background of and recent developments in this new turn in United States-Mexican relations occasioned by the prospects for a bilateral or even trilateral North American energy program (Fagen, 1978; Grayson, 1978, 1979; Kane, 1981; and Meyer, 1977). Meanwhile, other writings have clarified the emerging shape and character of Mexico's energy policy (Energy Policy Formulation, 1890; and U.S., Library of Congress, 1978).

The next section brings together references to the available literature on several energy topics bearing on the Brazilian situation. While Brazil does possess some exploit-able domestic oil deposits (Smith, 1977), its energy policy and future options (Goldenberg, 1978; Hammond, 1977b; Miller, 1978; and Schuh, 1978) are distinguished by its aggressive development of alcohol fuel for its transportation needs (Hammond, 1977a and d) and by its vigorous pursuit of civilian nuclear power (Gall, 1976; Hammond, 1977c; Krugmann, 1981; Louranc, 1976; Nuclear Energy, 1980; Perry and Kern, 1978; and Wonder, 1977) -- a quest which has raised fears over the possibility of nuclear proliferation.

Although Venezuela has shown interest in atomic energy development (see I.A.E.A., 1979), its energy context revolves primarily around its oil reserves (Baloyra, 1974; Betancourt, 1978; Fuad, 1973, 1974; Gall, 1975a and b; and Tugwell, 1974, 1975) and its role in OPEC (Vallenilla, 1975). Area specialists and policy analysts whose works are listed in this section have dwelled primarily upon the politics of Venezuelan oil (Avery, 1976; Blank, 1980; Gomez, 1976; Guy, 1979; and Nott, 1974, 1975), the history of its nationalization (Bye, 1979; Petras, et al., 1977; and Rossi-Guerrero, 1976), and its role in national economic development (Grove, 1976; Salazar-Carrillo, 1976; and von Lazar and Magid, 1975). The predominance of oil in the

Venezuelan policy-making arena is reflected in the works on its energy policy and future energy alternatives (Baloyra, 1974; Energy Policies, 1975; Fuad, 1975; Tugwell, 1974, 1975; Valdez, 1972; and von Lazar and Magid, 1975).

The Asian energy picture is highlighted in the succeeding chapter by separate sections on Japan, India, and China. The Japanese energy landscape is dominated by the extensive and pressing energy needs (Doernberg, 1978) born of Japan's post-war industrial superpower status. Lacking significant native energy resources (Yada, 1980), Japan has been extremely vulnerable to foreign supply disruptions and was particularly hard hit by the initial shocks of the oil crisis precipitated by the Arab oil embargo and by subsequent escalating petroleum prices (Japan, Institute of International Affairs, 1974; Sinha, 1974; Tsurumi, 1975; and U.S.-Japan Trade Council, 1974). Its energy import dependency has imposed constraints upon its foreign policy (Okita, 1974), stimulated a generously-funded energy R&D program (Goto, 1980), and led to an acceleration of its civilian atomic power program (Grey, 1979; Hodgetts, 1964; Huff, 1973; and Imai, 1975). It has also promoted fledgling projects in solar energy (Japan's Sunshine Project, 1975). Within this framework of energy dependency and industrial and technological prowess, Japan has attempted to fashion energy policies and long-run energy strategies capable of coping with near-term limitations and long-term possibilities (Eguchi, 1980; Levy, 1973; Matsui, 1977; Mazzanti, 1980; Menderhausen, 1980; O.E.C.D., 1970; Sakisaka, 1974, 1975, 1977; Sawhill, 1979; Surrey, 1974; Tisdell, 1975; Tsurumi, 1978; Turner, 1978; and Wu, 1977).

The energy situation in India illustrates the special problems and limited alternatives of a heavily-populated, energy-poor developing nation. With inadequate deposits of coal (Coal, 1975), and insufficient reserves of petroleum (Das Gupta, 1971), India in the past has had to forge an economic development strategy built around foreign oil imports. The energy crisis of the early seventies caused a serious setback to national economic programs in India, and in its wake Indian policy-makers and policy analysts have been compelled to review alternative development strategies in the light of their differing energy needs (Diwan, 1978; Mellor, 1976; Pachauri, 1977; Power, 1975; and Sankar, 1977). The temptation to emulate the capital-intensive/high-technology route of Western industrial nations, rather than the labor-intensive/low-technology route being advanced by soft path and appropriate technology schools, has been great. Symptomatic of this lure has been India's heightened interest in nuclear power development (Gillette, 1974; Tomar, 1980; and Walczak, 1974). Insofar as a coherent energy policy based upon long-range goals and objectives has crystallized out of the post-embargo era of policy flux in India, its main outlines and contradictory features are illuminated by Henderson (1975), Pachauri (1980), Parikh and Srivivasan (1980), Ramachandran and Gururaja (1977), Rudolph and Lenth (1978), Sankar (1977), and Venkataraman (1973).

In contrast to India, the energy picture in China appears far more promising. A highly-populated, disciplined nation with a vast expanse of untapped resource potential (Cheng, 1976, 1978; Dean, 1974; and Wu, 1975), China seems to have the necessary prerequisites for developing an energy and economic program adequate to bringing its inhabitants to a new level of economic development and social well-being. Analysts have taken special interest in China's oil reserves (Bartke, 1977; Carlson, 1979; Chen, 1976; and Harrison, 1975, 1977), its oil industry (Bartke, 1977, Chang, 1977; Cheng, 1976; Kambara, 1974; Ling, 1975; Sien-Chong, 1969; and Wolfe, 1976), and its uncertain potential (Foster, 1980; Hardy, 1978; U.S., C.I.A., 1977; and Williams,

212

1980). These concerns have prompted investigations of China's emergent energy policies (Chen, 1976; Goldman, 1980; Park, 1975; Smil, 1976a and b, 1977, 1978, and
1981; and Smil and Woodard, 1977), as well as the developing relationship between
China's energy resources and international affairs (Adie, 1976; Harrison, 1977; and
Woodard, 1975, 1980).

As with Latin America, the African energy scene is marked by great contrasts
and variations. The national energy programs and prospects delineated in some of the
references in this chapter (Adelman, 1975; Adeniji, 1977; Elmaihub,1977; Emembolu,
1975; and O'Keefe, 1980) present the reader with a virtual energy kaleidoscope. Some
analysts, however, have attempted to arrive at a comprehensive view of the African
energy context examining the energy situation in Africa as a whole (Arungu, 1977;
Wilson, 1978; and Wilson, et al., 1975), African energy resources (Fabrice, 1977), the
impact of the oil crisis on the continent (Green, 1975), and implications of energy for
economic development (Amann, 1969; Chibwe, 1976; and Howe, 1977). Other researchers have viewed energy issues relating to Africa with a more delimited focus, exploring such topics as coal reserves in the Union of South Africa (Friedland, 1976,
1979), and oil deposits in Nigeria (Adeniji, 1976; Emembolu, 1975; Madujibeya,
1975; Nwogugu, 1975; Pearson, 1970; Schatzl, 1969; and Usoro, 1972), Libya
(Elmaihub, 1977; Segal, 1972; and Waddams, 1980), and Algeria (Mansfield, 1978).

The last major chapter examines the literature on the energy plight of the developing nations. Many of these less developed countries (LDCs) are the worst victims of
the energy crisis in its real and contrived forms. For, not only were the halting efforts
of many of these nations at economic development shattered by the oil price spiral
following the oil embargo (Darmstadter and Hunter, 1973; Girvan, 1975; Grant, 1974;
Lunn, 1974; Powelson, 1977; and Shami, 1978), but some developing nations are suffering from an acute shortage of one of the most primitive energy sources of all – firewood (Eckholm, 1975). In the absence of external assistance or cooperative strategies
for energy development, most scholars and analysts doubt that in an era of resource
scarcity, autonomous economic and social development is possible for the overwhelm-
ing majority of the less developed nations of the world (Dasgupta, 1975; Friedman,
1976; Hartshorn, 1977; Hoffman, 1980; Hoffman and Johnson, 1980; Hoffmann and
Johnson,1979;   Howell and Morrow, 1974; and Mellor, 1976). The growing international debate over the inequality of North-South relations and over the need for a
new international economic order is indicative of the old economic and the new energy
realities thwarting the economic development of the developing nations.

General perspectives on energy in the developing world have been offered by a
number of authors (Aver, 1981; Center for Theoretical Studies, 1980; Dunn, 1976;
Hayes, 1977, 1978; MacKillop, 1980, 1981; Martin and Pinto, 1978; Palmedo, 1978;
Reddy, 1977, 1978; Smil and Knowland, 1980; and World Bank, 1980), while other
scholars (Baron, 1980, Berrie and Leslie, 1978; Cleveland, 1980; and Del Valle, 1979)
have directed their analytic talents to exploring the energy policies and options of the
developing nations. The importance of energy to agriculture in the less developed
world has also been probed (Energy for Agriculture, 1976; and Mikhijani and Poole,
1975). Finally, the hard path/soft path debate has been reintroduced, this time in the
context of appraising contending views regarding the most appropriate energy sources
for the economic development of the LDCs – nuclear power (Falls, 1973; Gottsetin,
1977; I.A.E.A., 1978; and Lopes, 1978) or solar energy and other decentralized renewable energy sources (Ashworth, 1979a and b, 1980; Biogas, 1978; Brown, 1980;
Brown and Howe, 1978; Datta, 1972; Eggers-Lura, 1979; Kettani, 1976; Kettani and

Soussou, 1976; Makhijani, 1976; Merriam, 1972; National Research Council, 1972, 1976; Parikh, 1979; and Siddiqi and Hein, 1977). The key issues in this pivotal policy dispute have been well-articulated by Hayes (1977), Hoffman (1980), and Reddy (1977, 1978).

1. An Introduction to Energy Policies in a Global Framework

Abrahamson, Bernard, ed. The Changing Economics of World Energy. Boulder, Colorado: Westview Press, 1976.

Adelman, Morris A. "American Import Policy and the World Oil Market." Energy Policy, 1 (September 1973), 91-99.

_____. "The Hinge of Energy Policy: Relations Between Energy Markets in the United States and Abroad." In Gary D. Eppen, ed. Energy: The Policy Issues. Chicago, Illinois: University of Chicago Press, 1975, pp. 71-81.

_____. "Politics, Economics, and World Oil." American Economic Review, 64 (May 1974), 58-67.

_____. The World Petroleum Market. Baltimore, Maryland: The Johns Hopkins University Press, 1972.

Akins, James E. "International Cooperative Efforts in Energy Supply." Annals of the American Academy of Political and Social Science, 410 (November 1973), 75-85.

_____ and Friis, Soren. "Changing Monopolies and European Oil Supplies: The Shifting Balance of Economic and Political Power in the World Oil Market." Energy Policy, 2 (December 1974), 275-292.

Allen, E. L.; Edmonds, J. A.; and Jeunne, R. E. "A Comparative Analysis of Global Energy Models." Energy Economics, 3 (January 1981), 2-13.

American Nuclear Society. Proceedings of the 1972 International Conference on Nuclear Solutions to World Energy Problems. Hinsdale, Illinois: American Nuclear Society, 1973.

Amuzegar, Jahangir. "The Oil Story: Facts, Fiction, and Fair Play." Foreign Affairs, 51 (July 1973), 676-689.

215

Barnes, Robert. "International Oil Companies Confront Governments: A Half-Century of Experience." International Studies Quarterly, 16 (December 1972), 454-471.

Basile, Paul S., ed. Energy Demand Studies: Major Consuming Countries. [First Technical Report of the Workshop on Alternative Energy Strategies (WAES).] Cambridge, Massachusetts: M.I.T. Press, 1976.

_____. Energy Supply/Demand Integrations to the Year 2000. [Workshop on Alternative Energy Strategies] Cambridge, Massachusetts: M.I.T. Press, 1977.

Beaujean, J. M.; Charpentier, J. P.; and Nakicenovic, N. "Global and International Energy Models: A Survey." In Jack M. Hollander, Melvin K. Simmons, and David O. Wood, eds. Annual Review of Energy. Volume 2. Palo Alto, California: Annual Reviews, Inc., 1977, pp. 153-170.

"The Big Five of World Energy: Coal, Petroleum, Natural Gas, Uranium, Hydropower." UNESCO Courier, 1 (January 1974), 6-13.

Blaney, Harry C. "The Energy Crisis: A Challenge to the International System." World Affairs, 136 (Winter 1973), 195-207.

Bolin, Bert. "The Impact of Production and Use of Energy on the Global Climate." In Jack M. Hollander, Melvin K. Simmons, and David O. Wood, eds. Annual Review of Energy. Volume 2. Palo Alto, California: Annual Reviews, Inc., 1977, pp. 197-226.

Brubaker, Sterling. "International Controls of Scarce Resources?" Current History, 69 (July-August 1975), 37-40.

Cassuto, A. "Competition in Nuclear Energy: International Issues." World Today, 20 (July 1964), 277-284.

Chandler, Geoffrey. "The Myth of Oil Power: International Groups and National Sovereignty." International Affairs, 46 (October 1970), 710-718.

Chayes, Abram and Lewis, W. Bennett, eds. International Arrangements for Nuclear Fuel Reprocessing. Cambridge, Massachusetts: Ballinger Publishing Company, 1977.

Choucri, Nazli. International Politics of Energy Interdependence. Lexington, Massachusetts: D. C. Heath and Company, 1976.

Committee for Economic Development. International Economic Consequences of High-Priced Energy. New York: Committee for Economic Development, 1975.

Corden, Max. "Implications of the Oil Price Rise." Journal of World Trade Law, 8 (March-April 1974), 133-143.

Cowen, Robert C. "Hanging Together: International Energy Strategies." Technology Review, 79 (July/August 1977), 8-9.

Crawford, Emmanuel J. Oil and the Changed Structure of the World at the Beginning of the 21st Century. Albuquerque, New Mexico: Institute for Economic and Political World Strategic Studies, 1980.

Darmstadter, Joel. "Intercountry Comparisons of Energy Use: Any Lessons for the United States?" In Bernard J. Abrahamson, ed. Conservation and the Changing Direction of Economic Growth. Boulder, Colorado: Westview Press, 1978, pp. 69-78.

_____. "World Energy Requirements." In Asher J. Finkel, ed. Energy, the Environment, and Human Health. Acton, Massachusetts: Publishing Sciences Group, 1973.

_____; Dunkerley, Joy; and Alterman, Jack. How Industrial Societies Use Energy: A Comparative Analysis. Baltimore, Maryland: The Johns Hopkins University Press, 1977.

_____; Tietlebaum, Perry D.; and Polach, Jaroslav. Energy in the World's Economy: A Statistical Review of Trends in Output, Trade, and Consumption Since 1925. Baltimore, Maryland: The Johns Hopkins University Press, 1971.

De Carmoy, Guy, et al. Cooperative Approaches to World Energy Problems. Washington, D. C.: The Brookings Institution, 1974.

De Leon, Daniel. A Comparative Analysis of High Technology Programs: The Development and Diffusion of the Nuclear Power Reactor in Six Nations. [U.S., U.S.S.R., Canada, Federal Republic of Germany, Great Britain, and France] Cambridge, Massachusetts: Ballinger Publishing Company, 1979.

De Winter, F. Description of the Solar Energy R & D Programs of Many Nations. J. W. De Winter, ed. San Mateo, California: Solar Energy Information Services, 1979.

Dickstein, H. L. "National Environmental Hazards and International Law." International and Comparative Law Quarterly, 23 (April 1974), 426-446.

Dreschhoff, Gisela; Saunders, D. F.; and Zeller, E. J. "International High Level Nuclear Waste Management." Bulletin of the Atomic Scientists, 30 (January 1974), 28-33.

Dunkerley, Joy, ed. International Comparisons of Energy Consumption. [Research Paper No. R-10] Washington, D.C.: Resources for the Future; Baltimore, Maryland: The Johns Hopkins University Press, 1978.

_____. International Energy Strategies. Cambridge, Massachusetts: Oelgeschlager, Gunn & Hain, 1980.

217

Ebinger, Charles K. International Politics of Nuclear Energy. Beverly Hills, California: Sage Publications, 1978.

Eckbo, Paul Leo. The Future of World Oil. Cambridge, Massachusetts: Ballinger Publishing Company, 1976.

Eklund, Sigvard. "The International Atom." Bulletin of the Atomic Scientists, 26 (June 1970), 56-61.

"Energy for the World's Technology." New Scientist, 44 (November 13, 1969), 1-24.

Energy: Global Prospects, 1985-2000: Report of the Workshop on Alternative Energy Strategies. Carroll L. Wilson, Project Director. New York: McGraw-Hill, 1977.

Erickson, Edward W. and Waverman, Leonard, eds. The Energy Question: An International Failure of Policy. Volume I: The World. Volume II: North America. Toronto, Ontario: University of Toronto Press, 1974.

Evan, Harry Z. "The Multinational Oil Company and the Nation State." Journal of World Trade Law, 4 (September-October 1970), 666-685.

Evans, Douglas. Western Energy Policy: The Case for Competition. New York: St. Martin's Press, 1979.

Feraru, A. T. "Transnational Political Interests and the Global Environment." International Organization, 28 (Winter 1974), 31-60.

Fermi, Laura. Atoms for the World: United States Participation in the Conference on the Peaceful Use of Atomic Energy. Chicago, Illinois: University of Chicago Press, 1974.

Flower, Andrew R. "World Oil Production." Scientific American, 238 (March 1978), 42-49.

Freymond, J. "New Dimensions in International Relations." Review of Politics, 37 (October 1975), 464-478.

Friedland, Edward; Seabury, Paul; and Wildavsky, Aaron. "Oil and the Decline of Western Power." Political Science Quarterly, 90 (Fall 1975), 437-450.

Gasteyger, Curt, et al. Energy, Inflation, and International Economic Relations: Atlantic Institute Studies - Two. New York: Praeger Publishers, 1975.

Ghobash, Saeed A. "Interdependence and the World Energy Picture." Natural Resources Lawyer, 9 (1976), 11-18.

Gray, John E. "Financing Free World Energy Supply and Use." Atlantic Community Quarterly, 13 (Spring 1975), 57-93.
218

Grenon, Michel. Ce Monde Affame D'Energie. Paris, France: Robert Laffont, 1973.

_____ . Energy R & D in the Industrialized Countries. New York: Ford Foundation, 1974.

_____ . "Global Energy Resources." In Jack M. Hollander, Melvin, K. Simmons, and David O. Wood, eds. Annual Review of Energy, Volume 2. Palo Alto, California: Annual Reviews, Inc., 1977, pp. 67-94.

Griffith, Edward D. and Clarke, Alan W. "World Coal Production." Scientific American, 240 (January 1979), 38-47.

Gustavson, Marvin. Dimensions of World Energy. Washington, D.C.: The MITRE Corporation, 1971.

Guyol, Nathaniel B. The World Electric Power Industry. Berkeley, California: University of California Press, 1969.

Hafele, Wolf, ed. Energy in a Finite World: Volume I: Paths to a Sustainable Future. Cambridge, Massachusetts: Ballinger Publishing Company, 1981a.

_____ , ed. Energy in a Finite World: Volume II: Global Systems Analysis. Cambridge, Massachusetts: Ballinger Publishing Company, 1981b.

_____ and Sassin, W. "The Global Energy System." In Jack M. Hollander, Melvin K. Simmons, and David O. Wood, eds. Annual Review of Energy. Volume 2. Palo Alto, California: Annual Reviews, Inc., 1977, pp. 1-30.

Halvorsen, Robert F. and Thornton, Judith A. "Comparative Responses to the Energy Crisis in Different Economic Systems: An Extensive Analysis." Journal of Comparative Economics, 2 (June 1978), 187-209.

Handl, G. "Territorial Sovereignty and the Problem of Transnational Pollution." American Journal of International Law, 69 (January 1975), 50-76.

Hansen, Joachim. "Movements in the International Oil Business." Aussen Politik, 23 (1972), 324-334.

Harris, William R. International Institutions for Nuclear Energy: Issues of Assessment and Design. Santa Monica, California: RAND Corporation, 1978.

Hartshorn, J. E. Politics and World Oil Economics: An Account of the International Oil Industry and Its Political Environment. New York: Praeger Publishers, 1967.

Hellman, Hal. Energy in the World of the Future. Philadelphia, Pennsylvania: M. Evans & Company; Distributed by J. B. Lippincott & Company, 1973.

Hodgetts, J. E. Administering the Atom for Peace. [U.S., Great Britain, France, Canada, Italy, Japan] New York: Atherton Press, 1964.

Hoffmann, Robert T. and Johnson, Brian. The World Energy Triangle: A Strategy for Cooperation. Cambridge, Massachusetts: Ballinger Publishing Company, 1981.

Hurewitz, C. J., ed. Oil, the Arab-Israeli Dispute, and the Industrial World: Horizons of Crisis. Boulder, Colorado: Westview Press, 1976.

International Atomic Energy Agency. Governmental Organization for the Regulation of Nuclear Power Plants. New York: Unipub, 1979.

_____. International Comparisons of Nuclear Power Costs. New York: Unipub, 1968.

_____. Nuclear Energy Costs and Economic Development. New York: Unipub, 1970.

_____. Nuclear Power and Its Fuel Cycle: Nuclear Power Prospects and Plans. Volume I. New York: Unipub, 1978.

_____. Nuclear Power and Its Fuel Cycle: The Nuclear Fuel Cycle, Part 1. Volume 2. New York: Unipub, 1978.

_____. Nuclear Power and Its Fuel Cycle: The Nuclear Fuel Cycle, Part 2. Volume 3. New York: Unipub, 1978.

_____. Nuclear Power and Its Fuel Cycle: Radioactivity Management. Volume 4. New York: Unipub, 1978.

_____. Power Reactors in Member States. New York: Unipub, 1975.

_____. Problems Associated with the Export of Nuclear Power Plants. Vienna, Austria: International Atomic Energy Agency, 1978; New York: Unipub, 1979.

_____. Reliability of Nuclear Power Plants. Vienna, Austria: International Atomic Energy Agency, 1975.

International Energy Agency/Organization for Economic Cooperation and Development. Energy Policies and Programs of IEA Countries, 1977 Review. Paris, France: Organization for Economic Cooperation and Development, 1978.

International Institute for Environmental Affairs. World Energy, the Environment, and Political Action. New York: International Institute for Environmental Affairs, 1973.

International Oil Symposium – Selected Papers. London, England: The Economist Intelligence Unit, 1973.

Ion, D. C. Availability of World Energy Resources. London, England: Graham and Trotman, 1976, 1978.

Jacoby, Neil Herman. Multinational Oil: A Study in Industrial Dynamics. New York: Macmillan, 1974.

Jensen, Walter G. W. Energy and the Economy of Nations. Henley-on-Thames, England: G. T. Foulis, 1970.

Johnson, Brian. "The Control of Nuclear Energy: New Aims for the Reform of International Institutions." Energy Policy, 5 (December 1977), 307-318.

_____. "Nuclear Power Proliferation: Problems of International Control." Energy Policy, 5 (September 1977), 179-194.

Joyner, Christopher C. "The Petrodollar Phenomenon and Changing International Economic Relations." World Affairs, 138 (Fall 1975), 152-176.

Katz, Julius L. "Energy and the World's Economy." U.S. Department of State Bulletin, 76 (January 24, 1977), 61-67.

Kelley, Donald R. ed. The Energy Crisis and the Environment: An International Perspective. New York: Praeger Publishers, 1977.

Kissinger, Henry. "Energy: The Necessity of Decision." Atlantic Community Quarterly, 13 (Spring 1975), 7-22.

Lantzke, Ulf. "International Cooperation on Energy – Problems and Prospects." World Today, 32 (March 1976), 84-94.

_____. "The Oil Crisis in Perspective: The OECD and Its International Energy Agency." Daedalus, 104 (Fall 1975), 217-227.

Lawrence, Robert M. and Heisler, Morris O., eds. International Energy Policy. Lexington, Massachusetts: Lexington Books, 1980.

Levy, Walter J. "An Atlantic-Japanese Energy Policy." Survey, 19 (Summer 1973), 50-73.

_____. "World Oil Cooperation or International Chaos." Foreign Affairs, 52 (July 1974), 690-713.

Lindberg, Leon N. "Comparing Energy Policies: Policy Implications." In Leon N. Lindberg. ed. The Energy Syndrome: Comparing National Responses to the Energy Crisis. Lexington, Massachusetts: Lexington Books, 1977a, pp. 357-382.

221

Lindberg, Leon N. "Comparing Energy Policies: Political Constraints and the Energy Syndrome." In Leon N. Lindberg, ed. The Energy Syndrome: Comparing National Responses to the Energy Crisis. Lexington, Massachusetts: Lexington Books, 1977b, pp, 325-356.

_____. "The Energy Crisis: A Political Economy Perspective." In Leon N. Lindberg, ed. The Energy Syndrome: Comparing National Responses to the Energy Crisis. Lexington, Massachusetts: Lexington Books, 1977c, pp. 1-32.

_____. "Energy Policy and the Politics of Economic Development" Comparative Political Studies, 10 (October 1977d), 355-382.

_____, ed. The Energy Syndrome: Comparing National Responses to the Energy Crisis. Lexington, Massachusetts: Lexington Books, 1977e.

Lovins, Amory B. "World Energy Strategies." Bulletin of the Atomic Scientists, 30 (May 1974), 14-32. [Also appeared in Science and Public Affairs, 30 (May 1974), 13-32]

_____. World Energy Strategies: Facts, Issues, and Options. Rev. ed. San Francisco, California: Friends of the Earth International; Cambridge, Massachusetts: Ballinger Publishing Company, 1975.

Lundqvist, Lennart J. Environmental Policies in Canada, Sweden, and the United States: A Comparative Overview. Beverly Hills, California: Sage Publications, 1974.

McKie, James W. "The Political Economy of World Petroleum." The American Economic Review, 64 (May 1974), 51-57.

Maddox, John Royden. Beyond the Energy Crisis: A Global Perspective. New York: McGraw-Hill, 1975.

Mangone, G. Energy Policies of the World. 2 volumes. New York: Elsevier-North Holland Publishing Company, 1977.

Merklein, H. A. "The Energy Crisis: Some Causes and Effects: International Monetary Considerations." World Oil (February 1, 1974), 33-42.

Mikdashi, Zuhayr. The International Politics of Natural Resources. Ithaca, New York: Cornell University Press, 1974.

Mitchell, Edward J., ed. Dialogue in World Oil; Proceedings. Washington, D.C.: American Enterprise Institute for Public Policy Research, 1974.

MITRE Corporation. Dimensions of World Energy. McLean, Virginia: The MITRE Corporation, November 1971.

Montbrial, Thierry de. Energy, the Countdown: A Report to the Club of Rome. Elmsford, New York: Pergamon Press, 1979.

Morganthau, Hans. J. "World Politics and the Politics of Oil." In Gary D. Eppen, ed. Energy: The Policy Issues. Chicago, Illinois: University of Chicago Press, 1975, pp. 43-51.

Muir, J. D. "Changing Legal Framework of International Energy Management." International Lawyer, 9 (Fall 1975), 605-614.

_____ . "Legal and Ecological Aspects of the International Energy Situation." International Lawyer, 8 (January 1974), 1-10.

Murray, A. H. and La Violette, P. A. "Assessing the Solar Transition." In E. Lazlo and J. Bierman, eds. Goals in a Global Community. Elmsford, New York: Pergamon Press, 1977.

Ninth World Energy Conference Transactions. New York: United States National Committee of the World Energy Conference, 1975.

Nordhaus, William D. and Goldstein, Ramy, eds. International Studies of the Demand for Energy. New York: Elsevier-North Holland Publishing Company, 1978.

The Nuclear Power Issue: A Guide to Who's Doing What in the U.S. and Abroad. Claremont, California: Center for California Public Affairs, 1979.

Odell, Peter R. Oil and World Power. 3rd ed. Harmondsworth, Middlesex, England: Penguin Books, 1974.

_____ . "The World of Oil Power in 1975." World Today, 31 (July 1975), 273-282.

Organization for Economic Cooperation and Development. Oil – The Present Situation and Future Prospects. Paris, France: Organization for Economic Cooperation and Development, 1973.

_____ . World Energy Outlook: A Reassessment of Long-Term Energy Developments and Related Policies. Paris, France: Organization for Economic Cooperation and Development, 1977.

Palz, Wolfgang. Solar Electricity: An Economic Approach to Solar Energy. Paris, France: UNESCO; London, England: Butterworths, 1978.

Park, Yoon S. Oil Money and the World Economy. Boulder, Colorado: Westview Press, 1976.

Parker, Albert. "World Energy Resources." Energy Policy, 3 (March 1975), 289-308.

Parvin, M. J. "Technology, Economics, and the Politics of Oil: A Global View." Journal of International Affairs, 30 (Spring 1976), 97-110.

Pindyck, Robert S. The Structure of World Energy Demand. Cambridge, Massachusetts: M.I.T. Press, 1979.

"Renewable Energy Prospects." [Special Issue] Energy - The International Journal, 4 (October 1979).

Ridgeway, James. The Last Play: The Struggle to Monopolize the World's Energy Resources. New York: E. P. Dutton and Company, 1973.

Rigin, Y. "The Energy Famine and the U.S. Monopolies." International Affairs, 5 (May 1974), 52-60.

Saltzman, Stephen, ed. Energy Technology and Global Policy: A Selection of Contributing Papers to the Conference on Energy Policies and the International System. Santa Barbara, California: American Bibliographical Center - Clio Press, 1977.

Sassin, Wolfgang. "Energy." [Special Issue on World Economic Development] Scientific American, 243 (September 1980), 118-132.

Scheinman, Lawrence and Curtis, Harold B. "The International Safeguards Problem." In Jack M. Hollander, Melvin K. Simmons, and David O. Wood, eds. Annual Review of Energy. Volume 2. Palo Alto, California: Annual Reviews, Inc., 1977, pp. 227-238.

Schmitt, R. W. and Stewart, P. J. "Role of Industry in International Energy Programs." Physics Today, 28 (March 1975), 40-44.

Schneider, Hans K. "International Energy Trade: Recent History and Prospects." In Jack M. Hollander, Melvin K. Simmons, and David O. Wood, eds. Annual Review of Energy. Volume 2. Palo Alto, California: Annual Reviews, Inc., 1977, pp. 31-66.

Schurr, Sam, et al. Middle Eastern Oil and the Western World: Prospects and Problems. New York: American Elsevier, 1971.

Sherbing, Naiem A. Arab Oil: Impact on the Arab Countries and Global Implications. New York: Praeger Publishers, 1976.

Shwadran, Benjamin. The Middle East, Oil, and the Great Powers. New York: Halsted Press, 1973, 1974.

Siddiqi, Toufiq. World Energy. New York: Holt, Rinehart and Winston, 1976.

Smart, Ian. "The Oil Crisis in Perspective: Uniqueness and Generality." Daedalus, 104 (Fall 1975), 259-281.

Solem, K. E. "Energy Resources and Global Strategic Planning." Impact of Science on Society, 26 (January-April 1976), 77-90.

Sporn, Philip. The Social Organization of Electric Power Supply in Modern Societies. Cambridge, Massachusetts: M.I.T. Press, 1971.

Steinberg, Eleanor B. "The Energy Needs of West Europe, Japan, and Australasia." Current History, 69 (July-August 1975), 15-18.

_____ and Yager, Joseph A. "Policy Alternatives of the Major Energy-Importing Nations." In Jack M. Hollander, Melvin K. Simmons, and David O. Wood, eds. Annual Review of Energy. Volume 2. Palo Alto: California: Annual Reviews, Inc., 1977, pp. 95-124.

Stunkel, Kenneth, ed. National Energy Profiles. New York: Praeger Publishers, 1979.

T. B. Taylor Associates. "A Preliminary Assessment of the Prospects for Worldwide Use of Solar Energy." New York: Rockefeller Foundation, 1978.

Tanzer, Michael. The Energy Crisis: World Struggle for Power and Wealth. New York: Monthly Review Press 1975.

_____ . "The International Oil Crisis: A Tightrope Between Depression and War." Social Policy, 5 (November 1974), 23-29.

Thomas, Trevor M. "World Energy Resources: Survey and Review." Geographical Review, 63 (April 1973), 246-258.

Thomson, E. Keith and Harbert, Richard R. "A Call for Leadership: An Approach Toward Resolution of Energy Problems Through International Coordination and Cooperation." Atlantic Community Monthly, 13 (Spring 1975), 46-56.

Thring, M. W. "Why the World Urgently Needs an Energy Policy." International Affairs, 4 (May 1973), 225-239.

Turner, Louis. "Politics of the Energy Crisis." International Affairs, 50 (July 1974), 404-415.

United Nations. Solar Energy: I-III. New York: United Nations, 1964.

_____ . "World Energy Requirements in 1975 and 2000." In Proceedings of the International Conference on the Peaceful Uses of Atomic Energy, Volume I. New York: United Nations, 1956, pp. 3-33.

United Nations. Department of Economic and Social Affairs. World Energy Supplies. [Statistical Papers, Series J] New York: United Nations. [Annual]

United Nations. Economic and Social Council. New Sources of Energy and Energy Development: Solar Energy, Wind Power, Geothermal Energy. New York: United Nations, 1972.

United States. Central Intelligence Agency. The International Energy Situation: Outlook to 1985. Washington, D.C.: Central Intelligence Agency, April 1977.

Vernon, Raymond, ed. The Oil Crisis. New York: W. W. Norton and Company, 1976.

_____. "The Oil Crisis in Perspective: The Distribution of Power." Daedalus, 104 (Fall 1975), 245-247.

"The Washington Energy Conference." Atlantic Community Quarterly, 12 (Spring 1974), 22-128.

Willrich, Mason. Energy and World Politics. New York: Free Press, 1975.

_____. "International Energy Issues and Options." In Jack M. Hollander and Melvin K. Simmons, eds. Annual Review of Energy. Volume I. Palo Alto, California: Annual Reviews, Inc., 1976, pp. 743-772.

_____ and Marston, P. M. "Prospects for a Uranium Cartel.""Orbis, 19 (1975), 166-174.

Wilson, Carroll, ed. Energy Supply, Global and National Studies. [Workshop on Alternative Energy Strategies Series] Cambridge, Massachusetts: M.I.T. Press, 1977.

Woodcliffe, J. C. "A New Dimension to International Cooperation: The OECD International Energy Agreement." International and Comparative Law Quarterly, 24 (July 1975), 525-541.

World Energy Conference. World Energy Resources, 1985-2020. Guildford, England: Published for the World Energy Conference by IPC Science and Technology Press, 1978.

Wyant, Frank R. "The Role of Multinational Oil Companies in the World Energy Trade." In Jack M. Hollander, Melvin K. Simmons, and David O. Wood, eds. Annual Review of Energy. Volume 2. Palo Alto, California: Annual Reviews, Inc., 1977, pp. 125-152.

2. Canada

Argue, Robert R.; Emanuel, Barbara; and Graham, Stephen. The Sun Builders: A People's Guide to Solar, Wind, and Wood Energy in Canada. Toronto, Ontario: Renewable Energy in Canada, 1978.

Berry, G. R. "The Oil Lobby and the Energy Crisis." Canadian Public Administration, 17 (Winter 1974), 600-635.

Biswas, Asit K. Energy and the Environment. [Report No. 10] Ottawa, Ontario: Environment Canada, Planning and Finance Service, 1974.

Boyd, F. C. "Nuclear Power in Canada: A Different Approach." Energy Policy, 2 (June 1974), 126-135.

Brooks, David. Economic Impact of Low Energy Growth in Canada; An Initial Analysis. [Discussion Paper No. 126] Ottawa, Ontario: Economic Council of Canada, 1978.

Campbell, C. and Reese, T. "The Energy Crisis and Tax Policy in Canada and the United States: Federal-Provincial Policy v. Congressional Lawmaking." Social Science Journal, 14 (January 1977), 17-32.

Canada. Energy Policy Sector. Ministry of Energy, Mines and Resources. Energy Conservation in Canada: Programs and Perspectives. Ottawa, Ontario: The Sector, 1977.

_____. An Energy Strategy for Canada: Policies for Self-Reliance. Ottawa, Ontario: The Sector, 1976.

Canada. Federal Government. An Energy Policy for Canada: Phase I. Ottawa, Ontario: The Queen's Printer, 1973.

Canada. National Energy Board. Energy Supply and Demand in Canada and Export Demand for Canadian Energy, 1966 to 1990. Ottawa, Ontario: Information Canada, 1969.

Chibuk, John. Energy and Urban Reform. Ottawa, Ontario: Ministry of State for Urban Affairs, Information Resource Service, June 1977.

Clark, R. E. "Energy from Fundy Tides." Canadian Geographical Journal, 85 (November 1972), 150-163.

Crane, David. "Canada's Energy Policies in Global Context." International Perspectives (August 1973a), 32-37.

_____. "The Pressing Need for Canada to Define Its Energy Policies: A New Dimension to Global Politics." International Perspectives [Canada] (July/August 1973b), 32-37.

Crommelin, Michael, et al. "Management of Oil and Gas Resources in Alberta: An Economic Evaluation of Public Policy." Natural Resources Journal, 18 (April 1978), 337-389.

Crowe, Marshall. "Canadian Energy Developments." Natural Resources Lawyer, 9 (1976), 1-10.

Daub, M. and Petersen, E. "The Accuracy of a Long-Term Forecast: Canadian Energy Requirements." International Journal of Energy Research, 5 (April-June 1981), 141-154.

De Leon, Daniel. A Comparative Analysis of High Technology Programs: The Development and Diffusion of the Nuclear Power Reactor in Six Nations. [U.S., U.S.S.R., Canada, Federal Republic of Germany, Great Britain, and France] Cambridge, Massachusetts: Ballinger Publishing Company, 1979.

Denny, M., et al. "The Demand for Energy in Canadian Manufacturing: Prologue to an Energy Policy." Canadian Journal of Economics [Ontario], 11 (May 1978), 300-313.

Downs, J. R. The Availability of Capital to Fund the Development of Canadian Energy Supplies. Calgary, Alberta: University of Calgary, Canadian Energy Research Institute, 1977.

Eggleston, Wilfred. Canada's Nuclear Story. Toronto, Ontario: Irwin Clarke and Company, 1965.

Erickson, Edward W. and Waverman, Leonard, eds. The Energy Question: An International Failure of Policy. Volume II: North America. Toronto, Ontario: University of Toronto Press, 1974.

Fischer, David W. and Keith, Robert F. "Canadian Energy Development: A Case Study of Policy Processes in Northern Petroleum Development." In Leon N. Lindberg, ed. The Energy Syndrome: Comparing National Responses to the Energy Crisis. Lexington, Massachusetts: Lexington Books, 1977, pp. 63-117.

Foster, John S. and Stewart, Gordon C. "Questions About Canadian Nuclear Power." In Ian E. Efford and Barbara M. Smith, eds. Energy and the Environment. Vancouver, British Columbia: Institute of Resource Ecology, University of British Columbia, 1972, pp. 90-106.

Galway, Michael A. "A Continental Energy Policy – An Examination of Some of the Current Issues." Case Western Reserve Journal of International Law, 5 (Winter 1972), 65-80.

Goldstein, Walter. "Canada's Constitutional Crisis: The Uncertain Development of Alberta's Energy Resources." Energy Policy, 9 (March 1981), 4-13.

Gordon, Richard L. Coal and Canada-U.S. Energy Relations. Washington, D.C.: Canadian-American Committee, 1976.

Gray, J. L. "Nuclear Power: An Energy Source for Canada." In Ian E. Efford and Barbara M. Smith, eds. Energy and the Environment. Vancouver, British Columbia: Institute of Resource Ecology, University of British Columbia, 1972, 45-68.

Greenwood, Ted. "Canadian-American Trade in Energy Resources." International Organization, 28 (Autumn 1974), 689-710.

Hamilton, Richard E. "Canada's 'Exportable Surplus' Natural Gas Policy: A Theoretical Analysis." Land Economics, 49 (August 1973a), 251-259.

_____. "A Marketing Board to Regulate Exports of Natural Gas?" Canadian Public Administration, 16 (Spring 1973b), 83-95.

Hartley, Karen. Energy R & D Decision Making for Canada. Brookfield, Vermont: Renouf USA, 1979.

Helliwell, John F. "Arctic Pipelines in the Context of Canadian Energy Requirements." Canadian Public Policy, 3 (Summer 1977), 344-354.

_____. "Canadian Energy Policy." In Jack M. Hollander, et al., eds. Annual Review of Energy. Volume 4. Palo Alto, California: Annual Reviews, Inc., 1979, 175-229.

Hodgetts, J. E. Administering the Atom for Peace. [U.S., Great Britain, France, Canada, Italy, Japan] New York: Atherton Press, 1964.

Hollands, K. G. T. and Orgill, J. F. Potential for Solar Heating in Canada. [Report 77-01] Waterloo, Ontario: University of Waterloo Research Institute, 1977.

Hooker, C. A., et al. Energy and the Quality of Life: Understanding Energy Policy in Canada. Toronto, Ontario: University of Toronto Press, 1980.

Jackson, E. L. "Perceptions of Energy Problems and the Adoption of Conservation Practices in Edmonton and Calgary." The Canadian Geographer, 24 (Summer 1980), 114-130.

Laxer, James. Canada's Energy Crisis. Toronto, Ontario: James Lewis & Samuel, 1974.

Lewis, W. B. "Canada's Steps Toward Nuclear Power." Progress in Nuclear Energy Economics. London, England: Pergamon Press, 1959, Volume 2, pp. 227-288.

Lundqvist, Lennart J. Environmental Policies in Canada, Sweden, and the United States: A Comparative Overview. [Sage Administrative and Policy Series] Beverly Hills, California: Sage Publications, 1974.

McLin, Jon B. Canada's Changing Policy, 1957-1963. Baltimore, Maryland: The Johns Hopkins University Press, 1967.

Macdonald, Robert. "Energy, Ecology, and Politics." In William Leiss, ed. Ecology versus Politics in Canada. Buffalo, New York: University of Toronto Press, 1979, 188-208.

MacKillop, A. "Canada's Energy Options." International Journal of Energy Research, 2 (October-December 1978), 307-336.

Maini, J. S. and Carlisle, A. Conservation in Canada: A Conspectus. New York: Unipub, 1977.

Martin, Fernand. "Effets de la Crise de l'Energie sur la Croissance Economique de Montreal et de Quebec." Actualite Economique, 50 (1974), 351-361.

Marx, H. "Energy Crisis and the Emergency Power in Canada." Dalhousie Law Journal, 2 (September 1975), 446-454.

Middleton Associates. Canada's Renewable Energy Resources: An Assessment of Potential. Toronto, Ontario: Middleton Associates, 1976.

Mitzman, Barry. "Canada's Energy Policy: No More Mr. Nice Guy." Environmental Action, 7 (February 26, 1976), 3-5.

Morrison, R. W. and Wonder, E. F. Canada's Nuclear Energy Policy. Ottawa, Ontario: The Norman Patterson School of International Affairs, Carleton University, 1978.

Powell, Wyley L., ed. The Canadian Energy Directory. Toronto, Ontario: Ontario Library Association, 1980.

"Power Reactors in Canada: NPD and CANDU Studied." Nuclear Engineering, 3 (August 1958), 334-338.

Rigin, Y. "Canada and the U.S. Oil Business." International Affairs [Moscow] (June 1975), 91-97.

Ritchie, Ronald S. "Canada's Energy Situation in a World Context." International Perspectives (March-April 1974), 13-17.

Robertson, J. A. L. "The CANDU Reactor System: An Appropriate Technology." Science, 199 (February 10, 1978), 657-664.

Rohmer, Richard H. The Arctic Imperative: An Overview of the Energy Crisis. Toronto, Ontario: McClelland and Stewart, 1973.

Scheinman, Lawrence, ed. North American Energy Policy. Cambridge, Massachusetts: Ballinger Publishing Company, 1977.

Science Council of Canada. Canada as a Conserver Society: Resource Uncertainties and the Need for New Technologies. [Report No. 27] Ottawa, Ontario: Science Council of Canada, 1977.

_____. Roads to Energy Self-Reliance: The Necessary National Demonstrations. [Report No. 30] Ottawa, Ontario: Science Council of Canada, 1979.

Shrum, Gordon M. "Meeting British Columbia's Energy Requirements." In Ian E. Efford and Barbara M. Smith, eds. Energy and the Environment. Vancouver, British Columbia: Institute of Resource Ecology, University of British Columbia, 1972, pp. 11-31.

Smil, Vaclav. Energy and the Environment: A Long-Range Forecasting Study. Winnipeg, Manitoba: Department of Geology, University of Manitoba, 1974.

Smith, Robert. "Canadian Gas Export Policy." Public Utilities Fortnightly, 84 (September 25, 1969), 26-34.

Swanick, Eric L. The Canadian Energy Crisis: First Supplementary Bibliography. [Exchange Bibliography No. 1188] Monticello, Illinois: Council of Planning Librarians, December 1976.

Sykes, Philip. Sellout: The Giveaway of Canada's Energy Resources. Edmonton, Ontario: Hurtig Publishers, 1973.

"Symposium: U.S.-Canada Energy Resource Development." Case Western Reserve Journal of International Law, 5 (Winter 1972), 36-64.

United States. Congress. Joint Economic Committee. Canadian Oil Policies and Northern Tier Energy Alternatives: Hearing, September 13, 1976. [94th Cong., 2nd sess.] Washington, D.C.: U.S. Government Printing Office, 1977.

Ziemba, W. T. and Schwartz, S. L. Energy Policy Modeling. Volume 2: United States and Canadian Experiences. The Hague, The Netherlands: Martins Nijhoff, 1980.

3. Western Europe

GENERAL

Adelman, Morris A. and Friis, Soren. "Changing Monopolies and European Oil Supplies: The Shifting Balance of Economic and Political Power in the World Oil Market." Energy Policy, 2 (December 1974), 275-292.

Alting von Geusau, Frans A. M., ed. Energy in the European Communities. Leyden, The Netherlands: A. W. Sijthoff, 1975.

Askari, Hossein and Cummings, John T. Oil, OECD, and the Third World: A Vicious Triangle? Austin, Texas: University of Texas Press, 1978.

Bailey, Richard. "Headings for an EEC Common Energy Policy." Energy Policy, 4 (December 1976), 308-321.

Brodman, J. R. and Hamilton, R. E. A Comparison of Energy Projections to 1985. [IEA Monograph No. 1] Paris, France: Organization for Economic Cooperation and Development, 1979.

Brondel, Georges and Morton, Noel. "The European Community - An Energy Perspective." In Jack M. Hollander, Melvin K. Simmons, and David O. Wood, eds. Annual Review of Energy. Volume 2. Palo Alto, California: Annual Reviews, Inc., 1977, pp. 343-364.

Burchard, Hans-Joachim. "Towards a Common European Energy Policy." Aussen Politik, 21 (1970), 76-82.

Burgess, W. Randolph. "After the Energy Hurricane." Atlantic Community Quarterly, 12 (Summer 1974), 162-170.

Cicco, John A. "The Atlantic Alliance and the Arab Challenge: The European Perspective." World Affairs, 137 (Spring 1975), 303-325.

"The Community and the Energy Crisis." Bulletin of the European Communities, 7 (1974), 10-16.

"Community Energy Policy: A New Strategy." Bulletin of the European Communities, 7 (1974), 13-19.

"Conservation of Energy: How Are Countries Performing?" OECD Observer (September-October 1976), 4-9.

Cooper, D. F. "The Changing Pattern of Energy Purchasing in Europe." Purchasing Journal, 27 (July 1970), 32-37.

De Carmoy, Guy. Energy for Europe: Economic and Political Implications. Washington, D.C.: American Enterprise Institute for Public Policy Research, 1977a.

_____. "Energy Policies of France, Great Britain, and Germany Compared." Revue de l'Energie, 28 (March 1977b), 160-164.

De Man, Reinier; De Vries, Bert; and Kommandeur, Jan. "Depletion Policy Options for Western Europe." Energy Policy, 5 (December 1977), 319-333.

De Vries, Bert and Kommandeur, Jan. "Gas for Western Europe: How Much for How Long?" Energy Policy, 3 (March 1975), 24-37.

Donnelly, Warren H. Commercial Nuclear Power in Europe: The Interaction of American Diplomacy with a New Technology. [Prepared for the Subcommittee on National Security Policy and Scientific Development of the Committee on Foreign Relations, United States Senate] Washington, D.C.: U.S. Government Printing Office, 1972.

Eggers-Lura, A. Survey of European Solar Energy Activities. San Mateo, California: Solar Energy Information Services, 1979.

Energy Policies in the European Community. Springfield, Virginia: National Technical Information Service, 1975.

"Energy Policy in the European Community." The OECD Observer, 58 (June 1972), 36-39.

Energy Research Unit, Queen Mary College, London. "World Energy Modelling: The Development of Western European Oil Prices." Energy Policy, 1 (June 1973), 21-34.

"Europe and the Energy Problem." Euro Cooperation (December 1973/March 1974), 7-94.

The European Community and the Energy Problem. London, England: Her Majesty's Stationery Office, 1980.

European Economic Community. Towards a New Energy Policy Strategy for the European Community. Brussels, Belgium: European Economic Community, 1974a.

European Economic Community. Trilateral Commission. Energy: A Strategy for International Action. [Report No. 6] Brussels, Belgium: European Economic Community, December 1974c.

_____. Energy: The Imperative for Trilateral Approach. [Report No. 5] Brussels, Belgium: European Economic Community, June 1974b.

Foratom. The Nuclear Power Industry in Europe. 2nd ed. Bonn, Federal Republic of Germany: Atomforum, 1974.

Galway, Michael A. "A Continental Energy Policy – An Examination of Some of the Current Issues." Case Western Reserve Journal of International Law, 5 (Winter 1972), 65-80.

Gordon, Richard L. The Evolution of Energy Policy in Western Europe: The Reluctant Retreat from Coal. New York: Praeger Publishers, 1971.

Griffin, James M. Energy Conservation in the OECD, 1980-2000. Cambridge, Massachusetts: Ballinger Publishing Company, 1979.

Gross, A. T. A. Wind Power Usage in Europe. Springfield, Virginia: National Technical Information Service, 1974.

Gueron, J. "Atomic Energy in Continental Western Europe." Bulletin of the Atomic Scientists, 26 (June 1970), 62-68.

Hafele, Wolf. "Energy Choices that Europe Faces: A European View of Energy." Science, 184 (April 19, 1974), 360-367.

_____ and Sassin Wolfgang. "Energy Options and Strategies for Western Europe." Science, 200 (April 14, 1978), 164-167.

Harris, D. J. and Davies, B. C. L. "European Energy Policy and Planning: The Role of Institutions. National Westminster Bank Quarterly Review, November 1980, 23-33.

Hawkes, Nigel. "Science in Europe: The Antinuclear Movement Takes Hold." Science, 197 (September 16, 1977), 1167-1169.

"Integration a l'Est et a l'Orient et Cooperation Paneuropeenne: L'Incidence du Facteur Energetique." Politique Etrangere, 39 (1974), 291-306.

International Energy Agency/Organization for Economic Cooperation and Development. Energy Research, Development, and Demonstration in the IEA Countries. Paris, France: Organization for Economic Cooperation and Development, 1980.

Jensen, W. G. Energy in Europe, 1945-1980. London, England: G. T. Foulis, 1967.

Kouris, George. "The Determinants of Energy in the EEC Area." Energy Policy, 4 (December 1976), 343-355.

_____ and Robinson, Colin. "EEC Demand for Imported Crude Oil, 1956-1985." Energy Policy, 5 (June 1977), 142-157.

Kuz, Tony and Smil, Vaclav. "European Energy Elasticities." Energy Policy, 4 (June 1976), 171-175.

Lantzke, Ulf. "Reducing OECD Countries' Dependence on Foreign Oil." OECD Observer (March-April 1976), 27-30.

Lovins, Amory B. Re-Examining the Nature of the E.C.E. Energy Problem. [Economic Commission of Europe Report No. ECE (XXXIII)/2/I.G.] Geneva, Switzerland: United Nations, 1978.

Lubell, Harold. Middle East Oil Crises and Western Europe's Energy Supplies. Baltimore, Maryland: The Johns Hopkins University Press, 1963.

Lucas, Nigel. Energy and the European Communities. London, England: Europa Publications; New York: International Publications Service, 1977.

_____. "Energy Carriers Within Europe: Electricity and SNG." Energy Policy, 5 (March 1977), 25-34.

_____. "Nuclear Power and the EEC: The Cost of Security." Energy Policy, 4 (June 1976), 98-108.

McLin, Jon. "Oil, Money, and the Common Market." Common Ground, 1 (January 1975), 9-20.

Mauther, Martin U. "The Politics of Energy." European Community, 174 (March 1974), 13-16.

Menderhausen, Horst. "Energy Prospects in Western Europe and Japan." In Hans H. Landsberg, ed. Selected Studies on Energy: Background Papers for Energy: The Next Twenty Years. Cambridge, Massachusetts: Ballinger Publishing Company, 1980, pp. 211-266.

_____. Europe's Changing Energy Relations. Santa Monica, California: RAND Corporation, 1976.

Metz, William D. "European Breeders (I): France Leads the Way." Science, 190 (December 26, 1975), 1279-1281.

_____. "European Breeders (II): The Nuclear Parts Are Not the Problem." Science, 191 (January 30, 1976a), 368-372.

236

Metz, William D. "European Breeders (III): Fuels and Fuel Cycle are Keys to Economy." Science, 191 (February 13, 1976b), 551-553.

Michaelis, Hans. Energiemarkt und Energiepolitik in einer Europaischen Union. Frankfurt am Main, Federal Republic of Germany: A. Metzner, 1976.

Nau, Henry R. National Politics and International Technology: Nuclear Reactor Development in Western Europe. Baltimore, Maryland: The Johns Hopkins University Press, 1974.

Nelkin, Dorothy. Technological Decisions and Democracy: European Experiments in Public Participation [Sweden, the Netherlands, and Austria]. Beverly Hills, California: Sage Publications, 1977.

_____ and Pollack, Michael. "Consensus and Conflict Resolution - The Politics of Assessing Risk." In M. Dierkes; S. Edwards; and R. Coppock, eds. Technological Risk: Its Perception and Handling in the European Community. Cambridge, England: Gunn & Hain, 1980.

"Nuclear Power in the European Prospect." Energia Nucleare, 24 (March 1977), 63-72.

Odell, Peter R. "The Energy Economy of Western Europe - A Return to the Use of Indigenous Resources." Geography, 66 (January 1981), 1-14.

_____. "Europe and the Cost of Energy: Nuclear Power or Oil and Gas?" Energy Policy, 4 (June 1976), 109-118.

_____. "Europe Sits on Its Own Energy: Oil for the 1980's." Geographical Magazine, 46 (March 1974), 241-245.

_____. "Europe's Oil." National Westminster Bank Quarterly Review, August 1972, 6-21.

_____. "Indigenous Oil and Gas Developments and Western Europe's Energy Policy Options." Energy Policy, 1 (June 1973), 47-64.

Organization for Economic Cooperation and Development. "Changing Role for OECD's Nuclear Energy Agency." OECD Observer, 66 (October 1973a), 19-26.

_____. Energy and the Environment – Methods to Analyse the Long-Term Relationship. Paris, France: Organization for Economic Cooperation and Development, 1974.

_____. Energy Balances of OECD Countries, 1960-1974. Paris, France: Organization for Economic Cooperation and Development, 1976a.

237

Organization for Economic Cooperation and Development. Energy Conservation in the International Energy Agency. Paris, France: Organization for Economic Cooperation and Development, 1976b.

_____. Energy Policy: Problems and Objectives. Paris, France: Organization for Economic Cooperation and Development, 1968.

_____. Energy Production and Environment: The Probable Impact on the Environment of Energy Production and Use in OECD Countries in the Years 1975-1985 As Projected in the "Energy Prospects to 1985." Paris, France: Organization for Economic Cooperation and Development, 1977.

_____. Energy R & D. Paris, France: Organization for Economic Cooperation and Development, 1975.

_____. Gaps in Technology: Comparisons Between Member Countries in Education, Research & Development, Technological Innovation. Paris, France: Organization for Economic Cooperation and Development, 1970.

_____. Oil, The Present Situation and Future Prospects. Paris, France: Organization for Economic Cooperation and Development, 1973b.

_____. Organization and General Regime Governing Nuclear Activities. Paris, France: European Nuclear Energy Agency, 1969.

_____. Reports by Member Countries on the Administrative Structure for Dealing with Energy Problems. Paris, France: Organization for Economic Cooperation and Development, 1969.

_____. Energy Committee. Energy Policy: Problems and Objectives. Paris, France: Organization for Economic Cooperation and Development, 1966.

_____. Secretary-General. Energy Prospects to 1985: An Assessment of Long-Term Energy Development and Related Policies. 2 volumes. Washington, D.C.: Organization for Economic Cooperation and Development, 1975.

Palz, W. and Steemers, T. C., eds. Solar Houses in Europe: How They Have Worked. Elmsford, New York: Pergamon Press, 1981.

Patterson, Walter C. "Harrisburg Ist Uberall." Bulletin of the Atomic Scientists, 35 (June 1979), 9-11.

Persson, Goran A. "Control of Sulfur Dioxide Emissions in Europe." Ambio, 5 (1976), 249-252.

Polach, Jaroslav. Euratom: Its Background, Issues, and Economic Implications. Dobbs Ferry, New York: Oceana Publications, 1964.

_____ . "Nuclear Power in Europe at the Crossroads." Bulletin of the Atomic Scientists, 25 (October 1969), 15-18, 20.

Prodi, Romano and Clo, Alberto. "The Oil Crisis in Perspective: Europe." Daedalus, 104 (Fall 1975), 91-112.

Ray, George. Western Europe and the Energy Crisis. London, England: Trade Policy Research Center, 1975.

_____ and Robinson, Colin. The European Energy Market in 1980. London, England: Staniland Hall Associates, 1975.

_____ . The European Energy Outlook to 1985. London, England: Staniland Hall Associates, 1978.

Roberts, Fred. "The Scope for Energy Conservation in the E.C.E." Energy Policy, 7 (June 1979), 117-130.

Simeons, Charles. Energy Research and Development Programmes in Western Europe. New York: Elsevier-North Holland, 1978.

Simonet, Henri. Energie et Europe. Brussels, Belgium: Presence et Action Culturelles, 1976.

_____ . "Energy and the Future of Europe." Foreign Affairs, 53 (April 1975), 450-463.

_____ . "European Energy Policy: Distant Mirage or Tomorrow's Reality." Revue de l'Energie, 28 (1977), 15-22.

"Some Implications of Nuclear Energy Policies in European Countries." British Nuclear Energy Society Journal, 4 (October 14, 1975), 303-309.

Spaak, Fernand. "An Energy Policy for the European Community." Energy Policy, 1 (June 1973), 35-37.

Steinberg, Eleanor B. "The Energy Needs of West Europe, Japan, and Australasia." Current History, 69 (July-August 1975), 15-18.

Stingelin, Peter. "Europe and the Oil Crisis." Current History, 68 (March 1975), 97-100.

239

Sweet, William. "The Opposition to Nuclear Power in Europe." Bulletin of the Atomic Scientists, 33 (December 1977), 40-47.

United Nations. Economic Commission for Europe. Coal: 1985 and Beyond – A Prospective Study. Elmsford, New York: Pergamon Press, 1978.

_____. Trends in the Formulation of Energy Research and Development Policy. New York: United Nations, August 27, 1974.

Van Victor, S. A. "Energy Conservation in the OECD: Progress and Results." Journal of Energy and Development, 3 (Spring 1978), 239-259.

Yergin, D. "How Europe Saves Energy." New Republic, 173 (December 13, 1975), 14-17.

Zaleski, Pierre, ed. Nuclear Energy Maturity: Proceedings of the European Nuclear Conference. 12 volumes. Elmsford, New York: Pergamon Press, 1976.

UNITED KINGDOM

Adamson, Colin. "Electric Energy Trends in the U.K." Power Engineering, 75 (February 1971), 26-32.

Arnold, Guy. Britain's Oil. London, England: Hamish Hamilton, 1978.

Bailey, Richard. "Britain and a Community Energy Policy." National Westminster Bank Quarterly Review, November 1973, 5-15.

_____. "The Changing Energy Problem." National Westminster Bank Quarterly Review, August 1975, 18-27.

_____. "Drafting Britain's First Energy Policy." National Westminster Bank Quarterly Review, November 1976, 20-31.

_____. "Energy Policy – The National Dilemma." National Westminster Bank Quarterly Review, May 1972, 7-19.

_____. "The U.K. Coal Industry – Recent Past and Future." Energy Policy, 2 (June 1974), 152-158.

Battelle Institute. Observations on Current American, British, and West German Underground Coal Mining Practices. [Battelle Energy Program Report] Columbus, Ohio: Battelle Institute, April 15, 1976.

Berkovitch, Israel. Coal on the Switchback: The Coal Industry Since Nationalisation. London, England: G. Allen & Unwin, 1977.

240

Blair, I. M., et al., eds. Aspects of Energy Conversion: Proceedings of the U.K. Science Research Council. Elmsford, New York: Pergamon Press, 1976.

Bossanyi, Ervin. "UK Primary Energy Consumption and the Changing Structure of Final Demand." Energy Policy, 7 (September 1979), 253-258.

British Industry Today: Energy. London, England: Her Majesty's Stationery Office, 1978.

Brookes, L. B. "Energy Policy, the Energy Price Fallacy, and the Role of Nuclear Energy in the U.K." Energy Policy, 6 (June 1978), 94-106.

Buckley, J. Natural Gas in the UK Energy Market. Cambridge, England: Cambridge Information and Research Services, 1979.

Bullock, Jim. Bowers Row – Recollections of a Mining Village. London, England: E. P. Group of Companies, 1976.

Burn, Duncan. Nuclear Power and the Energy Crisis: Politics and the Atomic Industry. New York: New York University Press, 1978.

_____ . The Political Economy of Nuclear Energy: An Economic Study of Contrasting Organizations in the U.K. and U.S.A., with Evaluation of Their Effectiveness. [Research Monograph No. 9] London, England: Institute of Economic Affairs, 1976.

Buxton, Neil K. The Economic Development of the British Coal Industry. London, England: Batsford, 1979.

Carl, Ann. "The Lloyd Harbor Study Group Intervention -- A Response." Bulletin of the Atomic Scientists, 28 (June 1972), 31-36.

Chapman, Peter. "The Economics of U.K. Solar Energy Schemes." Energy Policy, 5 (December 1977), 334-340.

_____ . Fuel's Paradise: Energy Options for Britain. Hammondsworth, Middlesex, England: Penguin Books, 1975.

_____ . "Three Energy Scenarios for the United Kingdom." Long Range Planning, 9 (April 1976), 2-18.

Chesshire, John and Buckley, Christopher. "Energy Use in U.K. Industry." Energy Policy, 4 (September 1976), 237-254.

_____ ; Friend, J. K.; Pollard, J. de B.; Stringer, J.; and Surrey, A. J. "Energy Policy in Britain: A Case Study of Adaptation and Change in a Policy System." In Leon N. Lindberg, ed. The Energy Syndrome: Comparing National Responses to the Energy Crisis. Lexington, Massachusetts: Lexington Books, 1977, pp. 33-62.

241

Cochrane, Stuart and Francis, John. "Offshore Petroleum Resources: A Review of U.K. Policy." Energy Policy, 5 (March 1977), 51-62.

Combined Heat and Power Group. Combined Heat and Electrical Power Generation in the U.K. [Energy Paper No. 35] London, England: Her Majesty's Stationery Office, 1979.

Connery, R. H. and Gilmour, R. S. The National Energy Problem. Farnborough, Hants, England: Saxon House, D. C. Heath, 1975.

Cook, Franklin H. "Public Ownership: United Kingdom versus United States." Public Utilities Fortnightly, 85 (February 1970), 20-30.

Cook, P. Lesley and Surrey, A. J. Energy Policy: Strategies for Uncertainty. London, England: Martin Robertson and Co., 1977.

Counter Information Services. The Nuclear Disaster. London, England: Counter Information Services, 1979.

Day, G. V.; Inston, H. H.; and Main, F. K. An Analysis of the Low Energy Strategy for the United Kingdom. London, England: Economics and Programmes Branch, United Kingdom Atomic Energy Authority, 1980.

De Carmoy, Guy. "Energy Policies of France, Great Britain, and Germany Compared." Revue de l'Energie, 28 (March 1977), 160-164.

De Leon, Daniel. A Comparative Analysis of High Technology Programs: The Development and Diffusion of the Nuclear Power Reactor in Six Nations. [U.S., U.S.S.R., Canada, Federal Republic of Germany, Great Britain, and France] Santa Monica, California: RAND Corporation, 1978.

The Development of Atomic Energy: Chronology of Events 1939-1978. Harwell, Didcot, England: United Kingdom Atomic Energy Authority, Authority Historians Office, 1979.

Energy Conservation: Scope for New Measures and Long-Term Strategy. [Energy Paper No. 33] London, England: Her Majesty's Stationery Office, 1979.

Energy Conservation: A Study by the Central Policy Review Staff. London, England: Her Majesty's Stationery Office, 1974.

Energy Policy: A Consultative Document. London, England: Her Majesty's Stationery Office, 1978.

Energy Technologies for the UK: An Appraisal of RD & D Planning. [Energy Paper No. 39] London, England: Her Majesty's Stationery Office, 1979.

Evans, Simon Caradoc, ed. Energy Options in the United Kingdom. London, England: Latimer, 1975.

Farmer, F. R. "Safety and Nuclear Power Plants: A British View." Bulletin of the Atomic Scientists, 27 (November 1971), 47-49.

Fernie, J. A Geography of Energy in the United Kingdom. New York: Longman, 1980.

Flood, M. Torness: Keep It Green. London, England: Friends of the Earth, 1979.

Flowers, Brian. "Nuclear Power and the Public Interest: A Watchdog's View." Bulletin of the Atomic Scientists, 32 (December 1976), 24-27.

Foley, Gerald. The Energy Question. London, England: Penguin Books, 1976.

_____ and Van Buren, Araine, eds. Nuclear or Not? Choices for Our Energy Future. London, England: Heinemann Educational Books, 1978.

Freedman, Lawrence. "British Oil: The Myth of Independence." World Today, 34 (August 1978), 287-295.

Galloway, Robert L. History of Coal Mining in Great Britain. New York: Augustus M. Kelley, Publishers, 1978.

Gowing, Margaret. Britain and Atomic Energy, 1939-1945. New York: Macmillan, 1964.

_____. Independence and Deterrence: Britain and Atomic Energy, 1945-1952. New York: St. Martin's Press, 1967.

Greater London Council. Energy Policy and London. London, England: Department of Planning and Transportation, Greater London Council, 1978.

Griffin, A. R. British Coal Mining Industry: Retrospect and Prospect. New York: State Mutual Book and Periodical Service, 1977.

Hannah, Leslie. Electricity Before Nationalisation: A Study of the Development of the Electricity Supply Industry in Britain to 1948. Baltimore, Maryland: The Johns Hopkins University Press, 1979.

Harris, D. J. and Davies, B. C. L. "The Coordination of UK Energy Planning." Futures, 11 (December 1979), 536-539.

Haynes, William W. Nationalization in Practice: The British Coal Industry. Cambridge, Massachusetts: Harvard University Press, 1953; Elmsford, New York: Pergamon Press, 1953 (reprint).

Heald, David. "UK Energy Policy: Economic and Financial Control of the Nationalized Energy Industries." Energy Policy, 9 (June 1981), 99-112.

Hill, Sir John. "The Energy Situation and the Role of Nuclear Power." Atom [Great Britain], January 1975, 2-7.

Hines, A. G.; Momferratos, P. J.; and Simpson, D. R. "Energy and the U.K. Economy." Journal of Industrial Economics, 24 (September 1975), 15-20.

Hodgetts, J. E. Administering the Atom for Peace. [U.S., Great Britain, France, Canada, Italy, Japan] New York: Atherton Press, 1964.

Holloman, J. Herbert. "United Kingdom." In Energy Research and Development. [Ford Foundation Energy Policy Project Report] Cambridge, Massachusetts: Ballinger Publishing Company, 1975, pp. 235-259.

_____ ; Raz, B.; and Treital, R. "Nuclear Power and Oil Imports: A Look at the Energy Balance." Energy Policy, 3 (December 1975), 299-305.

Hutcheson, A. Macgregor and Hogg, Alexander, eds. Scotland and Oil. New York: Oliver & Boyd, 1975.

Jackson, Michael Peart. The Price of Coal. London, England: Croom Helm, 1974.

Jay, Kenneth. Calder Hall. London, England: Methuen, 1961.

Jeffery, J. W. "The Real Costs of Nuclear Power in the UK." Energy Policy, 8 (December 1980), 344-346.

Jenkin, A. K. The Cornish Miner. North Pomfret, Vermont: David & Charles, 1971.

Jevons, W. S. The Coal Question. 3rd ed. London, England: Macmillan, 1906.

Joskow, Paul L. "Research and Development Strategies for Nuclear Power in the United Kingdom, France, and Germany." In Proceedings of the Workshop on Institutional Alternatives for LMFBR Development and Commercialization. McLean, Virginia: The MITRE Corporation, 1977, pp. 227-328.

Kent, Marian. Oil and Empire: British Policy and Mesopotamian Oil, 1900-1920. London, England: London School of Economics and Political Science, 1976.

Kirby, M. W. The British Coal Mining Industry, 1870-1946: A Political and Economic History. Hamden, Connecticut: Archon Books, 1977.

Kolbe, R. A. "Britain Out-Plans the U.S.: Energy Policies There and Here." Nation, 223 (September 18, 1976), 239-241.

Lalor, Eamon. Solar Energy for Ireland. Dublin, Ireland: National Science Council, 1975.

Leach, G.; Lewis, C.; Romig, F; van Buren, A.; and Foley, G. A Low Energy Strategy for the United Kingdom. New York: The Humanities Press, 1979.

Leifchild, John R. Our Coal and Our Coal Pits. London, England: Longman, Brown, Green, and Longmans, 1856, Fairfield, New Jersey: A. M. Kelly, 1968.

Leighton, L. H. "Energy Technology Activities of the British Government and Industry." In Energy Technology II. Washington, D.C.: Government Institutes, Inc., 1975, pp. 81-98.

Lewis, Chris. "A Low Energy Option for the U.K." Energy Policy, 7 (June 1979), 131-148.

Lock, Michael. "Trends in U.K. Energy Prices." Economic Trends, 277 (1976), 91-106.

Longhurst, Henry. Adventure in Oil: The Story of British Petroleum. London, England: Sedgwick and Jackson, 1959.

Marshall, Eileen. "Low Energy Strategies for the UK - An Economic Perspective." Energy Policy, 8 (December 1980), 339-343.

Motamen, Homa and Strange, Roger. "UK Oil Revenue: The Medium-Term Outlook." Energy Policy, 9 (March 1981), 14-19.

Nef, John Uric. The Rise of the British Coal Industry. 2 volumes. London, England: Frank Case, 1932; Hamden, Connecticut: Archon Books, 1966.

Nicholson, R. L. R. "The Nuclear Power Paradox in the U.K." Energy Policy, 1 (June 1973), 38-46.

"North Sea Oil and Gas – What Potential for the U.K.?" Engineering, 213 (March 1973), 149-159.

Owens, Susan E., ed. Energy: A Register of Research, Development, and Demonstration in the United Kingdom. London, England: Social Science Research Council, 1980.

Patterson, Walter C. "Conservation Cornucopia." Bulletin of the Atomic Scientists, 35 (May 1979), 42-44.

_____. The Fissile Society. London, England: Earth Resources Research, 1977.

_____. Red Alert – Nuclear Reactors. London, England: Earth Island, 1973.

_____. "The Windscale Report: A Nuclear Apologia." Bulletin of the Atomic Scientists, 34 (June 1978), 44-46.

Pearce, David. "Energy Conservation and Official UK Energy Forecasts." Energy Policy, 8 (September 1980), 245-248.

245

Pearce, David. "Nuclear Power and Britain's Energy Future." National Westminster Bank Quarterly Review, May 1980, 10-19.

_____; Edwards, Lynn; and Beuret, Geoff. Decision-Making for Energy Futures: A Case Study of the Windscale Inquiry. London, England: Macmillan, 1979.

Peierls, Sir Rudolf. "Britain in the Atomic Age." Bulletin of the Atomic Scientists, 26 (June 1970), 40-46.

Pocock, R. F. Nuclear Power: Its Development in the United Kingdom. Surrey, England: Gresham Books, Unwin Brothers, 1977.

Political and Economic Planning. A Fuel Policy for Britain. London, England: Political and Economic Planning, 1966.

Posner, Michael V. Fuel Policy: A Study in Applied Economics. New York: Macmillan, 1973.

Ray, G. F. and Uhlmann, L. The Innovation Process in the Energy Industries. London, England: Cambridge University Press, 1979.

Robinson, Colin. The Energy "Crisis" and British Coal: The Economics of the Fuel Market in the 1970's and Beyond. London, England: Institute of Economic Affairs, 1974; Levittown, New York: Transatlantic Arts, 1975.

Royal Institute. Solar Energy: A U.K. Assessment. London, England: Royal Institute, 1975.

Royal Society of Arts. Energy and the Environment. London, England: The Royal Society of Arts, July 1974.

Sailor, Vance L. "The Role of the Lloyd Harbor Study Group in the Shoreham Hearings." Bulletin of the Atomic Scientists, 28 (June 1972), 25-31.

Sassin, W.; Hoffman, F.; and Sadnicki, M., eds. Medium-Term Aspects of a Coal Revival: Two Case Studies. [Great Britain and the Federal Republic of Germany] Laxenburg, Austria: International Institute for Applied Systems Analysis, 1977.

"Scotland's Oil: Striking It Rich." Economist, 246 (January 20, 1973), 35-38.

Sell, George, ed. The Post-War Expansion of the U.K. Petroleum Industry. London, England: Institute of Petroleum, 1954.

Squires, Susan M. "The Impact of North Sea Oil in Norway and Scotland." Norsk Georgrafisk Tidsskrift, 29 (1975), 133-140.

Stott, Martin and Taylor, Peter. The Nuclear Controversy: A Guide to the Issues of the Windscale Inquiry. London, England: Political Ecology Research Group and the Town and Country Planning Association, 1980.

Stretch, K. L. A Power Policy for Britain. London, England: Ernest Benn, 1961.

"Structure of the British Nuclear Industry." Nuclear Engineering, 8 (March 1963), 96-99.

Surrey, John and Walker, William. "Energy R & D -- A U.K. Perspective." Energy Policy, 3 (June 1975), 90-115.

UK Atomic Energy Authority Annual Report, 1979/80. London, England: The Library and Information Centre, United Kingdom Atomic Energy Authority, 1980.

United Kingdom. Central Office of Information. Nuclear Energy in Britain. London, England: Her Majesty's Stationery Office, 1976.

United Kingdom. Department of Energy. Coal Industry Examination: Final Report. London, England: Department of Energy, 1974.

_____. Development of the Oil and Gas Resources of the United Kingdom. London, England: Her Majesty's Stationery Office, 1979a.

_____. District Heating Combined with Electricity Generation in the United Kingdom. [Energy Paper No. 20] London, England: Department of Energy, n.d.

_____. Energy Conservation: Scope for New Measures and Long-Term Strategy. London, England: Her Majesty's Stationery Office, 1979b.

_____. Energy Conservation Research, Development, and Demonstration: An Initial Strategy for Industry. London, England: Her Majesty's Stationery Office, 1978a.

_____. Energy Policy Review. London, England: Her Majesty's Stationery Office, 1977a.

_____. Energy Prospects. London, England: Her Majesty's Stationery Office, 1976a.

_____. Energy Research and Development in the United Kingdom: A Discussion Document. London, England: Her Majesty's Stationery Office, 1976b.

_____. Energy Technologies for the United Kingdom. Volumes 1 and 2. London, England: Her Majesty's Stationery Office, 1979c.

_____. National Energy Policy. London, England: Her Majesty's Stationery Office, 1979d.

United Kingdom. Department of Energy. Prospects for the Generation of Electricity from Wind Energy in the United Kingdom. London, England: Her Majesty's Stationery Office, 1977b.

_____. Reorganization of the Electricity Supply Industry in England and Wales. London, England: Her Majesty's Stationery Office, 1978b.

_____. Solar Energy: Its Potential Contribution within the United Kingdom. London, England: Her Majesty's Stationery Office, 1977c.

United Kingdom. National Economic Development Office. Energy Conservation in the United Kingdom: Achievements, Aims, Options. London, England: Her Majesty's Stationery Office, 1974.

Uri, N. D. "The Effects of Environmental Quality Standards on Pricing Electrical Energy." Environment and Planning, 8 (August 1976), 573-580.

Vielvoye, Roger, et al. Coal, Technology for Britain's Future. London, England: Macmillan, 1976.

Walsh, John. "Britain and Energy Policy: Problems of Interdependence." Science, 180 (June 29, 1973), 1343-1347.

The Watt Committee on Energy. Energy Development and Land in the UK. London, England: The Watt Committee on Energy, 1979.

Wayne, B. "Windscale - A Case History in the Political Art of Muddling Through." In T. O'Riordan and R. K. Turner, eds. Progress in Resource Management and Environmental Planning. Volume 2. New York: John Wiley & Sons, 1980, pp. 165-204.

What Price Energy? Surrey, England: Electrical Research Association, 1975.

Williams, Roger. Nuclear Power Decisions: British Policies, 1953-1978. London, England: Croom Helm, 1980.

Wonder, Edward F. "Decision-Making and Reorganization of the British Nuclear Power Industry." Research Policy, 5 (July 1976), 240-268.

FRANCE

Alternatives au Nucleaire. Grenoble-Cedex, France: L'Institut Economique et Juridique de l'Energie de Grenoble, 1975.

Bagarry, Alain. Energies Classiques, Energie Nucleaire: Evolution et Perspectives. Paris, France: Universite de Paris, Pantheon-Sorbonne, 1976.

Bauer, Etienne; Puiseux, Louis; and Teniere-Buchot, Pierre Frederic. "Nuclear Energy - A Fateful Choice for France." Bulletin of the Atomic Scientists, 27 (January 1976), 37-41.

Bupp, Irvin C. "The French Nuclear Harvest: Abundant Energy or Bitter Fruit?" Technology Review, 83 (November/December 1980), 30-39.

_____ and Derian, Claude. Light Water: How the Nuclear Dream Dissolved. New York: Basic Books, 1978. [Published in paperback as: The Failed Promise of Nuclear Power. New York: Harper & Row, 1980]

Chelet, Yves. L'Energie Nucleaire. Paris, France: Editions du Seuil, 1975.

Chevalier, Jean-Marie. Le Nouvel Enjeu Petrolier. [The New Oil Stakes] Paris, France: Calmann Levy, 1973.

De Carmoy, Guy. Energy in France: Planning, Politics, and Policy. London, England: Europa Publications, 1979.

_____ . "Energy Policies of France, Great Britain, and Germany Compared." Revue de l'Energie, 28 (March 1977), 160-164.

De Leon, Daniel. A Comparative Analysis of High Technology Programs: The Development and Diffusion of the Nuclear Power Reactor in Six Nations. [U.S., U.S.S.R., Canada, Federal Republic of Germany, Great Britain, and France] Santa Monica, California: RAND Corporation, 1978.

Desprairies, Pierre. "L'Evolution de la Crise Petroliere de 1970-1971." Revue de Defense Nationale, 28 (May 1972), 138-173.

Energie, Environnement: Rapport sur les Relations entre le Secteur de l'Energie et l'Environnement. Paris, France: Documentation Francaise, 1974.

Feldon, Merceau. Energie, le Defi Nucleaire. Paris, France: A. Leson, 1976.

Finon, Dominique. "Optimisation Model for the French Energy Sector." Energy Policy, 2 (June 1974), 136-151.

Gamblin, Andre. L'Energie en France, Etude de Geographie. Paris, France: Societe d' Edition d'Enseignement Superieur, 1968.

Gilpin, Robert. France in the Age of the Scientific State. Princeton, New Jersey: Princeton University Press, 1968.

Gilsbach, Albert. Die Konzeption der Franzosischen Energiepolitik. Munich, Federal Republic of Germany: Verlag R. Oldenbourg, 1965.

Goldschmidt, Bernard. The Atomic Adventure: Its Political and Technical Aspects. [Trans. by Peter Beer] London, England: Pergamon Press, 1964.

249

Grenon, Michel. Pour Une Politique de l'Energie. Paris, France: Marabout Universite, 1972.

Hodgetts, J. E. Administering the Atom for Peace. [U.S., Great Britain, France, Canada, Italy, Japan] New York: Atherton Press, 1964.

Hoffman, Stanley, et al. In Search of France. Cambridge, Massachusetts: Harvard University Press, 1963.

Holloman, J. Herbert. "France." In Energy Research and Development. [Ford Energy Policy Project Report] Cambridge, Massachusetts: Ballinger Publishing Company, 1975, pp. 179-204.

Joskow, Paul L. "Research and Development Strategies for Nuclear Power in the United Kingdom, France, and Germany." In Proceedings of the Workshop on Institutional Alternatives for LMFBR Development and Commercialization. McLean, Virginia: The MITRE Corporation, 1977, pp. 227-328.

Laponche, Bernard. "Many in France Oppose an 'All-Nuclear' Policy." Bulletin of the Atomic Scientists, 32 (December 1976), 44-45.

Ledoux, Alain. "Them I, Premiere Centrale Solaire." Science & Vie, June 1977, 108-112.

Lucas, Nigel. Energy in France: Planning, Politics, and Policy. London, England: Europa Publications, 1979.

_____. "Energy in the French Plans." International Relations (London), 6 (May 1978), 215-244.

_____. "The Role of Institutional Relationships in French Energy Policy." International Relations (London), 5 (November 1977), 87-121.

McArthur, John H. and Scott, Bruce R. Industrial Planning in France. Cambridge, Massachusetts: Division of Research, Graduate School of Business Administration, Harvard University, 1969.

Magnien, M. "Utilizing Alternative Energy Sources in France." International Journal of Energy Research, 1 (January-March 1977), 69-94.

Mainguy, Y. L'Economie de l'Energie. Paris, France: Dunod, 1967.

Mendershausen, Horst. Coping with the Oil Crisis: French and German Experiences. Baltimore, Maryland: The Johns Hopkins University Press, 1976.

Metz, William D. "European Breeders (I): France Leads the Way." Science, 190 (December 26, 1975), 1279-1281.

Nelkin, Dorothy and Pollak, Michael. The Atom Beseiged: Extraparliamentary Dissent in France and Germany. Cambridge, Massachusetts: The M.I.T. Press, 1980a.

_____ . "French and German Courts on Nuclear Power." Bulletin of the Atomic Scientists, 36 (May 1980b), 36-43.

"Nuclear Power Contrasted: France vs. the U.S." Technology Review, 81 (March/April 1979), 66-67.

Pheline, Jean. General Orientation on French Policy of Research and Development in the Field of Energy." In Energy Technology II. Washington, D.C.: Government Institutes, Inc., 1975.

Puiseux, Louis. L'Energie et le Desarroi Post-Industriel – Essai Sur la Croissance de l'Energie. [Energy and Post-Industrial Disorder – An Essay on the Growth of Energy] Paris, France: Collection Futuribles, Hachette Literature, 1973.

Rapport de la Commission de l'Energie sur les Orientations de le Politique Energetique. [Report of the Energy Commission on the Trends of Energy Policy] Paris, France: Documentation Francaise, July 1975.

Rapport sur les Relations entre le Secteur de l'Energie et l'Environnement Dans Une Perspective a Long Terme. Paris, France: Documentation Francaise, 1974.

Reflexions sur les Choix Energetiques Francais. Grenoble-Cedex, France: L'Institut Economique et Juridique de l'Energie de Grenoble, 1975.

Rondot, Jean. La Compagnie Francaise des Petroles: Du Franc-or au Petrole-franc. Paris, France: Librairie Plan, 1962; New York: Arno Press, 1976.

Roset, Claude. La Cinquieme Energie. Paris, France: Tema-Editions, 1976.

Saumon, Dominique and Puiseux, Louis. "Actors and Decisions in French Energy Policy." In Leon N. Lindberg, ed. The Energy Syndrome: Comparing National Responses to the Energy Crisis. Lexington, Massachusetts; Lexington Books, 1977, pp. 199-172.

Scargill, D. I. "Energy in France." Geography, 58 (April 1973), 159-162.

Scheinman, Lawrence. Atomic Energy Policy in France Under the Fourth Republic. Princeton, New Jersey: Princeton University Press, 1965.

Solar Energy from France. Paris, France: Delegation Aux Energies Nouvelles, Ministere de l'Industrie et de la Recherche, 1976.

Thiriet, Lucien. L'Energie Nucleaire: Quelles Politiques Pour Quel Avenir? Paris, France: Dunod, 1976.

Vandryes, Georges A. "Superphenix: A Full-Scale Breeder Reactor." Scientific American, 236 (March 1977), 26-35.

Vilain, Michel. La Politique de l'Energie en France. Paris, France: Editions Cujas, 1969.

Vizon, M. L'Evolution Recente de la Production Energetique Francaise. Paris, France: Librairie Larousse, 1973.

Wade, N. "France's All-Out Nuclear Program Takes Shape." Science, 209 (August 22, 1980), 884-886.

Wind Power Projects of the French Electrical Authority. [NTIS Report N75-13394] Springfield, Virginia: National Technical Information Service, December 1974.

Yergin, Daniel. "France's Tough Energy Program." Fortune Magazine, 98 (July 17, 1978), 101-108.

Zaleski, C. Pierre. "Breeder Reactors in France." Science, 208 (April 11, 1980), 137-144.

FEDERAL REPUBLIC OF GERMANY

Arntzen, R. and Schmitt, D. "Energy for West Germany Towards 1980—A Report from the Institute of Energy Economics, Cologne." Energy Policy, 1 (September 1973), 169-171.

Battelle Institute. Observations on Current American, British, and West German Underground Coal Mining Practices. [Battelle Energy Program Report] Columbus, Ohio: Battelle Institute, April 15, 1976.

Borkin, Joseph. The Crime and Punishment of I.G. Farben. New York: The Free Press, 1978.

Bossel, H., et al. Energie Richtig Genutzt. Karlsruhe, Federal Republic of Germany: C. F. Muller, 1976.

Braunthal, Gerald. "The Political Economy of West Germany." Current History, 68 (March 1975), 123-126.

Carl, M. H. "The Distribution of Competences in the Field of Energy Policy Between Federal and State Institutions in the Federal Republic of Germany." International Relations (London), 5 (November 1977), 130-136.

Cervenka, Zdenek and Rogers, Barbara. The Nuclear Axis: Secret Collaboration Between West Germany and South Africa. New York: Times Books, 1978.

De Carmoy, Guy. "Energy Policies of France, Great Britain, and Germany Compared." Revue de l'Energie, 28 (March 1977), 160-164.

De Leon, Daniel. A Comparative Analysis of High Technology Programs: The Development and Diffusion of the Nuclear Power Reactor in Six Nations. [U.S., U.S.S.R., Canada, Federal Republic of Germany, Great Britain, and France] Santa Monica, California: RAND Corporation, 1978.

Denton, Richard. "Energy Futures for the Federal Republic of Germany: Three Scenarios." Energy Policy, 5 (March 1977), 35-50.

Die Energiekrise: Episode Oder Ende einer Ara? Hamburg, Federal Republic of Germany: Hoffman und Campe, 1974.

Dolinski, Urs. and Ziesing, Hans-Joachim. "An Evaluation of the Availability of Crude Oil." Aussenpolitik, 27 (1976), 207-220.

Dollekes, Hans P. Planung der Energie und Umweltpolitik. Munster, Federal Republic of Germany: Institut fur Siedlungs-und Wohnungswesen der Universitat Munster, 1976.

"Energy Cost of Goods and Services in the Federal Republic of Germany." Energy Policy, 3 (December 1975), 279-284.

Fells, Ian. "The Energy Future of West Germany." Energy Policy, 5 (December 1977), 341-344.

Foster, C. R., ed. Comparative Public Policy and Citizen Participation: Energy, Education, Health, and Urban Issues in the U.S. and Germany. Elmsford, New York: Pergamon Press, 1980.

Gall, Norman. "Atoms for Brazil: Dangers for All." Bulletin of the Atomic Scientists, 32 (June 1976), 4-9, 41-48.

Gerwin, Robert. Atoms in Germany. Dusseldorf, Federal Republic of Germany: Econ-Verlag for the Bundesminister fur Wiessenschaftliche Forschung, 1964.

Goen, Richard. Comparison of Energy Consumption Between West Germany and the United States. Menlo Park, California: Stanford Research Institute, 1975.

Groner, H. "Problems of the German Energy Policy." German Economic Review, 12 (1974), 173-185.

Hoerster, H. Systems for Using Solar Energy and Rational Energy Application in Buildings. Nordrheim Westfall, Federal Republic of Germany: Federal Mineral Research and Technology, 1976.

Holloman, J. Herbert. "Federal Republic of Germany." In Energy Research and Development. [Ford Energy Policy Project Report] Cambridge, Massachusetts: Ballinger Publishing Company, 1975, 157-178.

Hütter, Ulrich. The Development of Wind Power Installations for Electrical Power Generation in Germany. Springfield, Virginia: National Technical Information Service, 1973.

Joskow, Paul L. "Research and Development Strategies for Nuclear Power in the United Kingdom, France, and Germany." In Proceedings of the Workshop on Institutional Alternatives for LMFBR Development and Commercialization. McLean, Virginia: The MITRE Corporation, 1977, pp. 227-328.

Keck, Otto. Policymaking in a Nuclear Program: The Case of the West German Fast Breeder. Lexington, Massachusetts: Lexington Books, 1981.

_____. "The West German Fast Breeder Programme: A Case Study in Governmental Decision Making." Energy Policy, 8 (December 1980), 277-292.

Krugmann, H. "West Germany's Efforts to Close the Nuclear Fuel Cycle: Strategies for Radioactive Waste Management." International Journal of Energy Research, 2 (April-June 1978), 107-122.

Lantzke, Ulf. "Germany's Energy Policy Before and After the 1973/1974 Energy Crisis." German Economic Review, 13 (1975), 246-256.

Louranc, William W. "Nuclear Futures for Sale: To Brazil from West Germany." International Security, 1 (Fall 1976), 147-166.

Mackenthun, W. "Survey of Germany: Electric Supply Industry." Nuclear Engineering, 13 (September 1968), 745-748.

Mendershausen, Horst. Coping with the Oil Crisis; French and German Experiences. Baltimore, Maryland: The Johns Hopkins University Press, 1976.

Nelkin, Dorothy and Pollak, Michael. The Atom Beseiged: Extraparliamentary Dissent in France and Germany. Cambridge, Massachusetts: The M.I.T. Press, 1980a.

_____. "French and German Courts on Nuclear Power." Bulletin of the Atomic Scientists, 36 (May 1980b), 36-43.

Nephew, E. A. Surface Mining and Land Reclamation in Germany. Oak Ridge, Tennessee: Oak Ridge National Laboratory, 1972.

"Nuclear Energy and Public Opinion in West Germany." Energie Nucleaire, 14 (March-April 1972), 134-139.

Sassin, W.; Hoffman, F.; and Sadnicki, M., eds. Medium-Term Aspects of a Coal Revival: Two Case Studies. [Great Britain and Germany] Laxenburg, Austria: International Institute for Applied Systems Analysis, 1977.

Schmitt, D. and Monig, W. "Energy in the Federal German Republic: The Electric Power Industry: Investment Requirements and Their Financing." Energy Policy, 3 (March 1975), 67-72.

Stocker, H. J. "Energy Research and Development Program of the Federal Republic of Germany." In Energy Technology II. Washington, D. C.: Government Institutes, Inc., 1975, pp. 122-126.

United States. Congress. Senate. Committee on International Insular Affairs. A Study of the Relationships Between the Government and the Petroleum Industry in Selected Foreign Countries: The Federal Republic of Germany. [Report No. 94] Washington, D.C.: U.S. Government Printing Office, 1975.

Weingart, J. M. Solar Energy Conversion and the Federal Republic of Germany -- Some Systems Considerations. [Draft] Laxenburg, Austria: International Institute for Applied Systems Analysis, 1975.

Wellhofer, E. Spencer. "The Politics of Energy Policy Choices: Germany, Sweden, and the United States." In Robert M. Lawrence and Morris O. Heisler, eds. International Energy Policy. Lexington, Massachusetts: Lexington Books, 1980, pp. 145-161.

Winnacker & Wirtz. Nuclear Energy in Germany. LaGrange Park, Illinois: American Nuclear Society, 1979.

Wonder, Edward F. "Nuclear Competition and Nuclear Proliferation: Germany and Brazil, 1975." Orbis, 21 (Summer 1977), 277-306.

SCANDINAVIA

Aamo, Bjorn Skogstad. "Norwegian Oil Policy -- Basic Objectives." Energy Policy, 4 (March 1976), 63-68.

Abrahamson, D.; Barnaby, W.; Johansson, T. B.; and Steen, P. "Sweden's Nuclear Debate -- Reply." Bulletin of the Atomic Scientists, 36 (September 1980), 61-62.

Ager-Hanssen, Henrik. "The Exploitation of Norwegian Oil and Gas." Energy Policy, 8 (June 1980), 153-164.

Almer, B.; Dickson, W.; and Miller, U. "Effects of Acidification on Swedish Lakes." Ambio, 3 (1974), 30-36.

Anderson, S. S. "Conflict Over New Technology -- The Case of Nuclear Power Planning in Norway 1972-74." Acta Sociologica, 23 (1980), 297-310.

255

Ausland, John C. "The Challenge of Oil to Norwegian Foreign Policy." Cooperation and Conflict, 10 (1975), 189-198.

Barnaby, W. "The Swedish Referendum – So Away With It But Not Yet." Bulletin of the Atomic Scientists, 36 (June 1980), 58-60.

Blegaa, S.; Josephsen, L.; Meyer, N. I.; and Sorensen, B. "Alternative Danish Energy Planning." Energy Policy, 5 (June 1977), 87-94.

"A Comparison of Residential and Commerical Energy Use in the United States and Sweden." Energy and Buildings, 1 (May 1977), 89-92.

Daleus, Lennart. "A Moratorium in Name Only." Bulletin of the Atomic Scientists, 31 (October 1975), 27-33.

Davidson, A. Energy Policy in Sweden. Stockholm, Sweden: Svenska Institutet, April 1977.

Doernberg, A. Comparative Analysis of Energy Use in Sweden and the United States. [BNL-20539] Upton, New York: Brookhaven National Laboratory, September 1975.

Elbek, Bent. "Is Nuclear Power the Answer?" Scandinavian Review, 66 (September 1978), 16-20.

Elemenius, Lars. "District Heating in Sweden." In Anton B. Schmalz, ed. Energy: Today's Choices, Tomorrow's Opportunities. Washington, D.C.: World Future Society, 1974.

Engström, S. "Wind Energy – From a Swedish Viewpoint." Ambio, 4 (1975), 75-79.

Exploring Wind Power for the Production of Electricity. [NTIS Report N75-13385] Springfield, Virginia: National Technical Information Service, December 1974.

Finland, Council of State. "The Finnish Energy Program." Energy - The International Journal, 6 (May 1981), 389-402.

Finland. Energy Department. Ministry of Trade and Industry. Features of the Finnish Energy Economy. Helsinki, Finland: Ministry of Trade and Industry, Energy Department, 1978.

Garris, Jerome H. "Sweden's Debate on Nuclear Weapons." Cooperation and Conflict, 3-4 (1973), 189-208.

Hambraeus, Gunnar and Stillesjö, Staffan. "Perspectives on Energy in Sweden." In Jack M. Hollander, Melvin K. Simmons, and David O. Wood, eds. Annual Review of Energy. Volume 2. Palo Alto, California: Annual Reviews, Inc., 1977, pp. 417-453.

256

Hinrichsen, D. "Scandinavian Energy Race." New Scientist, 76 (December 15, 1977), 713-717.

_____ and Cawood, P. "Fresh Breeze for Denmark's Windmills." New Scientist, 70 (1976), 567-570.

Holloman, Herbert J. "Sweden." In Energy Research and Development. [Ford Foundation Energy Policy Project Report] Cambridge, Massachusetts: Ballinger Publishing Company, 1975, pp. 223-233.

Institutt for Atomenergi, Kjeller [Norway]. Air Pollution Health Effects of Electric Power Generation: A Literature Survey. [NP-20649] Springfield, Virginia: National Technical Information Service, 1975.

Johansson, Thomas B. and Steen, Peter. Radioactive Waste from Nuclear Power Plants: Facing the Ringhals-3 Decision. Stockholm, Sweden: D.S.I., 1978a.

_____ . "Solar Sweden." Ambio, 7 (June 1978b), 70-74.

_____ . Solar Sweden: An Outline to a Renewable Energy System. Stockholm, Sweden: Secretariat for Future Studies, 1973c.

Kihlstedt, P. G. Samhallets Ravaruforsorjning under Energibrist. [IVA-Rapport No. 112] Stockholm, Sweden: Ingenjorsvetenskapsakademien, 1977.

Kjolberg, A. "Oljen og Norsk Utenrikspolitik-et Nytt Trumfkort?"["Oil and Norwegian Foreign Policy – A New Trumpcard?"] Internasjonal Politikk, October/December 1974, 823-850.

Kober, Kjell and Vinjar, Asbjorn. "Cascades to Kilowatts." Scandinavian Review, 66 (September 1978), 36-39.

Korsgaard, Vagn. "The Zero Energy House." Scandinavian Review, 66 (September 1978), 21-24.

Kristoferson, Lars. Energy in Society. [Special issue] Ambio, 2 (1973), 177-240.

_____ . "Sweden Boosts Alternative Energy R & D." Ambio, 7 (1978), 183-185.

Kvinnsland, Ole-Jacob. "Norwegian Oil – Economic and Political Problems in a European Context." Internasjonal Politikk, 4 (1972), 581-591.

Lehtonen, Martti. "The Outlook for Energy Demand, Supply, and Investment in Finland up to 1985." Bank of Finland Monthly Bulletin, 51 (August 1977), 20-25.

Lind, Carl-Erik. "District Heating in Sweden, 1972-77." Energy Policy, 7 (March 1979), 74-76.

Litell, Richard. "Wave Power 'Creats' in Norway." Ambio, 7 (1978), 129-130.

Lonnroth, Mans. "Swedish Energy Policy: Technology in the Political Process." In Leon N. Lindberg, ed. The Energy Syndrome: Comparing National Responses to the Energy Crisis. Lexington, Massachusetts: Lexington Books, 1977, pp. 255-283.

_____; Johansson, T. B.; and Steen, P. "Sweden Beyond Oil: Nuclear Commitments and Solar Options." Science, 208 (May 9, 1980), 557-563.

_____ ; Steen, P.; and Johansson, T. B. Energy in Transition. [Fack S-10310] Stockholm, Sweden: Secretariat for Future Studies, 1977.

Lucas, N. J. D. "The Role of Institutional Relations in Danish Energy Policy." International Relations (London), 6 (November 1978), 347-373.

Lundqvist, Lennart J. Environmental Policies in Canada, Sweden, and the United States: A Comparative Overview. [Sage Administrative and Policy Series] Beverly Hills, California: Sage Publications, 1974.

Lundsten, Henrik. "The Finnish Oil Problem." Kansallis-Osake-Pankki Economic Review, 1 (1974), 7-14.

Margen, Peter. "Wanted: Waste Heat." Scandinavian Review, 66 (September 1978), 30-33.

Miettinen, Jorma K. "The Nuclear Power Situation in Finland." Ambio, 5 (1976), 129-131.

Nathan, Ove. "Denmark: Power Options Still Open." Bulletin of the Atomic Scientists, 37 (March 1981), 29-33.

Nelkin, Dorothy. Technological Decisions and Democracy: European Experiments in Public Participation. [Sweden, the Netherlands, and Austria] Beverly Hills, California: Sage Publications, 1977.

Norgard, J. Husholdninger og Energi. [Report 4, DEMO-projekt. Chapter 13, Tekniske Elbesparelser] Lyngby, Denmark: Danmarks Tekniske Hojskole, 1977.

"The Nuclear Debate in Sweden." Nuclear Engineering International, 21 (January 1976), 50-52.

"Nuclear R & D in Sweden." Nuclear Engineering International, 21 (January 1976), 46-49.

Organization for Economic Cooperation and Development. "Sweden's Energy Policy." OECD Observer, 40 (June 1969), 43-45.

Pearson, Frederic S. and Nyden, Michael. "Energy Crisis and Government Regulations: Swedish and Dutch Responses in 1973." West European Politics, 3 (October 1980).

"Pulling Power Out of Thin Air." Audubon, 76 (May 1974), 81-88.

Rydberg, J. "Sweden's Nuclear Debate." Bulletin of the Atomic Scientists, 36 (September 1980), 59-60.

Saab-Scania Aerospace Division. Saab Wind Energy. Linkoping, Sweden: Saab-Scania Aerospace Division, 1977.

Sahr, Robert C. "The Politics of Energy Conservation in Sweden: With Applications to the American Experience." [Doctoral dissertation, Massachusetts Institute of Technology, 1979.]

_____. "The Politics of National Nuclear Policy Change: Sweden and the 1980 Nuclear Referendum." A paper prepared for delivery at the Northeast Political Science Association Conference, New Haven, Connecticut: November 1980.

Schipper, Lee. "Lessons from Scandinavia." Scandinavian Review, 66 (September 1978a), 7-15.

_____. "The Swedish-U.S. Energy Use Comparisons and Beyond: Summary." In Joy Dunkerley, ed. International Comparisons of Energy Consumption. [Research Paper R-10] Washington, D.C.: Resources for the Future, 1978b.

_____ and Lichtenberg, A. J. "Efficient Energy Use and Well-Being: The Swedish Example." Science, 194 (December 3, 1976), 1001-1013.

Sorensen, Bent. "Wonderful Wind Machines." Scandinavian Review, 66 (September 1978), 40-43.

Squires, Susan M. "The Impact of North Sea Oil in Norway and Scotland." Norsk Geografisk Tidsskrift, 29 (1975), 133-140.

Surrey, John and Huggett, Charlotte. "Opposition to Nuclear Power: A Review of International Experience." Energy Policy, 2 (December 1976), 286-307.

Svedin, Uno. "Sweden's Energy Debate: The Nuclear Controversy." Energy Policy, 3 (September 1975), 258-261.

Swiss, M. "Norway's Oil and Gas Wealth Stimulates Industrial Growth." Energy International, 15 (January 1978), 9-12.

Tinnin, David B. "Why Volvo is Staking Its Future on Norway's Oil." Fortune, 99 (February 12, 1979), 110-114+.

Wellhofer, E. Spencer. "The Politics of Energy Policy Choices: Germany, Sweden, and the United States." In Robert M. Lawrence and Morris O. Heisler, eds. International Energy Policy. Lexington, Massachusetts: Lexington Books, 1980, pp. 145-161.

ITALY

Angelini, Arnaldo M. "Electricity Generation and Distribution in Italy." Review of the Economic Conditions in Italy, 28 (January 1974), 14-27.

Baker, Steven J. Technology and Politics: The Italian Nuclear Program and Political Integration in Western Europe. [Ph.D. Dissertation, University of California, Los Angeles, 1973.]

Fogagnolo, Giorgio. "The Italian Energy Problem: A World View." Review of the Economic Conditions in Italy, 30 (September 1976), 385-402.

Frankel, P. H. Mattei: Oil and Power Politics. London, England: Faber and Faber, 1966.

Hodgetts, J. E. Administering the Atom for Peace. [U.S., Great Britain, France, Canada, Italy, Japan] New York: Atherton Press, 1964.

I Problemi dell'Energie in Italia. Milan, Italy: Franco Angeli, 1977.

"Italy Depends on Imported Energy." Petroleum Review, 30 (July 1976), 389-394.

"Italy's National Energy Plan." Petroleum Economist, 42 (November 1975), 413-415.

Jorio, Marco. Energia in Crisi? Il Ruole dell'Energia Nucleare Storia e Politica. Naples, Italy: Guida, 1976.

Mazzanti, G. "Importance and Evolution of the Energetic Systems in Italy and Japan." Rivista Internazionale di Scienze Economiche e Commerciali, 27 (July-August 1980), 683-689.

Pacione, M. "Italy and the Energy Crisis." Geography, 61 (April 1976), 99-102.

Robotti, A. C. "The Outlook for Solar Energy Exploitation in Italy." Review of the Economic Conditions in Italy, 29 (November 1975), 467-481.

"Some Considerations of Fast Breeder Reactor Penetration in Italy." Energia Nucleare, 23 (August-September 1976), 423-436.

"Some Considerations on the Italian Nuclear Development Strategy." Energia Nucleare, 23 (March 1976), 131-144.

Valtorta, M. "Electricity Development in Japan and Italy -- Common Features and Possible Strategies. Rivista Internazionale di Scienze Economiche e Commerciali, 27 (July-August 1980), 709-720.

NORTH SEA

Affolter, M. T. "North Sea Oil Development: Some Related Environmental and Planning Considerations for Scotland." Ambio, 5 (1976), 3-16.

Basnett, David. "North Sea Oil – A Chance to Tackle Unemployment." Lloyds Bank Review, October 1978, 1-17.

Brenscheidt, M. H. "Petroleum Legislation in the North Sea Countries." Texas International Law Journal, 11 (Spring 1976), 281-303.

Brittan, Samuel and Riley, Barry. "A People's Stake in North Sea Oil." Lloyds Bank Review, April 1978, 1-18.

Chapman, Keith. North Sea Oil and Gas. London, England: David & Charles, 1976.

Dam, Kenneth W. "The Evolution of North Sea Licensing Policy in Britain and Norway." Journal of Law and Economics, 17 (October 1974), 213-263.

Eckbo, Paul L. "Perspectives on North Sea Oil." In Jack M. Hollander, et al., eds. Annual Review of Energy. Volume 4. Palo Alto, California: Annual Reviews, Inc., 1979, 71-98.

"The Economics of North Sea Oil." Midland Bank Review (May 1975), 11-19.

Evensen, J. "Nordsjoolen-Folkerettslige og Utenrikspolitiske Problemer." ["North Sea Oil: Problems of International Law and International Politics"] Internasjonal Politikk, 3 (July/September 1973), 501-514.

_____ . "Nordsjoolen-Oljepolitik." ["North Sea Oil: Oil Policy"] Internasjonal Politikk, 2 (April/June 1973), 359-373.

Gore, Rick. "Striking It Rich in the North Sea." National Geographic Magazine, 151 (April 1977), 519-550.

Gurney, Judith. "North Sea Oil and Gas: Implications for Western Europe." World Today, 31 (October 1975), 415-424.

"How Much Oil Can the North Sea Produce?" Energy Policy, 5 (December 1977), 282-306.

Keto, David B. Law and Offshore Oil Development: The North Sea Experience. New York: Holt, Rinehart and Winston, 1978.

Lewis, T. M. and McNicoll, I. H. North Sea Oil and Scotland's Economic Prospects. London, England: Croom Helm, 1978.

MacKay, D. I. and MacKay, G. A. The Political Economy of North Sea Oil. London, England: M. Robertson, 1975.

Minford, Patrick. "North Sea Oil and the British Economy." Banker [London], 127 (December 1977), 23-27.

Noreng, Oystein. Oil Industry and Government Strategy in the North Sea. London, England: Croom Helm, 1980.

"The North Sea – A Major Oil Province." Petroleum Review, 29 (September 1975), 601-612.

"The North Sea -- A New Major Oil Province in a Changing World." Journal of Canadian Petroleum Technology, October-December 1975, 22-27.

"North Sea Oil and Gas – What Potential for the U.K.?" Engineering, 213 (March 1973), 149-159.

Odell, Peter R. and Rosing, Kenneth E. "The North Sea Oil Province: A Simulation Model of Development." Energy Policy, 2 (December 1974), 316-329.

_____. Optimal Development of the North Sea's Oil Fields: A Study in Divergent Government and Company Interests and Their Reconciliation. London, England: Kogan Page, 1976.

_____; and Beke-Vogelaar, H. "Optimising the Oil Pipeline System in the U.K. Sector of the North Sea." Energy Policy, 4 (March 1976), 50-55.

Page, S. A. B. "The Value and Distribution of the Benefits of North Sea Oil and Gas, 1970-1985." National Institute Economic Review, November 1977, 41-58.

Robinson, Colin and Morgan, Jon. "Depletion Control and Profitability: The Case of the U.K. North Sea." Energy Policy, 4 (September 1976), 255-267.

_____. North Sea Oil in the Future: Economic Analysis and Government Policy. London, England: Macmillan, 1978.

Saeter, Martin and Smart, Ian, eds. The Political Implications of North Sea Oil and Gas. Guilford, England: IPC Science and Technology Press, 1975.

Squires, Susan M. "The Impact of North Sea Oil in Norway and Scotland." Norsk Geografisk Tidsskrift, 29 (1975), 133-140.

Turner, Louis. "Oil and the North-Sea Dialogue." World Today, 33 (February 1977), 52-61.

Uhl, William C. North Sea Petroleum: An Investment and A Marketing Community. New York: McGraw-Hill, 1977.

White, Irvin L., et al. North Sea Oil and Gas. Norman,Oklahoma: University of Oklahoma Press, 1973.

OTHER

"Energy for the Environment in the Netherlands." Ekistics, 40 (October 1975), 273-280.

Gallagher, Charles F. "Spain, Development, and the Energy Crisis." Common Ground, 1 (January 1975), 31-40.

Patterson, Walter C. "Austria's Nuclear Referendum." Bulletin of the Atomic Scientists, 35 (January 1979), 6-7.

Petitpierre, G. and Giovannini, B. "From Switzerland: 1979 Initiative." Bulletin of the Atomic Scientists, 35 (May 1979), 49.

Quinlan, M. "A New Look at the Spanish Energy Sector." Petroleum Economist, 45 (1978), 157-159.

Smith, Philip B. and Spanhoff, Ruud. "The Nuclear Energy Debate in the Netherlands." Bulletin of the Atomic Scientists, 32 (February 1976), 41-44.

"Spain: Nuclear Power to Check Oil Use." Petroleum Economist, 43 (November 1976), 431-433.

Van Gool, W. "Energy Research and Development Program for the Netherlands." In Energy Technology II. Washington, D.C.: Government Institutes, Inc., 1975, pp. 135-142.

4. Union of Soviet Socialist Republics

Armstrong, Terence. The Russians in the Arctic: Aspects of Soviet Exploration and Exploitation of the Far North, 1937-1957. Westport, Connecticut: Greenwood Press, 1974.

Block, Herbert. "Energy Syndrome, Soviet Version." In Jack M. Hollander, Melvin K. Simmons, and David O. Wood, eds. Annual Review of Energy. Volume 2. Palo Alto, California: Annual Reviews, Inc., 1977, pp. 455-497.

Bush, K. "Environmental Problems in the USSR." Problems of Communism, 21 (July 1972), 21-31.

Campbell, Robert W. The Economics of Soviet Oil and Gas. Baltimore, Maryland: Published for Resources for the Future by the Johns Hopkins University Press, 1968.

_____ . Soviet Energy Balances. [R-2257-DOE] Santa Monica, California: RAND Corporation, 1978.

_____ . Soviet Energy R & D: Goals, Planning, and Organization. [R-2253-DOE] Santa Monica, California: RAND Corporation, 1978.

_____ . Trends in the Soviet Oil and Gas Industry. Baltimore, Maryland: Published for Resources for the Future by the Johns Hopkins University Press, 1976.

Carlson, Sevinc. "Responses to the Oil Crisis: The U.S.S.R. and Selected Asian Countries." Journal of Energy and Development, 1 (Autumn 1975), 84-92.

Chesshire, J. H. and Huggett, C. "Primary Energy Production in the Soviet Union: Problems and Prospects." Energy Policy, 3 (September 1975), 223-144.

De Leon, Daniel. A Comparative Analysis of High Technology Programs: The Development and Diffusion of the Nuclear Power Reactor in Six Nations. [U.S., U.S.S.R., France, Canada, Great Britain, and the Federal Republic of Germany] Santa Monica, California: RAND Corporation, 1978.

Dienes, Leslie. "Energy Self-Sufficiency in the Soviet Union." Current History, 69 (July-August 1975), 10-14.

_____. "Geographical Problems of Allocation in the Soviet Fuel Supply." Energy Policy, 1 (June 1973), 3-20.

_____. "Modernization and Energy Development in the Soviet Union." Soviet Geography, 21 (March 1980), 121-158.

_____. "Soviet Energy Resources and Prospects." Current History, 71 (October 1976), 114-118+.

_____. "Soviet Energy Resources and Prospects." Current History, 74 (March 1978), 117-120+.

_____. "The Soviet Union: An Energy Crunch Ahead." Problems of Communism, 26 (September/October 1977), 41-60.

_____ and Shabad, Theodore. The Soviet Energy System. New York: Halstead Press, 1979.

Dollezhal, N. A. The Role of the Nuclear Power System in the Fuel-Power Complex of the USSR. [NTIS Report PC-A02/MF-A01] Springfield, Virginia: National Technical Information Service, 1976.

_____ and Koryakin, Y. "Nuclear Power Engineering in the Soviet Union." Bulletin of the Atomic Scientists, 36 (January 1980), 33-37.

Duffy, Gloria. "Soviet Nuclear Exports." International Security, 3 (Summer 1978), 83-111.

Economic Intelligence Unit. "Soviet Oil to 1980." Quarterly Economic Review [London], Spec. No. 14 (June 1973), 49 p.

Elliot, Iain F. The Soviet Energy Balance: Natural Gas, Other Fossil Fuels, and Alternative Power Sources. New York: Praeger Publishers, 1975.

Emelyanov, V. S. "Nuclear Energy in the Soviet Union." Bulletin of the Atomic Scientists, 27 (November 1971), 38-41.

"Energy Economy - Energy Conservation." Energia es Atomtechnika, 29 (March 1976), 111-116.

"Fastreactor Progress in the Soviet Union." New Scientist, 68 (December 4, 1975), 570-572.

Fedorenko, N. and Gofman, K. "Problems of Optimization in the Planning and Control of the Environment." Soviet Review, 14 (Summer 1973), 24-38.

265

Fishkis, M. Soviet Advances in Solar Energy: Technology and Applications. Tel Aviv, Israel: Tel Aviv University, Interdisciplinary Center for Technological Analysis and Forecasting, 1976.

Gerasimov, I. P., ed. Natural Resources of the Soviet Union: Their Use and Renewal. San Francisco, California: W. H. Freeman and Company, 1971.

Gillette, Philip S. "American Capital in the Contest for Soviet Oil, 1920-23." Soviet Studies, 24 (April 1973), 477-490.

Goldman, Marshall I. "The Dilemmas of Soviet Oil Policy." Challenge, 20 (July/August 1977a), 20-28.

_____ . Enigma of Soviet Petroleum: Half Empty or Half Full? Winchester, Maine: Allen & Unwin, 1980.

_____ . "The Oil Crisis in Perspective: The Soviet Union." Daedalus, 104 (Fall 1975), 129-143.

_____ . "The Soviet Union as a World Oil Power." In Frank N. Trager. Oil, Divestiture, and National Security. New York: Crane, Russak and Company, 1977b, pp. 92-105.

Grenon, Michel. "A Few Aspects of the Soviet Energy Policy." Revue de l'Energie, 296 (August/September 1977), 40-49.

Grunbaum, Rolf. "Alternative Energy Sources in the USSR." Ambio, 7 (1978), 49-55.

_____ . "Nuclear Energy in the Soviet Union." Ambio, 5 (1976), 124-126.

Gumpel, W. Energy Policy of the Soviet Union. Stanford, California: Hoover International Press, Stanford University, 1979.

_____ . "UdSSR - Energiepolitik und Nahostkrise." ["USSR - Energy Policy and Middle East Crisis"] Aussenpolitik, 25 (1974), 32-41.

Hardt, John P. "West Siberia: The Quest for Energy." Problems of Communism, 22 (May-June 1973), 25-36.

Hodgkins, J. A. Soviet Power: Energy Resources, Production, and Potential. Englewood Cliffs, New Jersey: Prentice-Hall, 1961.

Hoffman, George W. "Energy Projections - Oil, Natural Gas, and Coal in the USSR and Eastern Europe." Energy Policy, 7 (September 1979), 232-241.

Holcomb, Robert W. "Fusion Power: Optimism and a Tokamak Gap at Dubna." Science, 166 (October 17, 1969), 363-364.

Hopkins, G. D.; Korens, N.; and Schmidt, R. A. Analysis of Energy Resources and Programs of the Soviet Union and Eastern Europe. [NTIS Report No. AD-A012-970] Menlo Park, California: Stanford Research Institute, 1973.

Horelick, Arnold L. "The Soviet Union, the Middle East, and the Evolving World Energy Situation." Policy Sciences, 6 (March 1975), 41-48.

Hunter, Robert E. The Soviet Dilemma in the Middle East. Part II: Oil and the Persian Gulf. [Adelphi Paper No. 60] London, England: Institute for Strategic Studies, 1969.

"Hydroelectric Power and Water Resources of the USSR." Power Engineering - USSR, 12 (1974), 33-44.

Iorysh, I. "Legal Regulation of Environmental Effects of Atomic Energy." Soviet Review, 15 (Summer 1974), 58-75.

Kirkpatrick, Meredith. Energy Resources and Energy Policies in Eastern Europe and the USSR: A Bibliography. Monticello, Illinois: Vance Bibliographies, 1979.

Klinghoffer, Arthur Jay. "Soviet Oil Politics and the Suez Canal." World Today, 31 (October 1975), 397-405.

_____. Soviet Union and International Oil Politics. New York: Columbia University Press, 1977.

Kramer, J. M. "Environmental Problems in the USSR: The Divergence of Theory and Practice." Journal of Politics, 36 (November 1974), 866-899.

Kramish, Arnold. Atomic Energy in the Soviet Union. Stanford, California: Stanford University Press, 1959a.

_____. "Atomic Energy in the USSR." Bulletin of the Atomic Scientists, 15 (October 1959b), 322-328.

Landis, Lincoln. Politics and Oil: Moscow in the Middle East. New York: Dunellen, 1973.

Maddock, R. "Energy and Integration - The Logic of Interdependence in the Soviet Union and Eastern Europe." Journal of Common Market Studies, 19 (September 1980), 21-34.

Mangone, G. J., ed. Energy Policies of the World: Indonesia, the North Sea Countries, the Soviet Union. Volume II. New York: Elsevier-North Holland, 1977.

Medvedev, Zhores. A Nuclear Disaster in the Urals. New York: W. W. Norton and Company, 1979.

"The Nuclear Power Controversy: Assessing the Risks." Atlas, 24 (April 1977), 31-39.

Olson, Russell U., Jr. and Berentsen, William H. "Regional Energy Assessibility in the USSR." Soviet Geography, 22 (March 1981), 135-154.

Papp, Daniel S. "Soviet Scarcity: The Response of a Socialist State." Social Science Quarterly, 57 (September 1976), 350-363.

Pervukhin, M. "Energy Resources of the USSR and Their Rational Utilization." Soviet and Eastern Europe Foreign Trade, 11 (Spring 1975), 91-104.

Petrosyants, Andranik. From Scientific Search to Atomic Industry. Danville, Illinois: The Interstate, 1975.

Pryde, Philip R. and Pryde, Lucy T. "Soviet Nuclear Power." Environment, 16 (April 1974), 26-34.

Roy, Rustum. "The Technology of Nuclear-Waste Management." Technology Review, 83 (April 1981), 38-51.

Russell, Jeremy. Energy as a Factor in Soviet Foreign Policy. Fainborough, England: Published for the Royal Institute of International Affairs by Saxon House/ Lexington Books, 1976.

Shabad, Theodore. "Energy in the Soviet Union." Energy Policy, 4 (June 1976), 177-179.

Slocum, Marianna P. "Soviet Energy: An Internal Assessment." Technology Review, 77 (October/November 1974), 16-33.

"The Soviet Energy Balance." Nature, 261 (May 6, 1976), 3-5.

Stein, Ellen L. "The Politics of Soviet Oil." Energy Policy, 8 (September 1980), 203-212.

Stern, Jonathan P. "Soviet Energy Prospects in the 1980's." The World Today, 36 (May 1980), 188-194.

Stowell, Christopher. Oil and Gas Development in the USSR. Tulsa, Oklahoma: Petroleum Publishing Company, 1974.

Surface Coal Mining Developments in the USSR. [Report CP-76-12] Laxenburg, Austria: International Institute for Applied Systems Analysis, December, 1976.

United States. Atomic Energy Commission. Division of Reactor Development and Technology. Soviet Power Reactors. [WASH-1175] Washington, D.C.: U.S. Atomic Energy Commission, August 1970.

United States. Central Intelligence Agency. Prospects for Soviet Oil Production. [ER-77-10270] Washington, D.C.: Central Intelligence Agency, April 1977.

_____. USSR: Development of the Gas Industry: A Research Paper. [ER-78-10393] Washington, D.C.: Central Intelligence Agency, July 1978.

United States. Congress. Office of Technology Assessment. Technology and Soviet Energy Availability. Washington, D.C.: Office of Technology Assessment, November, 1981.

United States. Congress. Senate. Committee on Intelligence. The Soviet Oil Situation: An Evaluation of CIA Analyses of Soviet Oil Production. [Staff Report] Washington, D.C.: U.S. Government Printing Office, 1978.

Wright, Arthur W. "Contrasts in Soviet and American Energy Policies." Energy Policy, 3 (March 1975), 38-46.

_____. "The Soviet Union in World Energy Markets." In E. W. Erickson and L. Waverman, eds. The Energy Question: An International Failure of Policy. Volume I: The World. Toronto, Ontario: University of Toronto Press, 1974, pp. 85-99.

Zybenko, Roman. "Fuel and Power Resources." Studies on the Soviet Union, 8 (1968), 9-13.

5. Eastern Europe

Carter, F. W. "Four Countries Develop Their Own Energy (Albania, Bulgaria, Rumania, Yugoslavia)." Geographical Magazine, 49 (October 1976), 10+.

"The Development of Nuclear Power in the People's Republic of Bulgaria." IAEA Bulletin, 16 (December 1974), 19-23.

Dienes, Leslie. "Energy Prospects for Eastern Europe." Energy Policy, 4 (June 1976), 119-129.

_____. "Environmental Disruption and Its Mechanism in East-Central Europe." Professional Geographer, 26 (November 1974), 375-381.

Dobozi, Istvan. "Energy Planning and the Energy Situation in a Socialist Planned Economy Lacking Energy: A Case Study of Hungary." In Leon N. Lindberg, ed. The Energy Syndrome: Comparing National Responses to the Energy Crisis. Lexington, Massachusetts: Lexington Books, 1977, pp. 173-204.

Froelich, Lech. "King Coal." Polish Perspectives, 4 (1975), 3-12.

Guha, A. "International Energy Crisis and the Socialist Countries." Alternative, 3 (August 1977), 109-135.

Haberstroh, John R. "The Case of Hungary: Liberal Socialism Under Stress." Journal of Comparative Economics, 2 (June 1978), 111-125.

Hoffmann, George W. "Energy Projections - Oil, Natural Gas, and Coal in the USSR and Eastern Europe." Energy Policy, 7 (September 1979), 232-241.

Hopkins, G. D.; Korens, N.; and Schmidt, R. A. Analysis of Energy Resources and Programs of the Soviet Union and Eastern Europe. [NTIS Report No. AD-A012-970] Menlo Park, California: Stanford Research Institute, 1973.

Houdek, Karel. "The Fuel and Energy Base: Concept, Plan, and Development." Czechoslovak Economic Digest (December 1978), 27-80.

Ilinich, Yuri V. "The Impact of Electric Power on Changes in the Spatial Structure of the Economies of European Socialist Countries: A Case Study of Poland." [In Russian] Voprosy Geografii, 97 (1974), 118-128.

Kopecki, K. "Forecasting the Demand for Electric Power in Poland up to 2000 and Conditions for Meeting this Demand." Przeglad Elektrotechniczny, 53 (July 1977), 285-289.

Korda, B. and Moravcik, I. "The Energy Problem in Eastern Europe and the Soviet Union." Canadian Slavonic Papers, 18 (March 1976), 1-14.

Kramer, John M. "Between Scylla and Charybdis: The Policies of Eastern Europe's Energy Problem." Orbis, 22 (Winter 1979), 929-950.

_____. "The Energy Gap in Eastern Europe." Survey, 21 (Winter/Spring 1975), 65-78.

Lee, J. R. "Petroleum Supply Problems in Eastern Europe." In United States. Congress. Joint Economic Committee. Reorientation and Commercial Relations of the Economies of Eastern Europe. Washington, D.C.: U.S. Government Printing Office, 1974, pp. 418-419.

Maddock, R. "Energy and Integration - The Logic of Interdependence in the Soviet Union and Eastern Europe." Journal of Common Market Studies, 19 (September 1980), 21-34.

Mathieson, R. S. "Nuclear Power in the Soviet Bloc." Annals of the Association of American Geographers, 70 (June 1980), 271-279.

Matusek, Maurice. "The Comprehensive Programme in the Power and Fuel Industry." Czechoslavak Economic Digest, September 1972, 40-62.

"New Nuclear Power Plants in Europe." Atomwirtschaft-Atomtechnik, 22 (May 1977), 281-285.

"Oil Drilling and Production: Eastern Europe." World Oil, 183 (August 15, 1976), 119-128.

Park, Daniel. Oil and Gas in Comecon Countries. New York: Nichols Publishing Company, 1979.

Pearton, Maurice. Oil and the Romanian State. London, England: Oxford Claredon Press, 1971.

Polach, Jaroslav G. "The Development of Energy in Eastern Europe." In U.S. Congress. Joint Economic Committee. Subcommittee on Foreign Policy. Economic Developments in Countries of Eastern Europe: A Compendium of Papers. Washington, D.C.: U.S. Government Printing Office, 1970, pp. 348-433.

271

Polach, Jaroslav G. "The Energy Gap in the Communist World."East Europe, 18 (April 1969), 19-26.

_____. "Nuclear Power in East Europe." East Europe, 17 (May 1968), 3-12.

"Quarter Century Hungarian Energetics Development and Perspective." Energia es Atomtechnika, 27 (April 1974), 145-160.

"Research and Development of Geothermal Energy Production in Hungary." Geothermics, 4 (March-December 1975), 44-57.

Russell, J. "Energy Considerations in Comecon Policies." World Today, 32 (February 1976), 39-48.

Schiele, Robin. "Inflation and the Energy Crisis, Czech Style." Canadian Business, 47 (June 1974), 51-52.

Suica, J. Salom. "Power Development in Yugoslavia." Bulletin of the Atomic Scientists, 27 (November 1971), 42-46.

"Survey of Poland's Energy Situation." Energy Policy, 2 (March 1974), 86-88.

Wasowski, Stanislaw. "The Fuel Situation in Eastern Europe." Soviet Studies, 21 (July 1969), 35-51.

Wilcyznski, Jozef. "Atomic Energy for Peaceful Purposes in the Warsaw Pact Countries." Soviet Studies, 26 (October 1974), 568-590.

6. Middle East/OPEC

Abir, Mordechal. Oil, Power, and Politics: Conflict in Arabia, the Red Sea, and the Gulf. London, England: Frank Cass, 1974.

Adelman, Morris A. "Is the Oil Shortage Real? Oil Companies as OPEC Tax-Collectors." Foreign Policy, 9 (Winter 1972-73), 69-107.

_____. "The World Oil Cartel: Scarcity, Economics, and Politics." Quarterly Review of Economics and Business, 16 (Summer 1976), 7-18.

Al-Chalabi, Fadhil J. OPEC and the International Oil Industry: A Changing Structure. Oxford, England: Oxford University Press for the Organization of Arab Petroleum Exporting Countries, 1980.

Ali, S. R. Saudi Arabia and Oil Diplomacy. New York: Praeger Publishers, 1976.

Allen, Loring. OPEC Oil. Cambridge, Massachusetts: Oelgeschlager, Gunn & Hain, Publishers, 1979.

Alnasraui, Abbas. Financing Economic Development in Iraq - The Role of Oil in a Middle Eastern Economy. New York: Praeger Publishers, 1967.

Al-Otaiba, Mana Saeed. OPEC and the Petroleum Industry. New York: Halsted Press, 1975.

_____. Petroleum and the Economy of the United Arab Emirates. London, England: Croom Helm, 1977.

Amiri, Kathleen B. Oil and Iran: A Systems Analysis of Policy and Organization, 1973-1976. [Ph.D. Dissertation, University of Illinois, 1976]

Amuzegar, Jahangir. Iran: An Economic Profile. Washington, D.C.: Middle East Institute, 1977.

Amuzegar, Jahangir. "Olpreis und Weltwirtschaftliches Gleichgewicht. Die Vorschlage des Schahs von Iran fur eine Internationale Entwicklungsund Hilfsorganisation." ["The Price of Oil and the World Economic Balance. The Shah of Iran's Proposals for an International Development and Assistance Organization"] Europa-Archiv, 29 (May 10, 1974), 277-284.

Anthony, John Duke, ed. The Middle East: Oil, Politics, and Development. Washington, D.C.: American Enterprise Institute for Public Policy Research 1975.

"Arab Oil Monies, 1974/75 and the United States." Middle East Review, Winter 1975/76, 17-27.

Arnaoot, Ghassan. "The Organization of Petroleum-Exporting Countries (OPEC)." Rocky Mountain Social Science Journal, 11 (April 1974), 11-18.

Askari, Houssein and Creasy, James. "Texas and OPEC: A Case of Economic Interdependence." Texas Business Review, 51 (September 1977), 193-197.

Bach, Christopher L. "OPEC (Organization of Petroleum Exporting Countries) Transactions in the U.S. International Accounts, 1972-1977." Survey of Current Business, 58 (April 1978), 21-32.

Bahadori, Mendi N. "Passive Cooling Systems in Iranian Architecture." Scientific American, 238 (February 1978), 144-154.

Barger, Thomas C. Energy Policies of the World: Arab States of the Persian Gulf. Newark, Delaware: Center for the Study of Marine Policy, College of Marine Studies, University of Delaware, 1975.

Brookes, L. G. "The Nuclear Power Implications of OPEC Prices." Energy Policy, 3 (June 1975), 124-135.

"Building a New Middle East: Private Enterprise and Petrodollars are Remaking the Area Despite Political Tensions." Business Week, May 26, 1975, 38-44.

Campbell, John C. "Oil Power in the Middle East." Foreign Affairs, 56 (October 1977), 89-110.

_____ and Caruso, Helen. The West and the Middle East. New York: Council on Foreign Relations, 1972.

Carey, Jane Perry Clarke. "Iran and Control of Its Oil Resources." Political Science Quarterly, 89 (March 1974), 147-174.

Caroe, O. K. Wells of Power: The Oilfields of South-Western Asia: A Regional and Global Study. Westport, Connecticut: Hyperion Press, 1976.

Choucri, Nazli. "OPEC and the World Oil Market." Technology Review, 83 (October 1980), 36-45.

Chubin, S. "Iran's Security in the 1980's." International Security, 2 (Winter 1978), 51-80.

Cleron, Jean Paul. Saudi Arabia 2000: A Strategy for Growth. London, England: Croom Helm, 1978.

Cochrane, S. McL. "The Price of OPEC Crude Oil." Energy Policy, 3 (September 1975), 181-191.

Comptroller General. Critical Factors Affecting Saudi Arabia's Oil Decisions. Washington, D.C.: General Accounting Office, May 12, 1978.

DeCarmoy, Guy. "Energy and Development Policies In Iran: A Western View." Energy Policy, 2 (December 1974), 293-306.

Edens, David G. Oil and Development in the Middle East. New York: Praeger Publishers, 1979.

El Mallakh, Ragaei. "OPEC: Issues of Supply and Demand." Current History, 74 (March 1978), 125-127+.

_____. OPEC: Twenty Years and Beyond. Boulder, Colorado: Westview Press, 1981.

_____. "Oil and the OPEC Members." Current History, 69 (July-August 1975), 6-9.

_____. "The Organization of the Arab Petroleum Exporting Countries: Objectives and Potential." In Jack M. Hollander, Melvin K. Simmons, and David O. Wood, eds. Annual Review of Energy. Volume 2. Palo Alto, California: Annual Reviews, Inc., 1977, pp. 399-415.

_____. Some Dimensions of Middle East Oil: The Producing Countries and the United States. New York: American-Arab Association for Commerce and Industry, 1970.

Elshafei, Alwalid N. "Energy Planning in the Arab World." Energy Policy, 7 (September 1979), 242-252.

Enders, Thomas O. "OPEC and the Industrial Countries: The Next Ten Years." Foreign Affairs, 53 (July 1975), 625-637.

Ezzati, Ali. World Energy Markets and OPEC Stability. Lexington, Massachusetts: Lexington Books, 1978.

Farmanfarmaian, Khodad, et al. "How Can the World Afford OPEC Oil?" Foreign Affairs, 53 (January 1975), 201-222.

Fesharaki, Fereidun. Development of the Iranian Oil Industry: International and Domestic Aspects. New York: Praeger Publishers, 1976.

_____. Revolution and Energy Policy in Iran. London, England: Economist Intelligence Unit, 1980.

Field, Michael. A Hundred Million Dollars a Day. New York: Praeger Publishers, 1976.

_____. "Oil: OPEC and Participation." World Today, 28 (January 1972), 5-13.

Finnie, D. H. Desert Enterprise: The Middle East Oil Industry in Its Local Environment. Cambridge, Massachusetts: Harvard University Press, 1958.

Ghazi Algosaibi, H. E. "The Strategy of Industrialization in Saudi Arabia." Journal of Energy and Development, 2 (Spring 1977), 218-223.

Gholamnezhad, A. "Critical Choices for OPEC Members and the United States." The Journal of Conflict Resolution, 25 (March 1981), 115-144.

Hamer, John. "Persian Gulf Oil." Editorial Research Reports, March 28, 1973, 231-248.

Hamilton, Charles W. Americans and Oil in the Middle East. Houston, Texas: Gulf Publishing Company, 1962.

Hansen, Herbert E. "Some Comments on Nationalization of Oil Properties by OPEC (Organization of Petroleum Exporting Countries) Members." Journal of Energy and Development, 1 (Spring 1976), 262-268.

Hartshorn, J. E. "OPEC and the Development of Fourth World Oil." Millennium, 6 (Autumn 1977), 162-174.

Hazleton, Jared E. Gold Rush Economics: Development Planning in the Persian/ Arabian Gulf. [Working Paper No. 4] Austin, Texas; Lyndon B. Johnson School of Public Affairs, University of Texas, 1976.

Hitti, Said H. and Abed, George T. "The Economy and Finances of Saudi Arabia." International Monetary Fund Staff Papers, 21 (July 1974), 247-306.

Horelick, Arnold L. "The Soviet Union, the Middle East, and the Evolving World Energy Situation." Policy Sciences, 6 (March 1975), 41-48.

Hurewitz, J. C., ed. Oil, the Arab-Israel Dispute, and the Industrial World: Horizons of Crisis. Boulder, Colorado: Westview Press, 1976.

Issawi, Charles P. Oil, the Middle East, and the World. Beverly Hills, California: Sage Publishing Company, 1972.

Issawi, Charles P. and Yeganeh, Mohammed. The Economics of Middle Eastern Oil. New York: Praeger Publishers, 1963.

_____. "Oil and Middle East Politics." Proceedings of the Academy of Political Science, 31 (December 1973), 111-122.

Itayim, F. "Arab Oil: The Political Dimension." Journal of Palestine Studies, 3 (Winter 1974), 84-97.

Johany, Ali D. The Myth of the OPEC Cartel - The Role of Saudi Arabia. New York: John Wiley & Sons, 1980.

Katouzian, M. A. "Oil versus Agriculture: A Case of Dual Resources Depletion in Iran." Journal of Peasant Studies, 5 (April 1978), 347-369.

Keiser, G. "Die Aussichten des Endolmonopols der OPEC." ["The Prospects for the OPEC Petrol Monopoly"] Europa-Archiv, 30 (March 10, 1975), 153-162.

Klebanoff, Shoshama. Middle East Oil and U.S. Foreign Policy with Special Reference to the U.S. Energy Crisis. New York: Praeger Publishers, 1974.

Knauerhase, R. The Saudi Arabian Economy. New York: Praeger Publishers, 1974.

Krasner, S. D. "The Great Oil Sheikdown." Foreign Policy, 13 (Winter 1973), 123-138.

Landis, Robin C. OPEC: Policy Implications for the United States. Ed. by Michael W. Class. New York: Praeger Publishers, 1980.

Lenczowski, George. Oil and State in the Middle East. Ithaca, New York: Cornell University Press, 1960.

_____ and Issawi, Charles. "Probing the Arab Motivations Behind Use of the 'Oil Wagon'; Checking on the Consequences of the Oil Squeeze by Arab States." International Perspective [Canada], March/April, 1974, 3-12.

Levy, Walter J. "The Years that the Locust Hath Eaten: Oil Policy and OPEC Development Prospects." Foreign Affairs, 57 (Winter 1978), 287-305.

Long, D. E. The Persian Gulf: An Introduction to Its People, Politics, and Economics. Boulder, Colorado: Westview Press, 1976.

Lubell, Harold. Middle East Oil Crises and Western Europe's Energy Supplies. Baltimore, Maryland: The Johns Hopkins University Press, 1963.

McLachlan, Keith and Ghorban, Narsi. Economic Development of the Middle East Oil Exporting States. London, England: The Economist Intelligence Unit, 1979.

Mabro, Robert and Monroe, Elizabeth. "Arab Wealth from Oil: Problems of Its Investment." International Affairs [London], 50 (January 1974), 15-27.

Magnus, Ralph H. "Middle East Oil." Current History, 68 (February 1975), 49-53+.

Mancke, Richard B. "The Future of OPEC." Journal of Business, 48 (January 1975), 11-19.

Mikdashi, Zuhayr. The Community of Oil Exporting Countries. Ithaca, New York: Cornell University Press, 1972.

_____. "Cooperation Among Oil Exporting Countries with Special Reference to Arab Countries: A Political Economy Analysis." International Organization, 28 (Winter 1974), 1-30.

_____. "The Oil Crisis in Perspective: The OPEC Process." Daedalus, 104 (Fall 1975), 203-215.

Miller, Aaron David. "The Influence of Middle East Oil on American Foreign Policy." Middle East Review, 9 (Spring 1977), 19-24.

Mingst, Karen A. "Regional Sectorial Economic Integration: The Case of OAPEC." Journal of Common Market Studies, 16 (December 1977), 95-113.

Moran, Theodore H. "Modelling OPEC Behavior - Economic and Political Alternatives." International Organization, 35 (Spring 1981), 241-272.

_____. Oil Prices and the Future of OPEC. Washington, D.C.: Resources for the Future, 1978.

_____. "Why Oil Prices Go Up? The Future: OPEC Wants Them." Foreign Policy, No. 25 (Winter 1976-1977), 58-79.

Mosely, Leonard. Power Play: Oil in the Middle East. New York: Random House, 1973.

Mossavar-Rahmani, Bijan. Revolution and Evolution of Energy Policies in Iran. [Discussion Paper No. 87] Santa Monica, California: California Seminar on Arms Control and Foreign Policy, July 1980.

Nelson, J. W.; Short, R.; and el Mallakh, R. The Arab Middle East: Economic Potential. Stanford, California: Stanford Research Institute, 1976.

OPEC Official Resolutions and Press Releases, 1960-1980. Elmsford, New York: Pergamon Press, 1980.

Oweiss, Ibrahim M. "Strategies for Arab Economic Development." Journal of Energy and Development, 3 (Autumn 1977), 103-114.

Paust, Jordan J. and Blaustein, Albert P. The Arab Oil Weapon. Leiden, The Netherlands: Sijthoff International, 1977.

Paust, Jordan J. and Blaustein, Albert P. "The Arab Oil Weapon -- A Threat to International Peace." American Journal of International Law, 68 (July 1974), 410-439.

Penrose, Edith. "The Oil Crisis in Perspective: The Development of a Crisis." Daedalus, 104 (Fall 1975), 39-57.

Price, David Lynn. Oil and Middle East Security. Washington, D.C.: Georgetown University, Center for Strategic and International Studies, 1976.

Rand, Christopher T. Making Democracy Safe for Oil -- Oilmen and the Islamic East. Boston, Massachusetts; Atlantic/Little Brown, 1975.

Remba, Oded. "Arab Oil Power, Western Responses and Israel's Future." Midstream, 21 (August/September 1975), 28-37.

Rouhani, Fuad. A History of O.P.E.C. New York: Praeger Publishers, 1971.

Rustow, Dankwart A. "Petroleum Politics 1951-1974." Dissent, 21 (Spring 1974), 144-153.

_____ and Mungo, John F. OPEC: Success and Prospects. New York: New York University Press, 1976.

Saleem Khan, M. A. "Oil Politics in the Persian Gulf Region." India Quarterly, 30 (January/March 1974), 25-41.

Sayigh, Yusif A. "Arab Oil Policies: Self-Interest versus International Responsibility." Journal of Palestine Studies, 4 (Spring 1975), 59-73.

Schmalensee, R. "Resource Exploitation Theory and the Behavior of the Oil Cartel." European Economic Review, 7 (April 1976), 257-279.

Schurr, Sam, et al. Middle Eastern Oil and the Western World: Prospects and Problems. New York: American Elsevier, 1971.

Scott, B. R. "OPEC, the American Scapegoat." Harvard Business Review, 59 (January-February 1981), 6-8.

Seifert, William W., et al. Energy and Development: A Case Study. Cambridge, Massachusetts: The M.I.T. Press, 1973.

Sharshar, A. M. "Trade Policy and Economic Development in Saudi Arabia." Virginia Social Science Journal, 13 (April 1978), 50-54.

Sherbing, Naiem A. Arab Oil: Impact on the Arab Countries and Global Implications. New York: Praeger Publishers, 1976.

Shihata, I. F. I. "Arab Oil Policies and the New International Order." Virginia Journal of International Law, 16 (Winter 1976), 261-288.

_____. "Destination Embargo of Arab Oil: Its Legality Under International Law." American Journal of International Law, 68 (October 1974), 591-627.

_____. "The OPEC Fund for International Development." Third World Quarterly, 3 (April 1981), 251-268.

Shwadran, Benjamin. The Middle East, Oil, and the Great Powers. New York: Halsted Press, 1973, 1974.

_____. Middle East Oil: Issues and Problems. Cambridge, Massachsuetts: Schenkman Publishing Company, 1977.

Sobhan, R. "Institutional Mechanisms for Channeling OPEC Surpluses Within the Third World." Third World Quarterly, 2 (October 1980), 721-745.

Stobaugh, Robert B. "The Evolution of Iranian Oil Policy, 1925-1975." In George Lenczowski, ed. Iran Under the Pahlavis. Stanford, California: Hoover Institution, 1978.

Stocking, George W. Middle East Oil: A Study in Political and Economic Controversy. Nashville, Tennessee: Vanderbilt University Press, 1970.

Stone, Russell A., ed. OPEC and the Middle East: The Impact of Oil on Societal Development. New York: Praeger Publishers, 1977.

Stork, Joe. Middle East Oil and the Energy Crisis. New York: Monthly Review Press, 1975.

Taher, Abdulhady H. "The Middle East Oil and Gas Policy." Journal of Energy and Development, 3 (Spring 1978), 260-269.

Tomeh, George. "OAPEC: Its Growing Role in Arab and World Affairs." Journal of Energy and Development, 3 (Autumn 1977), 26-36.

United States. Congress. Senate. Committee on Interior and Insular Affairs. United States - OPEC Relations: Selected Materials Pursuant to S. Res. 45, a National Fuels and Energy Policy Study. [Prepared by the Congressional Research Service at the Request of Henry M. Jackson, Chairman. 94th Cong., 2nd sess.] Washington, D.C.: U.S. Government Printing Office, 1976.

Vicker, Ray. The Kingdom of Oil -- The Middle East: Its People and Its Power. New York: Charles Scribner's Sons, 1974.

Willett, Thomas D. "Structure of OPEC and the Outlook for International Oil Prices." World Economy [London], 2 (January 1979), 51-64.

280

Williams, M. J. "The Aid Programs of the OPEC Countries." Foreign Affairs, 54 (January 1976), 257-279.

Willrich, M. and Mossavarrahmani, B. "Oil on Troubled Waters - Industrial World and OPEC Middle East." Orbis, 23 (Winter 1980), 859-874.

Zonis, Marvin. "Oil and Politics in the Middle East." In Gary D. Eppen, ed. Energy: The Policy Issues. Chicago, Illinois: University of Chicago Press, 1975, pp. 52-68.

7. Latin America

GENERAL

Balestrini, Cesar. La Industria Petrolera en America Latina. Caracas, Venezuela: Universidad Central de Venezuela, 1971.

Centro Nuclear de Puerto Rico. Simposio Sobre Energia Nuclear y el Desarrollo de Latinoamerica. Rio Piedras, Puerto Rico: University of Puerto Rico Press, n.d.

Eichner, Donald O. The Inter-American Nuclear Energy Commission: Its Goals and Achievements. Stuart Bruchey, ed. New York: Arno Press, 1979.

Hammond, Allen L. "Energy: Elements of a Latin American Strategy." Science, 200 (May 19, 1978), 753-754.

"Increased Latin American Solidarity on Oil Issue." [Report on the Latin American Energy Organization Conference, San Jose, Costa Rica, July 6-7, 1979] Comercio Exterior de Mexico, 25 (September 1979), 349-354.

Kottlowski, Frank E., et al. Coal Resources of the Americas: Selected Papers. Boulder, Colorado: Geological Society of America, 1979.

Mullen, J. W. Energy in Latin America: The Historical Record. Santiago, Chile: United Nations Economic Commission for Latin America, 1978.

_____ . World Oil Prices: Prospects and Implications for Energy Policy Makers in Latin America's Oil-Deficit Countries. Santiago, Chile: United Nations Economic Commission for Latin America, 1978.

Rubichek, E. Walter. "The Payments Impact of the Oil Crisis: The Case of Latin America." Finance and Development, 11 (December 1974), 12-17.

Sabato, Jorge and Frydman, Raul. "Latin America Goes Nuclear." Atlas, 24 (June 1977), 26-27+.

Street, James H. "Latin American Adjustments to the OPEC (Organization of Petroleum Exporting Countries) Crisis and the World Recession." Social Science Quarterly, 59 (June 1978), 60-76.

Strout, Alan M. "Energy and Economic Growth in Central America." In Jack M. Hollander, Melvin K. Simmons, and David O. Wood, eds. Annual Review of Energy. Volume 2. Palo Alto, California: Annual Reviews, Inc., 1977, pp. 291-305.

MEXICO

Bermudez, Antonio J. The Mexican National Petroleum Industry: A Case Study in Nationalization. Stanford, California: Institute of Hispanic and Luso-Brazilian Studies, 1963.

"Energy Policy Formulation: The Case of Mexican Gas." Inter-American Economic Affairs, 34 (Autumn 1980), 87-91.

Diaz Serrano, Jorge. "Mexico's Oil and Gas Reserves." Mexican-American Review, 45 (June 1977), 11-13.

Fagen, Richard. "El Petroleo Mexicano y la Seguridad Nacional de Estados Unidos." Foro Internacional, 19 (October-December 1978), 216-230.

Grayson, George W. "The Maple Leaf, the Cactus, and the Eagle: Energy Trilateralism." Inter-American Economic Affairs, 34 (Spring 1981a), 49-75.

_____. "Mexico and the United States: The Natural Gas Controversy." Inter-American Economic Affairs, 32 (Winter 1978), 3-27.

_____. "Mexico's Opportunity: The Oil Boom." Foreign Policy, 29 (Winter 1977), 65-89.

_____. "Mexico's Reluctant Oil Boom." Business Week, January 15, 1979, 64-75.

_____. "Oil and U.S. - Mexican Relations." Journal of Inter-American Studies and World Affairs, 21 (November 1979), 427-456.

_____. The Politics of Mexican Oil. Pittsburgh, Pennsylvania; University of Pittsburgh Press, 1981b.

Kane, N. Stephen. "The United States and the Development of the Mexican Petroleum Industry, 1945-1950: A Lost Opportunity." Inter-American Economic Affairs, 35 (Spring 1981), 45-72.

Mancke, Richard. Mexico Oil and Natural Gas: Political, Strategic, and Economic Implications. New York: Praeger Publishers, 1979a.

Mancke, Richard. "Mexico's Petroleum Resources." Current History, 76 (February 1979b), 74-77+.

Metz, William D. "Mexico: The Premier Oil Discovery in the Western Hemisphere." Science, 202 (December 22, 1978), 1261-1265.

Meyer, Lorenzo. Mexico and the United States in the Oil Controversy, 1917-1942. Austin, Texas: University of Texas Press, 1977.

Powell, J. Richard. The Mexican Petroleum Industry: 1938-1950. New York: Russell & Russell, 1972.

Sandeman, Hugh. "Pemex Comes Out of Its Shell." Fortune, 97 (April 10, 1978), 44-48.

Stewart-Gordon, T. J. "Mexico's Oil: Myth, Fact, and Future." World Oil, 188 (February 1, 1979), 35-41.

United States. Library of Congress. Congressional Research Service. Mexico's Oil and Gas Policy: An Analysis. Washington, D.C.: U.S. Government Printing Office, December 1978.

Williams, Edward J. The Rebirth of the Mexican Petroleum Industry: Developmental Directions and Policy Implications. Lexington, Massachusetts: Lexington Books, 1979.

BRAZIL

Brazil. Ministerio das Minas e Energia. National Energy Balance. Brasilia, Brazil: Ministerio das Minas e Energia, 1977.

Gall, Norman. "Atoms for Brazil: Dangers for All." Bulletin of the Atomic Scientists, 32 (June 1976), 4-9, 41-48.

Goldemberg, J. "Brazil: Energy Options and Current Outlook." Science, 200 (April 14, 1978), 158-164.

Hammond, Allen L. "Alcohol: A Brazilian Answer to the Energy Crisis." Science, 195 (February 11, 1977a), 564-566.

_____ . "Brazil's Nuclear Program: Carter's Nonproliferation Policy Backfires." Science, 195 (February 18, 1977c), 657-659.

_____ . "Energy: Brazil Seeks a Strategy Among Many Options." Science, 195 (February 11, 1977b), 566-567.

_____ . "Unconventional Energy Sources: Brazil Looks for Applications." Science, 195 (March 4, 1977d), 862-863.

Krugmann, Hartmut. "The German-Brazilian Nuclear Deal." Bulletin of the Atomic Scientists, 37 (February 1981), 32-37.

Louranc, William W. "Nuclear Futures for Sale: To Brazil from West Germany." International Security, 1 (Fall 1976), 147-166.

Miller, Jeffrey B. "Comparing Responses to the Energy Problem in Four Countries: A Synthesis." [Brazil, Hungary, Japan, and the United States.] Journal of Comparative Economics, 2 (June 1978), 177-186.

Nuclear Energy in Latin America: The Brazilian Case. New York: Unipub, 1980.

Perry, William and Kern, Sheila. "The Brazilian Nuclear Program in a Foreign Policy Context." Comparative Strategy, 1 (1978), 53-70.

Schuh, C. Edward. "The Case of Brazil: Import Substitution Revisited." Journal of Comparative Economics, 2 (June 1978), 97-110.

Smith, Peter Seaborn. "Brazilian Oil: From Myth to Reality?" Inter-American Economic Affairs, 30 (Spring 1977), 45-61.

Tendler, Judith. Electric Power in Brazil: Entrepreneurship in the Public Sector. Boston, Massachusetts: Harvard University Press, 1968.

Wonder, Edward F. "Nuclear Competition and Nuclear Proliferation: Germany and Brazil, 1975." Orbis, 21 (Summer 1977), 277-306.

VENEZUELA

Avery, William P. "Oil, Politics, and Economic Policymaking: Venezuela and the Andean Common Market." International Organization, 30 (Autumn 1976), 541-571.

Baloyra, Enrique A. "Oil Policies and Budgets in Venezuela, 1938-1968." Latin American Research Review, 9 (Summer 1974), 28-72.

Betancourt, Romulo. Venezuela's Oil. [Tr. by Donald Peck] London, England: Allen and Unwin, 1978.

Blank, Daivd Eugene. "Oil and Democracy in Venezuela." Current History, 78 (February 1980), 71-75+.

Bye, Vegard. "Nationalization of Oil in Venezuela: Re-Defined Dependence and Legitimization of Imperialism." Journal of Peace Research, 16 (1979), 57-78.

Energy Policies of the World: Venezuela. Newark, Delaware: University of Delaware Press, 1975.

285

Fuad, Kim. "Energy: Dotting the i's and Crossing the t's: The Pursuit of a Coherent Strategy in the Aftermath of Nationalization." Business Venezuela, September/October 1975, 26-30+.

_____. "Money Isn't Everything: Keeping Tight Rein on Its Colossal Windfall Wealth." Business Venezuela, July/August 1974, 6-10+.

_____. "Oil: The Point of Diminishing Returns." Business Venezuela, March/April 1973, 13-16.

Gall, Norman. "The Challenge of Venezuelan Oil." Foreign Policy, 19 (Spring 1975a), 44-67.

_____. "Oil and Democracy in Venezuela." Common Ground, 1 (January 1975b), 53-61.

Gomez, Gonzalo. "Venezuela: Oil and Politics." Inprecor, May 13, 1976, 27-31.

Grove, Neil. "Venezuela's Crisis of Wealth." The National Geographic Magazine, 150 (August 1976), 175-208.

Guy, James J. "Venezuela: Foreign Policy and Oil" World Today, 35 (December 1979), 508-512.

International Atomic Energy Agency. Estudio de Planificacion Nucleoelectrica para Venezuela. New York: Unipub, 1979.

Nott, David. "Venezuela: The Apportionment of Oil Wealth." Bank of London and South America Review, 9 (January 1975), 2-12.

_____. "Venezuela: The Oil Bonanza and the New Government [of President Carlos Andres Perez]." Bank of London and South America Review, 8 (April 1974), 196-204.

Petras, J. F.; Morley, M.; and Smith, S. Nationalization of Venezuelan Oil. New York: Praeger Publishers, 1977.

Rossi-Guerrero, Felix P. "The Transition from Private to Public Control in the Venezuelan Petroleum Industry." Vanderbilt Journal of International Law, 9 (Summer 1976), 475-488.

Salazar-Carrillo, Jorge. Oil in the Economic Development of Venezuela. New York: Praeger Publishers, 1976.

Tugwell, Franklin. "Petroleum Policy in Venezuela: Lessons in the Politics of Dependence Management." Studies in Comparative and International Development, 9 (Spring 1974), 84-120.

Tugwell, Franklin. The Politics of Oil in Venezuela. Stanford, California: Stanford University Press, 1975.

Valdez, M. A. The Petroleum Policies of the Venezuelan Government. New York: New York University Press, 1972.

Vallenilla, Luis. Oil: The Making of a New Economic Order: Venezuelan Oil and OPEC. New York: McGraw-Hill, 1975.

Von Lazar, Arpad and Magid, Bruce. "Energy Policy and Social Development: Notes on Trinidad, Tobago, and Venezuela." Energy Policy, 3 (1975), 201-210.

OTHER

"Argentina: The Energy Sector." Bank of London and South America Review, 9 (April 1975), 192-198.

"Ecuador: Petroleum and the Future." Bolsa Review, 7 (March 1973), 94-100.

Fitzsimmons, Allan K. and McIntosh, Terry L. "Energy Planning in Guatemala: Response to Crisis." Energy Policy, 6 (March 1978), 14-20.

Gillette, Robert. "India and Argentina: Developing a Nuclear Affinity." Science, 184 (June 28, 1974), 1351-1353.

Grayson, George W. "Populism, Petroleum, and Politics in Ecuador." Current History, 68 (January 1975), 15-19+.

Lozinov, D. "The Future of Ecuador's Oil." International Affairs [Moscow], July 1973, 40-44.

Trier, Alex. "Third World Energy Problems: The Chilean Experience." Energy Policy, 7 (March 1979), 72-74.

8. Asia

JAPAN

Bennett, J. W. "Japan: The Environmental Impact of Unrestricted Growth." Environment, 15 (December 1973), 6-13.

Doernberg, Andres. "Energy Use in Japan and the United States." In Joy Dunkerley, ed. International Comparisons of Energy Consumption. [Research Paper R-10] Washington, D.C.: Resources for the Future, 1978.

Eguchi, Yojiro. "Japanese Energy Policy." International Affairs, 56 (Spring 1980), 280-295.

Goto, K. "Energy R and D Scenario and Japan's Option." Rivista Internazionale di Scienze Economiche e Commerciali, 27 (July-August 1980), 690-708.

Grey, Peter. Japan: High Priced Energy: Inflation, Unemployment, and Nuclear Power. Melbourne, Australia: Committee for Economic Development of Australia, May 1977.

Hodgetts, J. E. Administering the Atom for Peace. [U.S., Great Britain, France, Canada, Italy, Japan] New York: Atherton Press, 1964.

Huff, Rodney L. Political Decision-Making in the Japanese Civilian Atomic Energy Program. [Ph.D. Dissertation, George Washington University, 1973]

Imai, Ryukichi. "Japan and the Nuclear Age." Bulletin of the Atomic Scientists, 26 (June 1970), 35-39.

_____. "The Political Outlook for Nuclear Power in Japan." Atlantic Community Quarterly, 13 (Summer 1975), 226-245.

Japan. Institute of International Affairs. The Oil Crisis -- Its Impact on Japan and Asia. Washington, D.C.: United States-Japan Trade Council, 1974.

Japan's Sunshine Project. Tokyo, Japan: MITI Agency of Industrial Science and Technology, 1975.

Levy, Walter J. "An Atlantic-Japanese Energy Policy." Survey, 19 (Summer 1973), 50-73.

Matsui, Ken-ichi. "Perspectives on Energy in Japan." In Jack M. Hollander, Melvin K. Simmons, and David O. Wood, eds. Annual Review of Energy. Volume 2. Palo Alto, California: Annual Reviews, Inc., 1977, pp. 387-397.

Mazzanti, G. "Importance and Evolution of the Energetic Systems in Italy and Japan." Rivista Internazionale di Scienze Economiche e Commerciali, 27 (July-August 1980), 683-689.

Mendershausen, Horst. "Energy Prospects in Western Europe and Japan." In Hans H. Landsberg, ed. Selected Studies on Energy: Background Papers for Energy: The Next Twenty Years. Cambridge, Massachusetts: Ballinger Publishing Company, 1980, pp. 211-266.

Okita, S. "Natural Resources Dependency and Japanese Foreign Policy." Foreign Affairs, 52 (July 1974), 714-724.

Organization for Economic Cooperation and Development. "The Energy Policy of Japan." OECD Observer, 48 (October 1970), 15-18.

Paone, R. M. "Energy Needs vs. Environment in Japan." Asia Pacific Community, No. 9 (Summer 1980), 107ff.

Patokallio, P. "Energy in Japanese-American Relations: A Structural View." Journal of Contemporary Asia, 5 (1975), 19-41.

Rees, Judith. "World Power Status at a Price." Geographical Magazine, 47 (March 1975), 386-390.

Sakisaka, Masao. "Japan's Energy Policy." Japanese Economic Studies, 4 (Summer 1975), 4-37.

_____. "Japan's Energy Policy Under Review." Energy Policy, 2 (December 1974), 346-350.

_____. "Options of Long-Term Energy Strategy for Japan." Revue de l'Energie, 296 (August/September 1977), 27-39.

_____. "World Energy Problems and Japan's International Role." Energy Policy, 1 (September 1973), 100-106.

Sawhill, John. "Some Considerations for Japan and the United States in Developing an Energy Strategy." In H. Passin, and A. Iriye, eds. Encounter at Shimoda. Boulder, Colorado: Westview Press, 1979, pp. 211-232.

Sinha, R. P. "Japan and the Oil Crisis." World Today, 30 (August 1974), 335-344.

Steinberg, Eleanor B. "The Energy Needs of West Europe, Japan, and Australasia." Current History, 69 (July-August 1975), 15-18.

Surrey, John. "Japan's Uncertain Energy Prospects: The Problem of Import Dependence." Energy Policy, 2 (September 1974), 204-230.

Suttmeier, Richard P. "Japanese Reactions to U.S. Nuclear Policy: Origins of an International Negotiating Position." Orbis, 22 (Fall 1978), 651-680.

Tisdell, Clem. "An Australian Review of Japanese Science and Energy Policy." Australian Quarterly, 47 (June 1975), 44-61.

Tsurumi, Yoshi. "The Case of Japan: Price Bargaining and Controls on Oil Products." Journal of Comparative Economics, 2 (June 1978), 126-143.

_____. "Die Folgen der Olkrise fur Japan. Innen und Aussen-Politische Schwierigkeiten." ["The Consequences of the Oil Crisis for Japan. Difficulties in Domestic Politics and Foreign Policy."] Europa-Archiv, 30 (October 25, 1975), 642-652.

Turner, Louis. "European and Japanese Energy Policies." Current History, 74 (March 1978), 104-108+.

United States-Japan Trade Council. How the Energy Crunch Affects Japan. Washington, D.C.: United States-Japan Trade Council, 1974.

Valtorta, M. "Electricity Development in Japan and Italy - Common Features and Possible Strategies." Rivista Internazionale di Scienze Economiche e Commerciali, 27 (July-August 1980), 709-720.

Wu, Y. L. Japan's Search for Oil: A Case Study on Economic Nationalism and International Security. Stanford, California: Hoover Institution Press, 1977.

Yada, T. "Energy." In K. Murata, ed. Industrial Geography of Japan. New York: St. Martin's Press, 1980, 130-138.

INDIA

Chitale, V. P. and Roy, Mrs. M. Energy Crisis in India. New Delhi, India: Economic and Scientific Research Foundation, Sapru House, 1975.

Chopra, Maharaj K. "India and the Energy Crisis." Military Review, 54 (August 1974), 30-40.

"Coal: Production and Policy in India." Energy Policy, 3 (March 1975), 81-82.

Dasgupta, Biplab. The Oil Industry in India: Some Economic Aspects. London, England: Frank Cass and Company, 1971.

Davar, G. R. "The Energy Crisis and Its Impact on Planning and Administration [in India]." Indian Journal of Public Administration, 21 (October-December 1975), 762-787.

Diwan, Romesh. "Energy Implications of Indian Economic Development: Decade of 1960-1970 and After." Journal of Energy and Development, 3 (Spring 1978), 318-337.

"Energy and the Third World: The Future of India - A Delphi Study." Energy Policy, 4 (March 1976), 69-73.

Gillette, Robert. "India and Argentina: Developing a Nuclear Affinity." Science, 184 (June 28, 1974), 1351-1353.

Henderson, Patrick David. India: The Energy Sector. New York: Published for the World Bank by Oxford University Press, 1975.

Honaver, R. M. "India and the Oil Problem." World Today, 30 (July 1974), 294-305.

Mahatme, D. B. "India and the World Oil Crisis." Journal of the Indian Institute of Bankers, 45 (January/March 1974), 17-32.

Mehta, Balraj. India and the World Oil Crisis. New Delhi, India: Sterling Publishers, 1974.

Mellor, John W. The New Economics of Growth: A Strategy for India and the Developing World. Ithaca, New York: Cornell University Press, 1976.

Pachauri, Rajendra K. Energy and Economic Development in India. New York: Praeger Publishers, 1977.

_____, ed. Energy Policy for India. Columbia, Missouri: South Asia Books, 1980.

Parikh, Kirit S. and Srivivasan, T. N. "Food and Energy Choices for India." Behavioral Science, 25 (September 1980), 367-386.

Power, P. F. "Energy Crisis and Indian Development." Asian Survey, 15 (April 1975), 328-345.

Ramachandran, A. and Gururaja, J. "Perspectives on Energy in India." In Jack M. Hollander, Melvin K. Simmons, and David O. Wood, eds. Annual Review of Energy. Volume 2. Palo Alto, California: Annual Reviews, Inc., 1977, pp. 365-386.

Revelle, Roger. "Energy Use in Rural India." Science, 192 (June 4, 1976), 969-975.

Rudolph, Lloyd I. and Lenth, Charles S. "Energy Options: Changing Views from India." Bulletin of the Atomic Scientists, 34 (June 1978), 6-9.

Sankar, T. L. "Alternative Development Strategies with a Low Energy Profile for a Low GNP/Capita Energy-Poor Country: The Case of India." In Leon N. Lindberg, ed. The Energy Syndrome: Comparing National Responses to the Energy Crisis. Lexington, Massachusetts: Lexington Books, 1977, pp. 205-254.

Tewari, Sharat K. "Economics of Wind Energy Use for Irrigation in India." Science, 202 (November 3, 1978), 481-486.

Tomar, Ravindra. "The Indian Nuclear Power Program: Myths and Mirages." Asian Survey, 20 (May 1980), 517-531.

Venkataraman, K. Power Development in India. New York: Halsted Press, 1973.

Walczak, James R. "Legal Implications of Indian Nuclear Development." Denver Journal of International Law and Policy, 4 (Fall 1974), 237-256.

PEOPLE'S REPUBLIC OF CHINA

Adie, W. A. C. "China's Oil: Some Domestic and International Implications." Rivista Internazionale di Scienze Economiche e Commerciali, 23 (October/November 1976), 999-1019.

Bartke, Wolfgang. Oil in the People's Republic of China: Industry Structure, Production, Exports. Montreal, Canada: McGill-Queens University Press, 1977.

Carlson, Sevinc. China's Oil: Problems and Prospects. Washington, D.C.: Georgetown University, Center for Strategic and International Studies, November 1979.

Chang, Raymond J. The Petroleum Industry in China: Its Planning in the Development of Energy Resources: A Bibliography with Selective Annotations. [Exchange Bibliography No. 1285] Monticello, Illinois: Council of Planning Librarians, May 1977.

Chen, King C. "China's Oil Policy." Yale Review, 66 (October 1976), 1-13.

Cheng, Chu-Yuan. "China's Energy Resources." Current History, 71 (September 1976), 73-76.

_____ . "China's Energy Resources." Current History, 74 (March 1978), 121-124+.

_____ . China's Petroleum Industry: Output Growth and Export Potential. New York: Frederick A. Praeger, Publishers, 1976.

Dean, Genevieve C. "Energy in the People's Republic of China." Energy Policy, 2 (March 1974), 33-54.

Foster, J. "Petroleum Prospects for the People's Republic of China." In Joy Dunkerley, ed. International Energy Strategies. Cambridge, Massachusetts: Oelgeschlager, Gunn & Hain, 1980, 381-392.

Goldman, Marshall I. "Energy Policy in the Soviet Union and China." In Hans J. Landsberg, ed. Selected Studies on Energy: Background Papers for Energy: The Next Twenty Years. Cambridge, Massachusetts: Ballinger Publishing Company, 1980, pp. 323-348.

Hardy, Randall W. China's Oil Future: A Case of Modest Expectations. Boulder, Colorado: Westview Press, 1978.

Harrison, Selig S. "China: The Next Oil Giant." Foreign Policy, No. 20 (Fall 1975), 3-49.

_____. China, Oil, and Asia: Conflict Ahead? New York: Columbia University Press, 1977.

Kambara, Tatsu. "The Petroleum Industry in China." China Quarterly, 60 (December 1974), 699-719.

Klott, W. "China's Food and Fuel Under New Management." International Affairs [London], 54 (January 1978), 60-74.

Lamarsh, J. R. "China Nuclear Power Program." Bulletin of the Atomic Scientists, 36 (May 1980), 28-32.

Ling, H. C. The Petroleum Industry in the People's Republic of China. Stanford, California: Hoover Institution Press, Stanford University, 1975.

Park, C. H. Energy Policies of the World: China. Newark, Delaware: Center for the Study of the Marine Policy, College of Marine Studies, University of Delaware, 1975.

Sien-Chong, Niu. "China's Petroleum Industry." Military Review, 49 (November 1969), 23-27.

Smil, Vaclav. "China Reveals Long-Term Energy Development Plans." Energy International, 15 (August 1978), 23-25, 29.

_____. China's Energy: Achievements, Problems, Prospects. New York: Praeger Publishers, 1976a.

_____. "China's Energy Performance." Current History, 73 (September 1977), 63-67.

293

Smil, Vaclav. "Energy Development in China: The Need for a Coherent Policy." Energy Policy, 9 (June 1981), 113-126.

_____ . "Energy in China: Achievements and Prospects." China Quarterly, 65 (March 1976b), 54-81.

_____ . "Intermediate Energy Technology in China." Bulletin of the Atomic Scientists, 33 (February 1977), 25-31.

_____ and Woodard, Kim. "Perspectives on Energy in the People's Republic of China." In Jack M. Hollander, Melvin K. Simmons, and David O. Wood, eds. Annual Review of Energy. Volume 2. Palo Alto, California: Annual Reviews, Inc., 1977, pp. 307-342.

United States. Central Intelligence Agency. China: Oil Production Prospects. Washington, D.C.: Central Intelligence Agency, June 1977.

Willums, J. O. "Projecting the Oil Supply Potential of China and Southeast Asia." In Joy Dunkerley, ed. International Energy Strategies. Cambridge, Massachusetts: Oelgeschlager, Gunn & Hain, 1980, 365-380.

Wolfe, Jessica Leatrice. "Political Implications of the Petroleum Industry in China." Asian Survey, 16 (1976), 525-539.

Woodard, Kim. The International Energy Relations of China. Stanford, California: Stanford University Press, 1980.

_____ . "People's China and the World Energy Crisis: The Chinese Attitude Toward Global Resource Distribution." Stanford Journal of International Studies, 10 (Spring 1975), 114-142.

Wu, Yuan-Li. "China's Energy Resources and Prospects." Current History, 69 (July-August 1975), 25-27.

_____ and Ling, H. C. Economic Development and the Use of Energy Resources in Communist China. New York: Praeger Publishers, 1963.

OTHER

Brown, Harrison and Smith, Kirk R. "Energy for the People of Asia and the Pacific." Annual Review of Energy, 5 (1980), 173-240.

Carlson, S. Indonesia's Oil. Boulder, Colorado: Westview Press, 1977.

Chang, K. S. "The Energy Situation in Taiwan, Republic of China: Past, Present, and Future." International Commercial Bank of China Economic Review, November/ December 1974, 1-13.

Energy Management in Selected Asian Countries. New York: Unipub, 1979.

International Atomic Energy Agency. Nuclear Power Planning for Hong Kong. New York: Unipub, 1977.

_____ . Nuclear Power Planning Study for Indonesia. New York: Unipub, 1977.

Johnson, Mark. "Oil: I. Recent Developments." Bulletin of Indonesian Economic Studies, 13 (November 1977), 34-48.

Kim, Nak Kwan. "Effects of the Oil Shocks on the South East Asia Economies." Asian Economies, September 1974, 50-60.

Mangone, G. J., ed. Energy Policies of the World: Indonesia, the North Sea Countries, and the Soviet Union. Volume II. New York: Elsevier-North Holland, 1977.

Reksohadiprodjo, Sukanto. "Oil and Other Energy Resources for Development: The Indonesian Case." The Journal of Energy and Development, 5 (1980), 289-325.

Swiss, Michael. "Pakistan Reviews Her Energy Options." Energy International, 16 (February 1979), 19-21+.

United Nations. Economic Commission for Asia and the Pacific. Proceedings of the Meeting of the Expert Working Group on the Use of Solar and Wind Energy. [Energy Resources Development Series, No. 16] New York: United Nations, 1976.

Wijarso. "Oil: II. A Doomsday Scenario - and Alternatives." Bulletin of the Indonesian Economic Studies, 13 (November 1977), 49-56.

9. Africa

Adelman, Kenneth L. "Energy Crisis Brightens Zaire's Future." Africa Today, 22 (October/December 1975), 49-55.

Adeniji, Kola. "State Participation in the Nigerian Petroleum Industry." Journal of World Trade Law, 11 (March/April 1977), 156-179.

Amann, Hans. Energy Supply and Economic Development in East Africa. New York: Humanities Press, 1969.

Arungu, Olenda S. "Africa Takes a Look at Its Energy Problems and Projects." Natural Resources Forum, 2 (July 1977), 361-367.

Cervenka, Zdenek and Rogers, Barbara. The Nuclear Axis: Secret Collaboration Between West Germany and South Africa. New York: Times Books, 1978.

Chibwe, E. C. Arab Dollars for Africa. London, England: Croom Helm, 1976.

Elmaihub, S. H. Public Investment in a Capital-Surplus Country: The Case of Libya. Fort Collins, Colorado: Colorado State University, 1977.

Emembolu, Gregory. "Future Prospects and the Role of Oil in Nigeria's Development." Journal of Energy and Development, 1 (Autumn 1975), 135-151.

Fabrice, Herve. "Energy Resources in Africa." Africa, December 1977, 80+.

Friedland, R. J. "South African Coal: Challenge to Petrick Findings." Energy Policy, 7 (March 1979), 71-72.

_____. "South African Coal: The Petrick Report." Energy Policy, 4 (December 1976), 359-362.

Green, R. H. "Petroleum Prices and African Development: Retrenchment or Reassessment?" International Journal, 30 (Summer 1975), 391-405.

Howe, J. W. Energy for the Villages of Africa. Washington, D.C.: Overseas Development Council, 1977.

Madujibeya, S. A. "Nigerian Oil: A Review of Nigeria's Petroleum Industry." Standard and Chartered Review, May 1975, 2-10.

Mansfield, David. "Sonatrach's Master Plan (Algeria's National Oil and Gas Company)." Petroleum Economist, 45 (November 1978), 468-470.

Nwogugu, E. I. "Law and Environment in the Nigerian Oil Industry." Earth Law Journal, 1 (May 1975), 91-104.

O'Keefe, Philip and Shakow, Don. "Facing Kenya's Energy Predicament." Energy Policy, 8 (June 1980), 173-175.

Pearson, S. R. Petroleum and the Nigerian Economy. Stanford, California: Stanford University Press, 1970.

Rivers, Bernard. Fuelling Apartheid: American Oil Interests in South Africa. New York: Council on Economic Priorities, December 4, 1978.

Schatzl, L. H. Petroleum in Nigeria. Ibadan, Nigeria: Oxford University Press, 1969.

Schliephake, Konrad. Oil and Regional Development: Examples from Algeria and Tunisia. New York: Praeger Publishers, 1977.

Segal. Aaron. "Libya's Economic Potential: Can Libya So Utilize Her Petroleum Resources Within a Generation That Other Sources of Growth Can Be Built Up Before the Oil Runs Out?" World Today, 28 (October 1972), 445-451.

Seifert, William W.; Bakr, Mohammet A.; and Kettani, M. Ali. Energy Development: A Case Study. Cambridge, Massachusetts: The M.I.T. Press, 1973.

Skeet, Trevor. "Oil in Africa." African Affairs, 70 (January 1971), 72-76.

Usoro, Eno J. "Foreign Oil Companies and Recent Nigerian Petroleum Oil Policies." Nigerian Journal of Economic and Social Studies, 14 (November 1972), 301-314.

Vahrman, Mark. "Energy in the Third World: Fuel and Power in Tanzania." Energy Policy, 2 (June 1974), 160-164.

Waddams, Frank C. The Libyan Oil Industry. London, England: Croom Helm, 1980.

Wilson, Ernest J., III. "Crisis of Resources: In Underdeveloped Africa." Black Scholar, 9 (March 1978), 2-15.

_____ . "Energy, Africa, and World Politics." Review of Black Political Economy, 3 (Summer 1973), 27-41.

Wilson, Ernest J., III; Emembolu, G. E.; and Pannu, S. S. "The Energy Crisis and African Underdevelopment: Africa - Oil and Development." Africa Today, 22 (October/ December 1975), 11-37, 39-47.

10. Developing Nations

Ashworth, John. "Renewable Energy for the World's Poor." Technology Review, 82 (November 1979b), 42-49.

_____. Renewable Energy Sources for the World's Poor: A Review of Current International Development Assistance Programs. [Report No. 1-195] Golden, Colorado: Solar Energy Research Institute, September 1979a.

_____. "Technology Diffusion Through Foreign Assistance: Making Renewable Energy Sources Available to the World's Poor." Policy Sciences, 11 (February 1980), 241-262.

Askari, Hossein and Cummings, John T. Oil, OECD, and the Third World: A Vicious Triangle? Austin, Texas: University of Texas Press, 1978.

Auer, Peter, ed. Energy and the Developing Nations. Elmsford, New York: Pergamon Press, 1981.

Baron, C. "Energy Policy and Social Progress in Developing Countries." International Labour Review, 119 (September-October 1980), 531-548.

Berrie, T. W. and Leslie, D. "Energy Policy in Developing Countries." Energy Policy, 6 (June 1978), 119-128.

Biogas Technology in the Third World: A Multidisciplinary Review. New York: Unipub, 1978.

Brown, Norman L. "Renewable Energy Resources for Developing Countries." Annual Review of Energy, 5 (1980), 389-413.

_____ and Howe, James W. "Solar Energy for Village Development." Science, 197 (February 10, 1978), 651-657.

Center for Economic and Social Information. "Oil and Gas for Poor Countries." Environment, 16 (March 1974), 10-14.

Center for Theoretical Studies. Energy for Developed and Developing Countries: Proceedings. Ed. by Behram Kursunogly and Arnold Perlmutter. Cambridge, Massachusetts: Ballinger Publishing Company, 1980.

Cleveland, Harlan, ed. Energy Futures of Developing Countries: The Neglected Victims of the Energy Crisis. New York: Praeger Publishers, 1980.

Darmstadter, Joel and Hunter, Robert F. "Energy in Crisis?" In The United States and the Developing World: Agenda for Action, 1973. Washington, D.C.: Overseas Development Council, 1973, pp. 90-100.

Dasgupta, B. "Soviet Oil and the Third World." World Development, 3 (May 1975), 345-360.

Datta, R. L. Solar Energy in Developing Countries: Perspectives and Prospects. Washington, D.C.: National Academy of Sciences, 1972.

Del Valle, Alfredo. Energy Policy Making in Developing Countries: A Survey of Literature. Philadelphia, Pennsylvania: Social Systems Sciences Program, the Wharton School, University of Pennsylvania, 1979.

Dunn, P. D. "Energy and the Developing Countries." In I. M. Blair, B. D. Jones, and A. J. Van Horn, eds. Aspects of Energy Conversion. Elmsford, New York: Pergamon Press, 1976, pp. 221-241.

Eckholm, Eric. The Other Energy Crisis: Firewood. Washington, D.C.: Worldwatch Institute, 1975.

Eggers-Lura, A. Solar Energy in Developing Countries. Elmsford, New York: Pergamon Press, 1979.

"Energy for Agriculture in the Developing Countries." Monthly Bulletin of Agricultural Economics and Statistics, 25 (February 1976), 1-8.

Falls, O. B. "A Survey of the Market for Nuclear Power in Developing Countries." Energy Policy, 1 (December 1973), 225-242.

Friedman, Efrain. "Financing Energy in Developing Countries." Energy Policy, 4 (March 1976), 37-49.

Girvan, Norman. "The Oil Crisis in Perspective: Economic Nationalism." Daedalus, 104 (Fall 1975), 145-158.

Gottsetin, Klaus. "Nuclear Energy for the Third World." Bulletin of the Atomic Scientists, 33 (June 1977), 45-48.

Grant, James P. "Energy Shock and the Development Prospect." International Development Review, 16 (1974), 2-8.

Hartshorn, J. E. "OPEC and the Development of Fourth World Oil." Millennium, 6 (Autumn 1977), 162-174.

Hayes, Denis. Energy for Development: Third World Options. Washington, D.C.: Worldwatch Institute, December 1977.

_____. "Priorities for the Third World." Bulletin of the Atomic Scientists, 34 (June 1978), 9-10.

Hoffman, Kurt. "Alternative Energy Technologies and Third World Rural Energy Needs: A Case of Emerging Technological Dependency." Development and Change, 11 (July 1980), 335-366.

Hoffman, Robert T. and Johnson, Brian. Energy Cooperation: A New Strategy Toward the Third World. Cambridge, Massachusetts: Ballinger Publishing Company, 1980.

Hoffmann, Thomas and Johnson, Brian. "Bypassing Oil and the Atom: The Politics of Aid and World Energy." Energy Policy, 7 (June 1979), 90-101.

Howell, Leon and Morrow, Michael. Asia, Oil, Politics, and the Energy Crisis: The Haves and the Have Nots. New York: IDOC/North America, Inc., 1974.

International Atomic Energy Agency. Nuclear Power and Its Fuel Cycle: Nuclear Power in Developing Countries. Volume 6. New York: Unipub, 1978.

Kettani, M. Ali. "Renewable Energy Resources: Heliotechnique and Development." Ambio, 5 (1976), 190-192.

_____ and Soussou, Joseph E., eds. Heliotechnique and Development. [Proceedings of International Conference on Heliotechnique and Development, az Zahran, Saudi Arabia, 1975] Cambridge, Massachusetts: Development Analysis Associates, 1976.

Lall, Sham. "The Energy Crisis: Symbol and Catalyst of a New Order." International Development Review, 17 (1975), 36-40.

Lewis, J. P. "Oil, Other Scarcities, and the Poor Countries." World Politics, 27 (October 1974), 63-86.

Lopes, J. Leite. "Atoms in the Developing Nations." Bulletin of the Atomic Scientists, 34 (April 1978), 31-34.

Low, Helen C. "The Oil-Dependent Developing Countries." Current History, 69 (July/August 1975), 19-24+.

Lunn, Jeremy. "Oil Price Rises and the Developing Countries." World Today, 30 (October 1974), 403-408.

MacKillop, Andrew. "Energy for the Developing World: A Critique of the New Wisdom." Energy Policy, 8 (December 1980), 260-276.

_____. "Energy for the Developing World: The 'New Wisdoms' are Wrong." International Journal of Energy Research, 5 (April-June 1981), 111-140.

Makhijani, Arjun. "Solar Energy and Rural Development for the Third World." Bulletin of the Atomic Scientists, 32 (June 1976), 14-24.

_____ and Poole, Alan. Energy and Agriculture in the Third World. Cambridge, Massachusetts: Ballinger Publishing Company, 1975.

Martin, William F. and Pinto, Frank J. "Energy for the Third World." Technology Review, 80 (June/July 1978), 48-56.

Mellor, John W. The New Economics of Growth: A Strategy for India and the Developing World. Ithaca, New York: Cornell University Press, 1976.

Merriam, Marshal F. "Decentralized Power Sources for Developing Countries." International Development Review, 14 (1972), 13-18.

National Research Council. Solar Energy in Developing Countries: Perspectives and Prospects. Washington, D.C.: National Academy of Sciences, 1972.

_____. Panel on Renewable Energy Sources. Energy for Rural Development: Renewable Resources and Alternative Technologies for Developing Countries. Washington, D.C.: National Academy of Sciences, 1976.

Palmedo, Phillip, et al. Energy Needs, Uses, and Resources in Developing Countries. Brookhaven, Illinois: Brookhaven National Laboratory, Center for the Analysis of Energy Systems, 1978.

Parikh, J. K. "Renewable Energy Options: What Could Developing Countries Expect from Them?" Energy - The International Journal, 4 (October 1979), 989-994.

Powelson, John P. "The Oil Price Increase: Impacts on Industrialized and Less Developed Countries." Journal of Energy and Development, 3 (Autumn 1977), 10-25.

Reddy, Amulya Kumai N. "Energy Options for the Third World." Bulletin of the Atomic Scientists, 34 (May 1978), 28-33.

_____. Energy Options for the Third World. [Earthscan Seminar, The Hague, April 18, 1977] Bangalore, India: Indian Institute of Science, 1977.

Revelle, R. "Energy Sources for Rural Development." Energy - The International Journal, 4 (October 1979), 969-988,

Rothstein, R. L. Weak in the World of the Strong: The Developing Countries in the International System. New York: Columbia University Press, 1977.

Shams B., Feraidoon. "Oil-Poor Developing Countries (Problems Created by the Sudden Rise in Oil Prices)." Current History, 74 (March 1978), 109-112+.

Siddiqi, Toufiq A. and Hein, Gerald F. "Energy Resources of the Developing Countries and Some Priority Markets for the Use of Solar Energy." Journal of Energy and Development, 3 (Autumn 1977), 164-189.

Smil, V. and Knowland, W. E., eds. Energy in the Developing World: The Real Energy Crisis. Cambridge, England: Oxford University Press, 1980.

United Nations. Department of Economic and Social Affairs. Petroleum Cooperation Among Developing Countries. New York: United Nations, 1977.

Usmani, I. H. "Energy Banks for Small Villages." Bulletin of the Atomic Scientists, 35 (June 1979), 40-44.

World Bank. Energy in the Developing Countries. Washington, D.C.: World Bank, 1980.

Zakariya, Hasan S. "New Directions in the Search for and Development of Petroleum Resources in the Developing Countries." Vanderbilt Journal of Transnational Law, 9 (Summer 1976), 545-577.

11. Energy Policies and Problems in Other Countries

Australia. Task Force on Energy of the Institution of Engineers. "An Energy Policy for Australia." [Extracts from the Summary Report and Recommendations of the Task Force on Energy of the Institution of Engineers] Energy - The International Journal, 5 (April 1980), 295-324.

Camilleri, Joseph. "Nuclear Controversy in Australia: The Uranium Campaign." Bulletin of the Atomic Scientists, 35 (April 1979), 40-44.

Conybeare, C. E. B. Oil Search in Australia. Trumbull, Connecticut: Books Australia, 1980.

Cook, C. Sharp. "A View of Iceland." Bulletin of the Atomic Scientists, 34 (March 1978), 34-36.

Friedman, Todd. "Israel's Nuclear Option." Bulletin of the Atomic Scientists, 30 (September 1974), 30-36.

James, D. E. "A System of Energy Accounts for Australia."The Economic Record, 56 (June 1980), 171-181.

Mazur, Allan. "Solar Heaters in Israel." Bulletin of the Atomic Scientists, 37 (February 1981), 56-58.

Morse, Roger N. "Solar Energy in Australia." Ambio, 6 (1977), 209-215.

Papadopoulos, Raphael. "The Greek Energy Debate." Energy Policy, 5 (September 1977), 254-255.

Saddler, Hugh. Energy in Australia. Winchester, Massachusetts: Allen & Unwin, 1981.

Saha, G. P. and Stephenson, J. "An Evaluation of Residential Conservation Strategies in New Zealand." Energy - The International Journal, 5 (May 1980), 445-450.

Smith, B. R. "Modelling New Zealand's Energy System." European Journal of Operational Research, 4 (March 1980), 173-184.

Sonnino, T. "National Energy Policy for Israel." Energy, 2 (June 1977), 141-148.

Winn, I. J. and Peranio, A. "Israel's Energy Dilemma." Bulletin of the Atomic Scientists, 36 (April 1980), 57-60.

Part VIII
Additional Information Sources

Because of the breadth and volume of the energy literature and the growing specialization in the field, no bibliographical guide on the subject would be complete without directing the reader to other sources of information, data, and research publications not covered in the bibliography proper. The following part consists of a list of additional bibliographies on various energy topics and issues, a set of basic sources of energy and energy-related statistics gathered by researchers, agencies, and organizations, and an enumeration of the growing number of journals and periodicals devoted to energy policy and politics, conventional and alternative technologies, and environmental and other consequences of energy development. While no compiler could hope to exhaust the rich and diverse fund of additional information sources available to social scientists and technical experts, we believe that collectively the materials listed in this part represent a mine to be exploited by scholars, students, and librarians interested in going beyond the references listed in previous parts of this bibliography.

Readers interested in pursuing further references to general and specialized energy issues will find a sizable number of bibliographies spanning the entire range of energy topics structured into this part. Among the most general and valuable to social scientists are those by Grissom and Thompson (1980), Hsieh (1976), Morrison, et al. (1975, 1977), Rycroft, et al. (1977), and Warren (1980). Somewhat more focused are bibliographies on specific energy sources - includng coal (Allison, Winter 1978; Bituminous Coal Research, Inc., 1975; Caldwell, 1978; Cook and Cook, June 1978; Frawley, 1971; Gifford, et al., 1972; Green, 1975, 1977; Keiffer, 1972; MITRE Corporation, 1975; Munn, 1965, 1973; Tompkins, 1973; U.S. A.E.C., 1974; U.S. E.R.D.A., 1976; and Yanarella, 1979), oil (Hodges, June 1979; Swanson, 1960; U.S. A.E.C., 1974; U.S. E.R.D.A, 1976), nuclear power (Brown, Spring 1971; Drayton, 1980; Drazon, June 1981; Hazelton, Spring 1976; and Rather, 1976), alternative energy sources generally (Dossong, 1978; Harrah and Harrah, 1975; Piasetzki, Spring 1976), solar (Arizona State University Library, 1980; Burg, April 1977; Guthrie and Riley, 1977; Liu, June 1976; National Solar Energy Education Campaign, 1977), conservation (Burg, April 1976, September 1977; Lehmann, 1975), and hydrogen (Cox and Natarajan, 1974).

Many bibliographies have been designed especially to aid social scientists in exploring issues, themes, and topics bearing on the interface between social inquiry and energy policy. Political scientists will find policy-oriented bibliographies focusing upon planning (Cook, January 1973; Gil, 1976; and Wilson, 1979), governmental structure

(Kavass and Bieber, 1974; Klema and West, 1977), public attitudes and citizen participation (Howell and Olsen, July 1981; Cunningham and Lopreato, 1977; and Frankena, July 1977), land use (Heitz, 1976; LeFaver and Brutoco, 1979), transportation (Jackson, 1976), and international relations (Averitt and Carter, 1970; Kirkpatrick, 1979, January 1979). The concerns of sociologists, too, are represented in bibliographies devoted to energy and agriculture (Buttel, December 1977; Frankena and Powers, April 1979), boom-town phenomena (Cortez and Cortez, June 1978; Cortez and Jones, 1976; Freudenberg, 1976; and Gould, Winter 1978), and environmental issues (Environment Information, 1977; Layton, 1975; Peters, September 1976; and Vance, 1975). More broadly, bibliographies centering on socio-political impacts of various energy developments (Chiang and Snead, October 1976; Coleman, 1974; Schnell and Krannich, October 1977), as well as those organized around references to energy statistics (Balachandran, June 1976, March 1977) and energy modeling (Charpentier, 1975), should prove useful to all scholars in the social science community interested in the social science/energy nexus. Nor have the topical interests of economists been slighted by compilers, as illustrated by bibliographies on energy and economy by Burg (September 1977), Chiang and Snead (October 1976), Environment Information (1973), Frankena (July 1978), Lehmann (July 1974), and Little and Lovejoy (June 1977).

The references collected under the rubric, basic sources, include a wide variety of materials of interest and use to social scientists of an empirical bent. Two general and regularly updated statistical surveys with pertinent energy sections are the American Statistics Index and the U.S. Bureau of Census' Pocket Data Book: U.S.A., published annually and biennially, respectively. More directly tailored to the needs of technical experts and policy analysts are a number of energy information guides and locators -- including Averitt and Carter (1970), Bloch (1979), Mencke and Horton (1977), and U.S., F.E.A. (annual). Also helpful to energy specialists seeking data bases and sources on energy topics are several indexes and bibliographies on energy statistics (Balachandran, June 1976, March 1977, and 1980; Parker, 1981; and Pronin, 1981). Hans Landesberg and his associates (1974) have put together an important work for the social science community on unmet research needs in the energy field.

For energy policy students interested in the location of data on specific energy sources, a host of individual researchers, industries and trade organizations, investment firms, and government agencies have compiled statistical reports on individual energy industries and sources (American Petroleum Institute, annual, 1971; British Petroleum, 1973; Edison Electric Institute, annual; Moody's Investor Service, annual; Lieberman, 1976; U.S., A.E.C., 1973; U.S., Bureau of Mines, annual; U.S., J.C.A.E., February 1968; and U.S., D.O.E., June 1978; and passim). Then, too, there are a multitude of empirically-based interpretive studies relating to energy prices (Foster Associates, 1974), taxes and subsidies (Brannon, 1974), energy capital requirements (Bankers Trust Company, 1978), environmental issues (Council on Environmental Quality, annual), transportation use (Motor Vehicles Manufacturers Association, annual; Mutch, 1973; Society of Automotive Engineers, October 1975), and economic concentration and trends in the energy industry (Mulholland and Webbink, March 1974).

Energy economists will find especially valuable the many studies devoted to energy supply and demand statistics and trends, past, present, and future. Among these studies, the most numerous are those of an international or cross-national character (Basile, 1976; C.I.A., annual, 1978; Darmstadter, 1971, 1977, 1978; Dunkerley, 1977;

E.E.C., annual; Grenon, 1975, 1977; Hubbert, September 1971; McMullan, et al., 1976; O.E.C.D., annual; Pindyck, 1979; Thomas, April 1973; U.N., annual, 1975, 1976, 1979; and U.S. Office of Federal Statistical Policy and Standards, 1973, 1976). Works directed to energy supply and demand trends in the United States are similarly plentiful, and include: Center for Advanced Computation (1976), Lehmann (July 1974), Maadah and Maddox (1976), Newman and Day (1975), Schurr and Netschert (1972), Stanford Research Institute (1971, 1972), U.S., Congress, House, Committee on Interstate and Foreign Commerce (1971a and b), and U.S., D.O.E. (passim). Finally, energy studies and statistics bearing on energy production and use in the Soviet Union (Campbell, 1978; C.I.A., annual; Dienes, March 1974; and Elliot, 1975), Eastern Europe (Haberstroh, 1977), and the People's Republic of China (C.I.A., 1977, annual) provide a statistical foundation for assessing energy supply and demand problems and potentials in Communist nations.

The concluding section of this part presents a comprehensive, but hardly exhaustive, list of scholarly journals, trade magazines, technical reviews, professional newsletters, research abstracts, and other periodicals serving as frequent outlets for energy research information, interpretation, and communications. The serious energy policy analyst would wish to have at least a passing familiarity with most of these periodicals, although some journals inspire more interest and deserve greater attention than others. For social scientists and students interested in energy issues, several journals should be consistently surveyed for important research on social science issues. Among them, the Annual Review of Energy (annual), Energy Policy (quarterly), and the Journal of Energy and Development (semi-annual) are perhaps the best, with Energy -- The International Journal (monthly), the International Journal of Energy Research (quarterly), and Energy Communications (bi-monthly) being also worthy of periodic scrutiny. Periodicals and acquisitions librarians should give serious thought to subscriptions to several or all of these scholarly journals.

Because they oftentimes publish technical studies on energy issues geared to a general scholarly audience, the Bulletin of the Atomic Scientists (10 issues/year), Technology Review (8 issues/year), Scientific American (monthly), and Science (weekly) ought to be on the reading list of social scientists and policy analysts, and should be reviewed by undergraduates and graduate students studying or doing research papers on social and public dimensions of energy technologies. At a more general and popular level, journals sponsored by various research institutes and public-oriented organizations – such as Not Man Apart (monthly), Critical Mass (monthly), Energy Action (monthly), People and Energy (bi-monthly), The Power Line (monthly), Resources (3 issues/year), and Self Reliance (bi-monthly) – are both important sources of information on current energy developments and valuable sources of views and activities of segments of the alternative energy movement.

1. Bibliographical Sources

Allison, Peter B. "A Bibliography of Bibliographies Connecting Energy and the Social Sciences." Social Science Energy Review, 1 (Spring 1978), 61-70.

_____. "Men and Coal in Appalachia: A Survey of the Academic Literature." Social Science Energy Review, 1 (Winter 1978), 31-47.

Arizona State University Library. Solar Energy Index. Elmsford, New York: Pergamon Press, 1980.

Averitt, Paul and Carter, M. Devereux. Selected Sources of Information on United States and World Energy Resources: An Annotated Bibliography. [U.S. Geological Survey Circular No. 641] Washington, D.C.: U.S. Government Printing Office, 1970.

Balachandran, Sarojini. Energy Statistics: A Guide to Sources. [Exchange Bibliography No. 1065] Monticello, Illinois: Council of Planning Librarians, June 1976.

_____. Energy Statistics: An Update to No. 1065. [Exchange Bibliography No. 1247] Monticello, Illinois: Council of Planning Librarians, March 1977.

A Bibliography of Selected Rand Publications: Energy. [SB-1052] Santa Monica, California: RAND Corporation, February 1981.

Bituminous Coal Research, Inc. Reclamation of Coal-Mined Land: A Bibliography with Abstracts. Monroeville, Pennsylvania: Bituminous Coal Research, Inc., 1975.

Bossong, Ken. Bibliography of Materials on State/Local Conservation/Alternative Energy Programs. [Citizens Energy Project] Washington, D.C.: Center for Science in the Public Interest, 1978.

Browne, M. "Legal Aspects of Atomic Power Plant Development: A Selected Bibliography." Atomic Energy Law Journal, 13 (Spring 1971), 50-75.

Burg, Nan C. An Annotated Bibliography of Solar Energy Research and Technology Applicable to Community Buildings and Other Non-Residential Construction. [Exchange Bibliography No. 1263] Monticello, Illinois: Council of Planning Librarians, April 1977.

_____. Energy Conservation and Economy Through Marginal Cost and Peak Load Pricing of Electric Utilities. [Exchange Bibliography No. 1356] Monticello, Illinois: Council of Planning Librarians, September 1977.

_____. Energy Crisis in the United States. [Exchange Bibliography No. 550] Monticello, Illinois: Council of Planning Librarians, March 1974.

_____. Energy for the Future: A Selected Bibliography on Conservation. [Exchange Bibliography No. 776] Monticello, Illinois: Council of Planning Librarians, April 1975.

_____. Energy for the Future: An Update to Exchange Bibliography No. 776. [Exchange Bibliography No. 946] Monticello, Illinois: Council of Planning Librarians, December 1975.

_____. Reclamation of Energy from Solid Wastes: Theory and Practice: A Selected Annotated Bibliography for Municipal Officers. [Exchange Bibliography No. 1228] Monticello, Illinois: Council of Planning Librarians, February 1977.

Buttel, Frederick H. Energy and Agriculture: A Bibliography of Social Science Literature. [Exchange Bibliography No. 1430] Monticello, Illinois: Council of Planning Librarians, December 1977.

Caldwell, Nellie B., comp. An Annotated Bibliography of Surface-Mined Area Reclamation Research. Berea, Kentucky: Northeast Forest Experiment Station, 1978.

Charpentier, J. P. Review of Energy Models, No. 2. Laxenburg, Austria: International Institute for Applied Systems Analysis, 1975.

Chiang, S. S. and Snead, R. N. Social and Economic Factors Associated with Electric Power Generating Stations: An Annotated Bibliography. Washington, D.C.: Inforum, October 1976.

Coleman, Brian. Bibliography of the Social Effects of Nuclear Power: 1945-1973. Vancouver, British Columbia: British Columbia Hydro and Power Authority, 1974.

Cook, C. S. "Energy: Planning for the Future." American Scientist, 61 (January 1973), 61-65.

Cook, Earleen H. and Cook, Joseph Lee. Coal Slurry Lines: A Bibliography, 1967-1977. [Exchange Bibliography No. 1554] Monticello, Illinois: Council of Planning Librarians, June 1978.

Cortese, Charles F. and Cortese, Jane Archer. The Social Effects of Energy Boomtowns in the West: A Partially Annotated Bibliography. [Exchange Bibliography No. 1557] Monticello, Illinois: Council of Planning Librarians, June 1978.

_____. and Jones, Bernie. Boom Towns: A Social Impact Model with Propositions and a Bibliography. Denver, Colorado: Social Change Systems, Inc., 1976.

Cox, Kenneth E. and Natarajan, M., eds. Hydrogen Energy: A Bibliography. Albuquerque, New Mexico, 1974. [Updated quarterly]

Cunningham, William H. and Lopreato, Sally Cook. "Annotated Bibliography on Energy Attitude Surveys." Appendix B in Energy Use and Conservation Incentives: A Study of the Southwestern United States. New York: Praeger Publishers, 1977, pp. 121-162.

Dalstead, Norman L. and Leistritz, F. Larrry. A Selected Bibliography on Coal-Energy Development of Particular Interest to the Western States. [Miscellaneous Report No. 16] Fargo, North Dakota: Department of Agricultural Economics, North Dakota Agricultural Experiment Station, North Dakota State University, April 1974.

Darmstadter, Joel. An Energy Library: A Selected Annotated List of Useful Publications. Washington, D.C.: Resources for the Future, March 1976.

Drayton, John. Nuclear Energy and Public Safety: A Bibliography. Mankato, Minnesota: Minnesota Scholarly Press, 1980.

Drazan, Joseph Gerald. Three Mile Island: A Preliminary Checklist. Monticello, Ilinois: Vance Bibliographies, June 1981.

Environment Information. Energy/Environment/Economy: An Annotated Bibliography of Selected U.S. Government Publications Concerning United States Energy Policy. Greenbay, Wisconsin: Environment Information, 1973.

Farhar, Barbara C.; Unseld, Charles T.; Vories, Rebecca; and Crews, Robin. "Public Opinion About Energy." In Jack M. Hollander, et al., eds. Annual Review of Energy. Volume 5. Palo Alto, California: Annual Reviews, Inc., 1980, pp. 141-172.

_____; Weiss, Patricia; Unseld, Charles; and Burns, Barbara. Public Opinion About Energy: An Energy Search. [SERI/TR-53-155] Golden, Colorado: Solar Energy Research Institute, June 1979.

Frankena, Frederick. Behavioral Experiments in Energy Conservation: An Annotated Bibliography. [Exchange Bibliography No. 1330] Monticello, Illinois: Council of Planning Librarians, August 1977.

311

Frankena, Frederick. Cities, Regions, and the Energy Problem: A Bibliography and Index. [Public Administration Series, No. P-157] Monticello, Illinois: Vance Bibliographies, January 1979.

_____. Energy Analysis/Energy Accounting: A Bibliography. [Public Administration Series, No. P-24] Monticello, Illinois: Vance Bibliographies, July 1978.

_____. Energy and the Poor: An Annotated Bibliography of Social Research. [Exchange Bibliography No. 1307] Monticello, Illinois: Council of Planning Librarians, July 1977.

_____. Energy Intensity: A Selected Annotated Bibliography. [Exchange Bibliography No. 1306] Monticello, Illinois: Council of Planning Librarians, July 1977.

_____. Survey Research on Energy: An Annotated Bibliography. [Exchange Bibliography No. 1305] Monticello, Illinois: Council of Planning Librarians, July 1977.

_____; Buttel, Frederick H.; and Morrison, Denton E. "Energy/Society Annotations: A Comprehensive Bibliography of Behavioral-Empirical Studies." In Charles T. Unseld, et al., eds. Sociopolitical Effects of Energy Use and Policy. Washington, D.C.: National Academy of Sciences, 1979.

_____ and Powers, Sharon. Energy Use in Agriculture and the Food Systems: A Bibliography of Energy Analysis Applications and Critiques. [Public Administration Series, No. P-226] Monticello, Illinois: Vance Bibliographies, April 1979.

Frawley, Margaret. Surface-Mined Areas: Control and Reclamation of Environmental Damage: A Bibliography. Springfield, Virginia: National Technical Information Service, 1971.

Freudenburg, William R. Social Science Perspectives on the Energy Boomtown. [C00-4287-3] New Haven, Connecticut: Mapping Project on Energy and the Social Sciences, Yale University, 1976.

Gifford, Gerald F.; Dwyer, Don D.; and Norton, Brien E. A Bibliography of Literature Pertinent to Mining Reclamation in Arid and Semi-Arid Environments. Logan, Utah: The Environment and Man Program, Utah State University, 1972.

Gil, Efraim. Energy-Efficient Planning: An Annotated Bibliography. [Planning Advisory Service Report No. 315] Chicago, Illinois: American Society of Planning Officials, 1976.

Gould, Leroy C. "Social Science Research on 'The Energy Boomtown'." Social Science Energy Review, 1 (Winter 1978), 8-31.

312

Green, Jerry E. Selected Materials for Planning the Reclamation of Mined Land. Monticello, Illinois: Council of Planning Librarians, 1975.

_____. The Underground Gasification of Coal. Monticello, Illinois: Council of Planning Librarians, 1977.

Grissom, M. C. and Thompson, L. M. Energy: Social and Economic Aspects: A Bibliography. [DOE/TIC-3383] Oak Ridge, Tennessee: Technical Information Center, July 1980.

Guthrie, David L. and Riley, Robert A., eds. A Solar Energy Bibliography. Silver Springs, Maryland: Information Transfer, 1977.

Guthrie, M. P.; Huber, E. E.; and Norwood, G. A., eds. Energy Research and Development: A Selected Reading List. [A Bibliography prepared by the Environmental Information System Office, Oak Ridge National Laboratory, Report No. ORNL-EIS-73-65] Springfield, Virginia: National Technical Information Service, 1973.

Hamilton, Michael S. Power Plant Siting (With Special Emphasis on Western United States). [Exchange Bibliography No. 1359-1360] Monticello, Illinois: Council of Planning Librarians, September 1977.

Harrah, Barbara K. and Harrah, David. Alternative Sources of Energy: A Bibliography of Solar, Geothermal, Wind, and Tidal Energy and Environmental Architecture. Metuchen, New Jersey: Scarecrow Press, 1975.

Hazelton, P. "Literature Labyrinth of Nuclear Power: A Bibliography." Environmental Law, 6 (Spring 1976), 921-943.

Heitz, Michael S. Land Use and Energy Bibliography. Sacramento, California: Energy Resources Conservation and Development Commission, State of California, 1976.

Hodges, Michael. The Development of the International Oil Industry. Monticello, Illinois: Vance Bibliographies, June 1979.

Howell, Robert E. and Olsen, Darryll. Who Will Decide? The Role of Citizen Participation in Controversial Natural Resources and Energy Decisions. Monticello, Illinois: Vance Bibliographies, July 1981.

Hsieh, Kitty. Energy - A Scientific, Technical, and Socioeconomic Bibliography. Corvallis, Oregon: Oregon State University Press, 1976.

Institutt for Atomenergi, Kjeller (Norway). Air Pollution Health Effects of Electric Power Generation: A Literature Survey. [NP-20649] Springfield, Virginia: National Technical Information Service, 1975.

Jackson, Cynthia. Transportation and Energy: 1970 - March 1974, Updated March 1976. Evanston, Illinois: Northwestern University Transportation Center Library, 1976.

Kavass, Igor I. and Bieber, Doris M. Energy and Congress: An Annotated Bibliography of Congressional Hearings and Reports, 1971-1973. [Library Bibliographies and Occasional Papers Service] Durham, North Carolina: Duke University Law School, 1974.

Keiffer, F. V. A Bibliography of Surface Coal Mining in the United States. Columbus, Ohio: Forum Associates, 1972.

Kirkpatrick, Meredith. Energy Resources and Energy Policies in Eastern Europe and the USSR: A Bibliography. Monticello, Illinois: Vance Bibliographies, 1979.

_____. Energy Resources and Energy Policies in Western Europe: A Bibliography. [Public Administration Series, No. P-158] Monticello, Illinois: Vance Bibliographies, January 1979.

Klema, Ernest D. and West, Robert L. Public Regulation of Site Selection for Nuclear Power Plants: Present Procedures and Reform Proposals: An Annotated Bibliography. Baltimore, Maryland: Published for Resources for the Future by the Johns Hopkins University Press, 1977.

Layton, Mark. Energy and the Environment - A Bibliography. London, England: Department of the Environment Headquarters Library, 1975.

LeFaver, Scott and Brutoco, Sheila. Annotated Bibliography on Energy and Land Use. [Public Administration Series, No. P-159] Monticello, Illinois: Vance Bibliographies, January 1979.

Lehmann, Edward J. Energy Conservation: A Bibliography with Abstracts. [PS-75/214/7WE] Springfield, Virginia: National Technical Information Service, February 1975.

_____. Energy Supply and Demand and the Availability of Energy Sources: A Bibliography with Abstracts. Springfield, Virginia: National Technical Information Service, July 1974.

Little, Ronald L. and Lovejoy, Stephen B. Western Energy Development as a Type of Rural Industrialization: A Partially Annotated Bibliography. [Exchange Bibliography No. 1298] Monticello, Illinois: Council of Planning Librarians, June 1977.

Liu, Robert C. Solar Energy Utilization - A Bibliographic Guide. [ARS-NE-74] Beltsville, Maryland: United States Department of Agriculture, Agricultural Research Service, June 1976.

Mills, Madolia Massey. Energy-Conscious Design Techniques: Associated Economic, Environmental, Legal, and Social Considerations. Monticello, Illinois: Council of Planning Librarians, 1978.

MITRE Corporation. A Preliminary Bibliography on Energy, Resources, and the Environment (Key-Word-Out-Of-Context Listing). McLean, Virginia: MITRE Corporation, October 1972.

_____. Western Coal Development and Utilization: A Policy-Oriented Selected Bibliography with Abstracts. [PB-244-271] Springfield, Virginia: National Technical Information Service, 1975.

Morrison, Denton E., et al. Energy: A Bibliography of Social Science and Related Literature. New York: Garland Press, 1975.

_____. Energy II: A Bibliography of 1975-76 Social Science and Related Literature. New York: Garland Press, 1977.

Munn, Robert F. The Coal Industry in America: A Bibliography and Guide to Studies. Morgantown, West Virginia: West Virginia University Library, 1965.

_____. Strip Mining: An Annotated Bibliography. Morgantown, West Virginia: West Virginia University Library, 1973.

National Solar Energy Education Campaign. Solar Energy Bibliography/Book Catalog. Beltsville, Maryland: International Compendium, 1977.

Peters, Elizabeth. The Energy-Environment Dilemma: A Selective Bibliographic Guide with Annotations. [Exchange Bibliography No. 1111] Monticello, Illinois: Council of Planning Librarians, September 1976.

Piasetzki, J. P. Energy and Environmentally Appropriate Technologies: A Selectively Annotated Bibliography. [Exchange Bibliography No. 1182] Monticello, Illinois: Council of Planning Librarians, September 1976.

Quarterly Literature Review of Hydrogen Energy: A Bibliography with Abstracts. Albuquerque, New Mexico: Technology Application Center, New Mexico University, 1974.

RAND Corporation. Energy: A Bibliography of Selected RAND Publications. Santa Monica, California: RAND Corporation, May 1974.

Rather, Barbara L. "Nuclear Safeguards and Proliferation: A Selected Bibliography." In United States. Congress. Senate. Committee on Governmental Operations. Hearings on the Export Reorganization Act of 1976. Washington, D.C.: U.S. Government Printing Office, 1976, pp. 2013-2032.

Rycroft, Robert W., et al. Energy Policy-Making: A Selected Bibliography. Norman, Oklahoma: University of Oklahoma Press, 1977.

Saskatchewan Provincial Library. Energy Crisis: A Bibliography. Regina, Saskatchewan: Provincial Library, Bibliographic Services Division, 1973.

Schnell, John F. and Krannich, Richard S. Social and Economic Impacts of Energy Development Projects: A Working Bibliography. [Exchange Bibliography No. 1366] Monticello, Illinois: Council of Planning Librarians, October 1977.

Swanick, Eric L. The Energy Situation: Crisis and Outlook; An Introductory Non-Technical Bibliography. [Exchange Bibliography No. 742] Monticello, Illinois: Council of Planning Librarians, December 1976.

Swanson, E. B. A Century of Oil and Gas in Books. New York: Appleton-Century-Crofts, 1960.

Texas A&M University Library. Energy Bibliography and Index. Volumes 1-3. Houston, Texas: Gulf Publishing Company, 1980.

Tompkins, Dorothy Campbell, comp. Strip Mining for Coal. [Public Policy Bibliographies No. 4] Berkeley, California: Institute of Governmental Studies, University of California, 1973.

United States. Atomic Energy Commission. Technical Information Center. Coal Processing: Gasification, Liquefaction, Desulfurization: A Bibliography, 1930-1974. Oak Ridge, Tennessee: U.S. Atomic Energy Commission, Office of Information Services, Technical Information Center, 1974.

_____. Solar Energy: A Bibliography. Oak Ridge, Tennessee: U.S. Atomic Energy Commission, Office of Information Services, Technical Information Center, December 1974.

United States. Energy Research and Development Administration. Technical Information Center. Coal Processing, Production, and Properties: A Bibliography. Oak Ridge, Tennessee: Technical Information Center, Energy Research and Development Administration, 1976.

_____. Solar Energy: A Bibliography. Oak Ridge, Tennessee: Technical Information Center, Energy Research and Development Administration, 1976.

United States. General Accounting Office. GAO Energy Digest. Washington, D.C.: U.S. Government Printing Office, 1977.

United States. Library of Congress. Legislative Reference Service. Science Policy Research Division. An Inventory of Energy Research. Washington, D.C.: U.S. Government Printing Office, 1972.

Vance, Mary. Energy and the Environment: New Publications for Planning Librarians. [List No. 14] Monticello, Illinois: Council of Planning Librarians, 1975.

Warren, Betty. The Energy and Environment Checklist. San Francisco, California: Friends of the Earth Books, April 1980.

Wilkie, B. Energy in Canada: A Selective Bibliographic Review. [Occasional Paper No. 8] Waterloo, Ontario: Faculty on Environment Studies, University of Waterloo, n.d.

Wilson, David E. National Planning in the United States: An Annotated Bibliography. Boulder, Colorado: Westview Press, 1979.

Xerox University Microfilms. Energy: A Key-Phrase Dissertation Index. Ann Arbor, Michigan: University Microfilms, International, 1976.

Yanarella, Ann-Marie, ed. Coal and the Social Sciences: A Bibliographical Guide to the Literature. [Compiled by Curtis Harvey, Herbert G. Reid, David S. Walls, and Ann-Marie Yanarella] Lexington, Kentucky: Social Science/Technology Development Group, University of Kentucky, November 1979.

2. Basic Sources

American Petroleum Institute. Monthly Statistical Report. Washington, D.C.: American Petroleum Institute. [Monthly]

_____. Petroleum Facts and Figures. Washington, D.C.: American Petroleum Institute, 1971.

American Statistics Index: A Comprehensive Guide and Index to the Statistical Publications of the U.S. Government. [Energy Topics] Washington, D.C.: Congressional Information Service, Inc. [Annual]

Averitt, Paul. Coal Resources of the United States: January 1, 1974. [Geological Survey Bulletin 1412] Washington, D.C.: U.S. Government Printing Office, 1975.

_____ and Carter, M. Devereux. Selected Sources of Information on United States and World Energy Resources: An Annotated Bibliography. [U.S. Geological Survey Circular No. 641] Washington, D.C.: U.S. Government Printing Office, 1970.

Balachandran, Sarojini, ed. Energy Statistics: A Guide to Information Sources. Detroit, Michigan: Gale Research Co., 1980.

_____. Energy Statistics: A Guide to Sources. [Exchange Bibliography No. 1065] Monticello, Illinois: Council of Planning Librarians, June 1976.

_____. Energy Statistics: An Update to No. 1065. [Exchange Bibliography No. 1247] Monticello, Illinois: Council of Planning Librarians, March 1977.

Bankers Trust Company. U.S. Energy and Capital. New York: Bankers Trust Company, 1978.

Basile, Paul S., ed. Energy Demand Studies: Major Consuming Countries. [First Technical Report of the Workshop on Alternative Energy Studies] Cambridge, Massachusetts: M.I.T. Press, 1976.

Bemis, Virginia. Energy Guide: A Directory of Information Resources. New York: Garland Press, 1977.

Benson, David C. and Doyle, Frank J. Projects to Expand Fuel Sources in Eastern States - An Update of Information Circular 8725. [Bureau of Mines Information Circular 8765] Washington, D.C.: U.S. Government Printing Office, 1978.

Bloch, Carolyn C. Federal Energy Information Sources and Data Bases. Park Ridge, New Jersey: Noyes Data Corporation, 1979.

Brannon, Gerard M. Energy Taxes and Subsidies. Cambridge, Massachusetts: Ballinger Publishing Company, 1974.

British Petroleum. BP Statistical Review of the World Oil Industry, 1973. London, England: British Petroleum, 1973.

Campbell, Robert. Soviet Energy Balances. [R-2257-DOE] Santa Monica, California: RAND Corporation, 1978.

Center for Advanced Computation. Energy Flow Through the United States Economy. Urbana, Illinois: University of Illinois Press, 1976.

Central Intelligence Agency. China: Oil Production Prospects. Washington, D.C.: U.S. Library of Congress, 1977.

_____. "Energy Production and Consumption (Communist, U.S., OECD, and Selected Other Countries, Regarding Energy Reserves, Production, and Consumption, and Oil Trade)." Handbook of Economic Statistics. Washington, D.C.: U.S. Library of Congress. [Annual]

_____. International Energy Statistical Review. Washington, D.C.: Central Intelligence Agency, April 19, 1978.

Council of State Governments. Book of the States. (Energy Sections) Lexington, Kentucky: Council of State Governments. [Biennial]

Council on Environmental Quality. Annual Report. Washington, D.C.: Council on Environmental Quality. [Annual]

Darmstadter, Joel. "Intercountry Comparisons of Energy Use: Any Lessons for the United States?" In Bernard J. Abrahamson, ed. Conservation and the Changing Direction of Economic Growth. Boulder, Colorado: Westview Press, 1978, pp. 69-78.

_____, et al. Energy in the World Economy: A Statistical Review of Trends in Output, Trade, and Consumption Since 1925. Washington, D.C.: Resources for the Future, 1971.

Darmstadter, Joel, et al. How Industrial Societies Use Energy: A Comparative Analysis. Baltimore, Maryland: The Johns Hopkins University Press, 1977.

Dienes, Leslie. "Soviet Energy Resources and Prospects." Current History, 74 (March 1978), 117-120+.

Dunkerley, Joy, ed. International Comparisons of Energy Consumption. [Report No. R-10] Washington, D.C.: Resources for the Future; Baltimore, Maryland: The Johns Hopkins University Press, 1977.

Edison Electric Institute. Statistical Yearbook of the Electric Utility Industry. New York: Edison Electric Institute. [Annual]

Elliot, Iain F. The Soviet Energy Balance: Natural Gas, Other Fossil Fuels, and Alternative Power Sources. New York: Praeger Publishers, 1975.

European Economic Community. Energy Statistics. Brussels, Belgium: European Economic Community. [Annual]

Foster Associates. Energy Prices, 1960-1973. Cambridge, Massachusetts: Ballinger Publishing Company, 1974.

Grenon, Michel. Energy Resources. Laxenburg, Austria: International Institute for Applied Systems Analysis, 1975.

_____. "Global Energy Resources." In Jack M. Hollander, Melvin K. Simmons, and David O. Wood, eds. Annual Review of Energy. Volume 2. Palo Alto, California: Annual Reviews, Inc., 1977, pp. 67-94.

Griffith, Edward D. and Clarke, Alan W. "World Coal Production." Scientific American, 240 (January 1979), 38-47.

Haberstroh, John R. "East Europe: Energy Problems." [East European energy supply, demand, and import statistics] In East European Economics, Post-Helsinki. Joint Economic Committee Print. Washington, D.C.: U.S. Government Printing Office, August 25, 1977, pp. 379-395.

Hirst, Eric. Energy Intensiveness of Passenger and Freight Transport Modes: 1950-1970. Oak Ridge, Tennessee: Oak Ridge National Laboratory, 1973.

Hubbert, M. King. "The Energy Resources of the Earth." Scientific American, 225 (September 1971), 61-70.

International Atomic Energy Agency. International Comparisons of Nuclear Power Costs. New York: Unipub, 1968.

Kottlowski, Frank E., et al. Coal Resources of the Americas: Selected Papers. Boulder, Colorado: Geological Society of America, 1979.

Landsberg, Hans H., et al. Energy and the Social Sciences: An Examination of Research Needs. Baltimore, Maryland: The Johns Hopkins University Press, 1974.

Lehmann, Edward J. Energy Supply and Demand and the Availability of Energy Sources: A Bibliography with Abstracts. Springfield, Virginia: National Technical Information Service, July 1974.

Lieberman, M. A. "United States Uranium Resources -- An Analysis of Historical Data." Science, 192 (April 30, 1976), 431-436.

McMullan, J. T., et al. Energy Resources and Supply. New York: Wiley Interscience, 1976.

Maadah, Ali G. and Maddox, R. N. "Energy Consumption in a Typical American Home." Energy Communications, 2 (1976), 237-261.

Mann, Charles E. and Heller, James N. Coal and Profitability: An Investor's Guide. New York: McGraw-Hill, 1979.

Mencke, Claire, ed. Solar Energy Update: A Select Guide to Federal and State Government Agencies, Trade and Professional Associations Information Systems, Centers, and Publications. New York: Environmental Information Center, 1977.

_____ and Horton, Craig, eds. Energy Information Locator: A Select Guide to Information Centers, Systems, Data Bases, Abstracting Services, Directories, Newsletters, Binder Services, and Journals. New York: Environmental Information Center, 1977.

Moody's Investors Service. Moody's Public Utility Manual. New York: Moody's Investors Service. [Annual]

Motor Vehicles Manufacturers Association. Motor Vehicles Facts & Figures. Detroit, Michigan: Motor Vehicles Manufacturers Association. [Annual]

Mulholland, Joseph P. and Webbink, Douglas W. Concentration Levels and Trends in the Energy Sector of the U.S. Economy. [Staff Report to the Federal Trade Commission] Washington, D.C.: U.S. Government Printing Office, March 1974.

Mutch, James J. Transportation Use in the United States: A Statistical History, 1955-1971. Santa Monica, California: RAND Corporation, December 1973.

National Petroleum Council. Committee on U.S. Energy Outlook. Other Energy Resources Subcommittee. Coal Task Group. U.S. Energy Outlook: Coal Availability; A Report. Washington, D.C.: National Petroleum Council, 1973.

Newman, Dorothy and Day, Dawn. The American Energy Consumer. Cambridge, Massachusetts: Ballinger Publishing Company, 1975.

Noyes, R., ed. Coal Resources: Characteristics and Ownership in the U.S.A. Park Ridge, New Jersey: Noyes Data Corporation, 1978.

Organization for Economic Cooperation and Development. Statistics of Energy. Paris, France: Organization for Economic Cooperation and Development. [Annual]

Parker, Albert. "World Energy Resources." Energy Policy, 3 (March 1975), 289-305.

Parker, Sybil P., ed. McGraw-Hill Encyclopedia of Energy. 2nd. ed. New York: McGraw-Hill, 1981.

Pindyck, Robert S. The Structure of World Energy Demand. Cambridge, Massachusetts: M.I.T. Press, 1979.

Pronin, Monica, ed. Energy Index, Nineteen Eighty: A Guide to Energy Documents, Laws & Statistics. New York: Environmental Information Center, 1981.

Schurr, Sam H. and Netschert, Bruce C., et al. Energy in the American Economy, 1850-1975: An Economic Study of Its History and Prospects. Baltimore, Maryland: The Johns Hopkins University Press, 1972.

Society of Automotive Engineers. Passenger Car Fuel Economy Trends Through 1976. [Paper No. 750957] Warrendale, Pennsylvania: Society of Automotive Engineers, October 1975.

Stanford Research Institute. End Uses of Energy. Menlo Park, California: Stanford Research Institute, 1971.

_____. Patterns of Energy Consumption in the United States. Washington, D.C.: U.S. Government Printing Office, 1972.

Thomas, Trevor. "World Energy Resources: Survey and Review." Geographical Review, 63 (April 1973), 246-255.

United Nations. Department of International Economic and Social Affairs. Statistical Office. "Energy." Statistical Yearbook. New York: United Nations. [Annual]

_____. World Energy Supplies, 1950-1974. New York: United Nations, 1976.

_____. World Energy Supplies, 1970-1973. New York: United Nations, 1975.

_____. World Energy Supplies, 1973-1978. New York: United Nations, 1979.

United Nations. Economic Commission for Europe. Annual Bulletin of Coal Statistics for Europe. New York: United Nations. [Annual]

United Nations. Economic Commission for Europe. Annual Bulletin of Electrical Energy Statistics for Europe. New York: United Nations. [Annual]

_____ Annual Bulletin of General Energy Statistics for Europe. New York: United Nations. [Annual]

United States. Atomic Energy Commission. The Nuclear Industry, 1973. Washington, D.C.: U.S. Government Printing Office, 1973.

United States. Bureau of Census. Science and Energy Section. Pocket Data Book: USA. Washington, D.C.: U.S. Department of Commerce. [Biennial]

United States. Bureau of Mines. International Petroleum Annual. Washington, D.C.: U.S. Government Printing Office. [Annual]

_____ . Minerals Yearbook. Washington, D.C.: U.S. Government Printing Office. [Annual]

United States. Congress. House. Committee on Interstate and Foreign Commerce. Subcommittee on Energy and Power. Energy Information Digest: Basic Data on Energy Resources, Production, Consumption, and Prices. [95th Congress, 1st session] Washington, D.C.: U.S. Government Printing Office, 1977a.

_____ . Energy Information Handbook. [95th Congress, 1st session] Washington, D.C.: U.S. Government Printing Office, 1977b.

United States. Congress. Joint Committee on Atomic Energy. Nuclear Power Economics – 1962 through 1967. [90th Congress, 2nd session] Washington, D.C.: U.S. Government Printing Office, February 1968.

United States. Department of Energy. Solar Energy: A Status Report. [DOE/ET-0062] Washington, D.C.: U.S. Department of Energy, June 1978.

_____ . Statistical Data on the Uranium Industry. [GJO-100] Grand Junction, Colorado: U.S. Department of Energy, 1978.

United States. Department of Energy. Economic Regulatory Commission. Division of Power Supply and Reliability. Additions to Generating Capacity 1978-1987 for the Contiguous United States. Washington, D.C.: U.S. Government Printing Office. [Annual ten-year update]

_____ . Bulk Electric Power Load and Supply Projections 1988-1997 for the Contiguous United States. Washington, D.C.: U.S. Government Printing Office. [Annual update]

323

United States. Department of Energy. Economic Regulatory Commission. Division of Power Supply and Reliability. Electric Power Supply and Demand 1978-1987 for the Contiguous United States. Washington, D.C.: U.S. Government Printing Office. [Annual ten-year update]

United States. Department of Energy. Energy Information Administration. Annual Report to the Congress. Washington, D.C.: U.S. Government Printing Office. [Annual]

_____. Bituminous Coal, Sub-Bituminous Coal, and Lignite Distribution. Washington, D.C.: U.S. Government Printing Office. [Quarterly and annual]

_____. Coal - Bituminous and Lignite. Washington, D.C.: U.S. Government Printing Office. [Annual]

_____. Coke and Coal Chemicals. Washington, D.C.: U.S. Government Printing Office. [Quarterly and annual]

_____. Energy Supply and Demand in the Midterm, 1990 and 1995. Washington, D.C.: U.S. Government Printing Office, April 1979.

_____. Energy Supply and Demand in the Short Term, 1979 and 1980. Washington, D.C.: U.S. Government Printing Office, June 1979.

_____. Monthly Energy Review. Springfield, Virginia: National Technical Information Service. [Monthly]

_____. Weekly Coal Report. Washington, D.C.: U.S. Government Printing Office. [Weekly]

United States. Department of Energy. Federal Energy Regulatory Commission. Office of Electric Power Regulation. Annual Summary of Cost and Quality of Electric Utility Plant Fuels. Washington, D.C.: U.S. Government Printing Office. [Annual]

_____. Status of Coal Supply Contracts for New Electrical Generating Units. Washington, D.C.: U.S. Government Printing Office.

United States. Federal Energy Administration. End Uses of Petroleum Products in the U.S., 1965-1975. 2 volumes. Washington, D.C.: U.S. Government Printing Office, 1975.

United States. Federal Energy Administration. <u>Federal Energy Information Locator</u> <u>System: Energy Information in the Federal Government</u>. Washington, D.C.: Federal Energy Administration. [Annual]

_____. <u>Monthly Energy Review</u>. Washington, D.C.: U.S. Government Printing Office. [Monthly]

United States. Office of Federal Statistical Policy and Standards. "International Comparison (of estimated populations and per capita GNP, and per capita energy consumption, selected countries)." <u>Social Indicators</u>. Washington, D.C.: U.S. Government Printing Office, 1973, 1976.

3. Journals and Other Periodicals

Alternative Sources of Energy. Milaca, Minnesota: Alternative Sources of Energy. [6 issues/year]

Annual Review of Energy. Palo Alto, California: Annual Reviews, Inc. [Annual]

Annual Review of Nuclear Science. Palo Alto, California: Annual Reviews, Inc. [Annual]

Annual Review of Solar Energy. Golden, Colorado: Solar Energy Research Institute. [Annual]

Atomic Energy Review. New York: Unipub. [Quarterly]

Bulletin of the Atomic Scientists. Chicago, Illinois: Educational Foundation for Nuclear Science. [10 issues/year]

Coal Abstracts. London, England: International Energy Agency Coal Research, Technical Information Service. [Monthly]

Coal Age. New York: McGraw-Hill. [Monthly]

Coal Data. Washington, D.C.: National Coal Association. [Annual]

Coal Facts. Washington, D.C.: National Coal Association. [Biennial]

Coal International. Washington, D.C.: Zinder-Neris. [Monthly]

Coal News. Washington, D.C.: National Coal Association. [Weekly]

The Coal Observer. New York: Dean Witter Reynolds, Inc. [Monthly]

Coal Week. New York: McGraw-Hill. [Weekly]

Conservation & Recycling. Elmsford, New York: Pergamon Press. [Quarterly]

Critical Mass. Washington, D.C.: Citizen's Movement for Safe and Efficient Energy. [Monthly]

EPRI Journal. Palo Alto, California: Electrical Power Research Institute. [Bi-Monthly]

The Elements. Washington, D.C.: Public Resource Center. [11 issues/year]

Energy - The International Journal. Elmsford, New York: Pergamon Press. [Monthly]

Energy Abstracts for Policy Analysis. Oak Ridge, Tennessee: U.S. Department of Energy, Technical Information Center. [Monthly]

Energy Action. Palos Verdes Peninsula, California: Technology News Center. [Monthly]

Energy Communications. New York: Marcel Dekker. [Bi-Monthly]

Energy Conservation Update. Oak Ridge, Tennessee: U.S. Department of Energy, Technical Information Center. [Monthly]

Energy Consumer. Washington, D.C.: U.S. Department of Energy, Office of Consumer Affairs. [Monthly]

Energy Conversion and Management. Elmsford, New York: Pergamon Press. [Quarterly]

The Energy Daily. Washington, D.C.: Llewellyn King, Publisher. [Daily]

Energy Economics Newsletter. Menlo Park, California: Stanford Research Institute. [Periodically]

Energy Information Abstracts. New York: Environmental Information Center. [Monthly]

Energy Perspectives. Columbus, Ohio: Battelle Institute Energy Program. [Monthly]

Energy Policy. Surrey, Guilford, England: IPC Science and Technology Press. [Quarterly]

Energy Regulation Update. New York: Environmental Information Center. [Monthly]

Energy Report to the States. Denver, Colorado: National Conference of State Legislatures. [Bi-weekly]

Energy Systems and Policy. New York: Crane, Russak & Company. [Quarterly]

Energy User News. New York: Fairchild Publications. [Weekly]

Hydrogen Energy. Elmsford, New York: Pergamon Press. [Quarterly]

327

INIS Atom Index. New York: Unipub (for the International Atomic Energy Agency). [Bi-weekly]

International Coal. Washington, D.C.: National Coal Association. [Annual]

International Journal of Energy Research. New York: John Wiley & Sons. [Quarterly]

International Journal of Hydrogen Energy. Elmsford, New York: Pergamon Press. [Quarterly]

Journal of Energy. New York: American Institute of Aeronautics and Astronautics. [Bi-monthly]

Journal of Energy and Development. Boulder, Colorado: University of Colorado, International Research Center for Energy and Economic Development. [Semi-Annually]

Keystone Coal Industry Manual. New York: McGraw-Hill. [Annual]

Natural Gas. Washington, D.C.: U.S. Energy Information Administration. [Monthly]

Not Man Apart. San Fancisco, California: Friends of the Earth. [Monthly]

Nuclear Fusion. New York: Unipub. [Monthly]

Nuclear Industry. Washington, D.C.: Atomic Industrial Forum. [Monthly]

Nuclear Report. LaGrange Park, Illinois: American Nuclear Society. [Monthly]

Nuclear Safety. Oak Ridge, Tennessee: U.S. Nuclear Regulatory Agency. [Bi-monthly]

Nucleonics Week. New York: McGraw-Hill. [Weekly]

Oil and Gas Journal. Tulsa, Oklahoma: The Petroleum Publishing Company. [Weekly]

People and Energy. Washington, D.C.: Institute for Ecological Policies. [Bi-monthly]

Power Line. Washington, D.C.: Environmental Action Foundation. [Monthly]

Public Utilities Fortnightly. Washington, D.C.: Public Utilities Reports, Inc. [Bi-weekly]

Resources. Washington, D.C.: Resources for the Future. [3 issues/year]

Science. Washington, D.C.: American Association for the Advancement of Science. [Weekly]

Scientific American. New York: Scientific American, Inc. [Monthly]

Self-Reliance. Washington, D.C.: Institute for Local Self-Reliance. [Bi-monthly]

Soft Energy Notes. San Francisco, California: Friends of the Earth. [Bi-monthly]

Solar Age. Port Jervis, New York: Kurt Wasserman, Publisher. [Monthly]

Solar Energy. Elmsford, New York: Pergamon Press. [Monthly]

Solar Energy Digest. San Diego, California: William & Lillian Edmundson Publishers. [Monthly]

Solar Energy Intelligence Report. Silver Springs, Maryland: Business Publishers. [Bi-weekly]

Solar Energy Update. Oak Ridge, Tennessee: Energy Research and Development Administration. [Monthly]

Solar Law Reporter. Golden, Colorado: Solar Energy Research Institute. [Bi-monthly]

Solar Utilization News. Philadelphia, Pennsylvania: Solar Utilization Network. [Quarterly]

Sun Times. Washington, D.C.: Solar Lobby. [Monthly]

Synfuels Week. Washington, D.C.: Pasha Publications. [Weekly]

Technology Review. Cambridge, Massachusetts: Alumni Association of the Massachusetts Institute of Technology. [8 issue/year]

United Mine Workers Journal. Indianapolis, Indiana: United Mine Workers of America. [Monthly]

WISE [World Information Service on Energy]. Washington, D.C.: Wise Network. [6 issues/year]

Weekly Government Abstracts: Energy. Springfield, Virginia: U.S. Department of Energy, National Technical Information Service. [Weekly]

World Coal. San Francisco, California: Miller Freeman Publications. [Monthly]

Part IX
Building an Energy Library
Collection: An Annotated List
of Essential Books

The burgeoning literature on energy and energy-related topics undoubtedly presents the professional librarian interested in building a library collection or in adding to an existing one, with a monumental task of sifting through the ever-expanding list of recently-published books. Moreover, students new to the field, concerned with developing a broad knowledge of methodological and substantive issues informing energy policy analysis, can easily become overwhelmed by the voluminous nature of the available literature. The concluding section of this work seeks to assist acquisitions librarians and students of the energy literature by offering a broad-ranging, balanced, and, we hope, judicious selection of 75 "must" books for any serious energy collection. As an additional feature, each of these choices has been annotated in order to give the reader some understanding of the scope of the work as well as some sense of its distinctiveness or special contribution.

The major criteria used in selecting these books spring from our conviction that, in order to promote general knowledge and generate serious research, an energy library collection must incorporate materials of divergent and conflicting ideological perspectives, works of varying technical complexity, and writings of wide-ranging themes and foci. Precisely because the energy crisis is a fundamental matter of public policy, an informed and literate populace and a knowledgeable and broadly-grounded group of energy researchers are absolute prerequisites to its solution.

While a collection such as the one offered below neither exhausts our knowledge of the energy crisis and policy recommendations for transcending it, nor provides a consensus around which such a solution can be forged, it does – we would argue – provide a set of important and recognized works which helps to establish the foundations for bringing public consciousness of the multiple problems of energy to a new level of general understanding and for bringing professional comprehension of the complexity and multidimensionality of the energy crisis to a new plateau of specialized awareness.

Adelman, Morris, et al. No Time to Confuse. San Francisco, California: Institute for Contemporary Studies, 1975.

The deficiencies of the corporate liberal policy recommendations of the Ford Foundation Energy Study (A Time to Choose) and the superiority of "free market conservative" energy policies domestically and internationally are the main topics of this edited collection. While critics will indict this energy program as based upon a pseudo-free market economic philosophy covertly supportive of the corporate interests of giant energy industries, proponents will praise it for prefiguring the main outlines of the Reagan Administration's energy program.

Barnet, Richard J. The Lean Years: Politics in the Age of Scarcity. New York: Simon & Schuster, 1980.

  This timely book by a respected economic and political analyst of the left places the energy crisis and the transition to a post-petroleum world in the larger framework of the international political economy within which the struggle for control of the world's energy reserves and other increasingly scarce natural resources is unfolding. Seeing the control of these global resources and the development of world unemployment as the key causes of the new international politics and the developing domestic conflicts of the Age of Scarcity, Barnet explores the transformations which are taking place in the world's political, economic, and military order in reaction to these problems and then charts an alternative democratic course for a politics of survival.

Behrman, Daniel. Solar Energy: The Awakening Science. Boston, Massachusetts: Little, Brown and Company, 1976.

  The present reality and future promise of solar energy are examined by a noted science writer in this general, non-technical book. It recounts the history of solar energy from its use in ancient Rome and Greece to its employment in passive and active designs in the United States, Western Europe, and the Third World. It forecasts a bright future for solar technologies in meeting the energy needs of nations throughout the world.

Bernard, Harold W., Jr. The Greenhouse Effect. Cambridge, Massachusetts: Ballinger Publishing Company, 1980.

  A matter of growing concern to the international scientific community, the possible creation of a "greenhouse effect" resulting from continued dependence upon fossil fuels for the world's energy supply is the subject of this book. Noting the uncertain, but possibly cataclysmic, effects of the warming of the earth's atmosphere even by only a few degrees, the author analyzes the complex relationships between the energy system and the globe's ecosystem and proposes a rather novel middle-range energy strategy involving use of nuclear power and natural gas as the major bridge fuels to a sustainable, environmentally-beneficent energy future based upon alternative energy sources.

Blair, John M. The Control of Oil. New York: Pantheon Books, 1976.

  This impassioned, well-researched book on the international oil system carefully documents the early history of the evolution of the international oil industry and the economic consequences of its oligopolistic practices. Seeing control of energy resources as at least as important to the multinational oil corporations as profit, Blair reveals the strategies designed by Big Oil -- from secret cartel agreements to vertical integration – to establish and then to extend such control over most of the globe's richest oil reserves.

Bupp, Irvin C., Jr., and Derian, Jean-Claude. Light Water: How the Nuclear Dream Dissolved. New York: Basic Books, 1978. [Published in paperback as: The Failed Promise of Nuclear Power. New York: Harper & Row, 1980.]

This excellent comparative study of the light water reactor in the United States and France contributes much to our understanding of the reasons why the Nuclear Dream has proven chimerical. In the process, the authors, knowledgeable students of their respective nations' atomic energy programs, illuminate both the institutional relationships and larger environmental contexts within which the mythical claims of civilian nuclear power were propagated and then de-mystified.

Burton, Dudley J. The Governance of Energy: Problems, Prospects, and Underlying Issues. New York: Praeger Publishers, 1980.

Rejecting technocratic and corporate solutions to the energy crisis, the author explores the political problems, ideological beliefs, and institutional barriers which must be overcome if energy governance is to be infused with genuine democratic participation by the citizenry. Inspired by the democratic tradition in Western political thought, Burton's analysis provides less a blueprint for instituting organizational reforms in the energy arena than a critique of past patterns and a prospectus on the future social design for democratic governance. In so doing, he opens up new vistas on our energy future and brings fresh perspectives to our continuing national dialogue.

Casper, Barry M., and Wellstone, Paul David. Powerline: The First Battle of America's Energy War. Amherst, Massachusetts: The University of Massachusetts Press, 1981.

This dramatic case study of the struggle of Minnesota farmers to reverse a utility decision to run a powerline across prime farm land is recounted by Casper and Wellstone with grace, sensitivity, and passion. Beyond its rich and authentic description of this clash of social forces and cultural values in the heart of rural America is its claim that this event anticipates an inevitable confrontation between the Country and the City over the shape of our energy future and the values which will animate it.

Cesaretti, C.A., ed. The Prometheus Question: A Moral and Theological Perspective on the Energy Crisis. New York: Seabury Press, 1980.

Prepared by the Public Issues Officer of the Episcopal Church, this study guide is designed for citizens interested in grappling with the religious and ethical dimensions of key energy policy issues. It combines sets of discussion topics and questions with resource materials which unfold theological and moral perspectives on energy production and use. The twin responsibilities of the social sciences to promote political education and to advance value clarification make this work a valuable reference for college teaching and community education projects.

Clark, Wilson. Energy for Survival: An Alternative to Extinction. New York: Anchor/ Doubleday, 1975.

This extensively-documented, yet non-technical, overview of the spectrum of conventional and alternative energy resources blends historical insight with a sensitivity to the socio-economic, political, and environmental facets of energy. Until the appearance of Lovins' Soft Energy Paths in 1977, Clark's book was the only serious introductory guide for energy activists and students sympathetic to a solar-based energy future.

Coates, Gary J., ed. Resettling America: Energy, Ecology, and Community. Andover, Massachusetts: Brick House Publishing Company, 1981.
    Inspired by the growing movement toward community self-reliance, this beautifully illustrated and well-organized collection of essays seeks to weave together perspectives from energy, ecology, and community into a coherent vision of a decentralized and resettled America in tune with humankind's social existence and with nature's forms and rhythms. Concrete examples of the successful efforts of diverse communities to overcome problems of food, shelter, and energy through strategies of local self-reliance, as well as theoretical essays seeking to break down the divisions between the city and the country, humankind and nature, industry and agriculture, make this work an essential guide to all those social scientists, planners, architects, and other futurists struggling to transcend the fragmented views and artificial divisions in their own specialties in search of a holistic vision of America in the dawning age of physical and material limits and of human and aesthetic possibilities.

Commoner, Barry. The Politics of Energy. New York: Alfred A. Knopf, 1979.
    Commoner extends his critical perspectives on energy – begun in his earlier study, The Poverty of Power – to the context of the first Carter National Energy Plan. Arguing that policy selection of alternative energy technologies for our collective energy future reduces to a fateful decision to adopt solar energy or to embrace breeder reactors, Commoner attempts to dispel many myths about the solar option by detailing a concrete blueprint for making the transition to the Solar Age. His noted and partially camouflaged advocacy of socialism under the code word "social governance" has the virtue of widening public discussion of the social and political dimensions of the energy debate beyond the mainstream ideologies of American political thought.

Commoner, Barry. The Poverty of Power: Energy and the Economic Crisis. New York: Alfred A. Knopf, 1976.
    Eschewing a narrow definition of the energy crisis, noted biologist and public interest scientist Barry Commoner seeks to relate our energy woes to a larger economic crisis of our advanced capitalist society. Exhibiting the full scope of his interdisciplinary erudition, the author demonstrates the irrationality of our present energy/economic course and pari passu sheds light on the import of the second law of thermodynamics for energy analysis, the environmental dangers of coal and nuclear power, the destructive facets of the petrochemical industry, and the need for structural changes in our political economy in order to resolve our energy crisis.

Cook, Earl. Man, Energy, Society. San Francisco, California: W. H. Freeman and Company, 1976.
    This broad introduction to energy and human affairs remains the best

general college text to date on the subject matter. Written by a renowned geographer, this clearly-written and nicely-illustrated work addresses a wide variety of energy topics bearing on the technical, socio-economic, political, and ethical aspects of humankind's use of energy. A briefly annotated bibliography supplements each chapter.

Daneke, Gregory A., and Lagassa, George K., eds. Energy Policy and Public Administration. Lexington, Massachusetts: Lexington Books, 1980.
This volume by two young, prolific scholars draws together the writings of public administration theorists and practitioners bearing on energy policy-making and government administration. Animated by a desire to deal realistically with soft energy path options in a hard energy technology world, the editors and the other contributors lend considerable insight into the supports and impediments to realizing the possibilities of the soft energy paths future. For a work compiled from two conference panels, this volume as a whole is surprisingly coherent in its architecture and convergent in its analysis.

Darmstadter, Joel; Dunkerley, Joy; and Alterman, Jack. How Industrial Societies Use Energy: A Comparative Analysis. Baltimore, Maryland: The Johns Hopkins University Press, 1977.
This well-documented research monograph sets into comparative relief the energy extravagance of the United States. Employing a variety of analytic techniques, the energy team of Resources for the Future examines energy consumption and other patterns in the U.S., Canada, France, West Germany, Italy, the Netherlands, Britain, Sweden, and Japan. The indigenous factors supporting American energy profligacy and the long-term merits of significant conservation strategies for reducing American energy consumption are well-argued and vigorously defended.

Davis, David H. Energy Politics. New York: St. Martin's Press, 1974.
Both historical and contemporary in its analysis, this rather spare book examines the politics and policy developments of major energy sources -- including coal, oil, natural gas, electricity, and atomic energy. Chief among its distinctions is the model of energy policy-making which guides its analysis -- a conceptual scheme which is predominant in most energy policy analysis today.

Diesendorf, Mark, ed. Energy and People: Social Implications of Different Energy Futures. Forest Grove, Oregon: International Scholarly Book Service, 1980.
The broad appeal of this outstanding collection of conference papers springs from the generally high quality of the contributions and the consistency of the main argument. In sharp contrast to the essays in the Erickson and Waverman volume, the reflective writings by international energy experts in this work view the world's energy woes as basically institutional, political, and ethical in nature and call for serious scrutiny by energy policy-makers and citizens alike of the social ramifications of divergent and competing energy paths and scenarios.

Doran, Charles F. Myth, Oil, and Politics: Introduction to the Political Economy of Petroleum. Riverside, New Jersey: The Free Press, 1977.
Mustering available empirical and historical data, the author attempts to

sift through six myths about the international political economy of petroleum in order to develop a theory of "co-dependence" among consumer nations, OPEC, and the oil industry appropriate to new domestic and international realities. Since one or more of these supposed myths is shared by each major party in the energy debate, Doran's book is bound to remain hotly disputed and a source of controversy for some time to come.

Engler, Robert. The Brotherhood of Oil: Energy Policy and the Public Interest. Chicago, Illinois: University of Chicago Press, 1977.

Robert Engler builds upon his early and highly-acclaimed study, The Politics of Oil, to expose the degree to which the political economy of oil continues to elude public accountability and to pervert the public interest in the creation of an equitable and democratic energy policy system. Unimpressed with claims that the rise of OPEC has turned multinational oil corporations into mere pawns of the Arab-dominated oil cartel, the author seeks to show how, internationally, OPEC and Big Oil have formed a duopoly and how, domestically, the oil majors have increasingly trained their sights on exploitation of America's indigenous and largely publicly-owned energy reserves.

Erickson, Edward W., and Waverman, Leonard, eds. The Energy Question: An International Failure of Policy. 2 volumes. Buffalo, New York: University of Toronto Press, 1974.

It is the consensus of the authors of this work's eighteen essays on past and future trends in energy exploration, production, and government policy that the energy crisis is really a policy-induced crisis -- that is, an international failure of policy -- rather than a true physical shortage of energy. Examining these issues in the context of the larger global community (Volume 1) and of North America (Volume 2), these energy analysts appeal to old nostrums derived from the law of supply and demand and the market, allege the competitiveness of the international petroleum industry, and generally attribute energy policy failures to unwarranted government intervention.

Freeman, S. David, et al. A Time to Choose. [Final Report of the Ford Foundation Energy Policy Project] Cambridge, Massachusetts: Ballinger Publishing Company, 1974.

Plausible energy scenarios and recommendations for future U.S. energy policy are offered by a panel of influential political, economic, and academic figures in this sophisticated and controversial study. Drawing upon extensive data compiled by researchers in studies commissioned by the chairman and his committee, the authors disclose the benefits of energy conservation, the merits of renewable energy resources, and the potentials and risks of coal and nuclear power. The influence of this report upon the energy policies of the Carter Administration will be evident to any reader.

Fuller, John. We Almost Lost Detroit. New York: Reader's Digest Press, 1975.

The near catastrophic consequences of the mishap at the Enrico Fermi Breeder Reactor plant located close to Detroit are disclosed in this journalistic account by John Fuller. Observing that the incident should serve as a warning about the dangers of other complex energy systems, the author also documents other accidents which have occurred at nuclear power plants.

Funigiello, Phillip J. Toward a National Power Policy: The New Deal and the Electric Utility Industry, 1933-1941. Pittsburgh, Pennsylvania: University of Pittsburgh Press, 1973.

Funigiello's work demonstrates definitively that the course of American energy policy since World War II -- especially its bias toward central power station electric generation and its ambivalent planning imperatives -- cannot be fully comprehended without a sensitive and thorough understanding of the effort to develop a national power policy by Franklin Roosevelt and his fellow New Dealers. With grace, style, and insight, the author explores the corporate and bureaucratic-political factors which derailed this first attempt to forge a comprehensive national energy plan -- one which sought to assert a sense of the public interest and a planning vision as significant dimensions of policy-making in this area.

Gaines, Linda, and Berry, K. Stephen. TOSCA: The Total Social Cost of Coal and Nuclear Power. Cambridge, Massachusetts: Ballinger Publishing Company, 1979.

The sophistication of the authors' analysis of the role of coal and/or nuclear power as the bridge fuel(s) to an energy future of alternative, sustainable energy sources makes this somewhat technical study of the comparative direct and indirect social costs of these two energy forms indispensable to the policy-maker and energy analyst alike. Using a demand model as their chief methodological tool, these researchers seek to establish the optimum mix of coal-fired and nuclear power plants which will achieve the maximum energy production with the minimum social cost.

Georgescu-Roegen, Nicholas. Energy and Economic Myths. Elmsford, New York: Pergamon Press, 1977.

In his fundamental challenge to orthodox views on the relationship between energy and economic activity held by mainstream economists, the Rumanian-born author of this volume has sought to resituate economic processes within a new theoretical framework built upon certain laws of biology and thermodynamics (particularly the law of entropy). The work of this seminal thinker has caused a veritable paradigmatic revolution in resource economics and has influenced the views of such energy analysts as Barry Commoner, Herman Daly, and Hazel Henderson. This volume collects some of Georgescu-Roegen's most weighty and thought-provoking essays on energy, economics, and entropy.

Goodwin, Craufurd D., ed. Energy Policy in Perspective: Today's Problems, Yesterday's Solutions. Washington, D.C.: Brookings Institution, 1981.

Written under the auspices of the Brookings Institution, this somewhat deceptively titled book is the fruit of a team of economists dedicated to clarifying the formulation of energy policy in the United States from the end of World War II through 1979. Each successive chapter sifts through public documents and major studies to delineate the changing forms and vicissitudes of energy policy in each Presidential administration from Truman to Carter.

Gordon, Richard L. Coal in the U.S. Energy Market. Lexington, Massachusetts: Lexington Books, 1978.

This study of the U.S. coal industry examines the many economic dimensions of this key energy source and enterprise, including consumption patterns, the economics of fuel choice, and government rules and regulations influencing coal production and use. It is a work valuable not only to specialists in energy economics but also to students of energy policy interested in understanding the policies and constraints affecting coal expansion and utilization.

Grossman, Richard, and Daneker, Gail. Energy, Jobs, and the Economy. Boston, Massachusetts: Carrier Pigeon Press, 1979.

Prepared by two members of Environmentalists for Full Employment, this volume is a revision and update of their earlier study, Jobs and Energy. By exhibiting that an energy economy based upon conservation and renewable energy technologies provides more jobs and a cleaner and healthier environment than one based upon nuclear and other hard energy technologies, the authors demonstrate the imperative need for coalition-building strategies to unite American labor and the environmental movement behind such an energy program.

Gyorgy, Anna, and Friends. No Nukes: Everyone's Guide to Nuclear Power. Boston, Massachusetts: South End Press, 1979.

Written by a loose network of committed anti-nuclear activists, this comprehensive guide to nuclear power and its dangers is refreshing in its thoroughness and its non-polemical tone. While implacably opposed to a nuclear future, the authors eschew emotional rhetoric and instead recount a devastating critique of nuclear power fashioned out of public exposure of the cover-ups and deceits of the profit-hungry and bureaucratic authoritarian elements of the atomic-industrial complex. Copiously illustrated and extensively referenced, it is unmatched in its quality and breadth by any work coming from the pro-nuclear camp.

Hafele, Wolf, ed. Energy in a Finite World. [The International Institute for Applied Systems Analysis Energy Group] 2 volumes. Cambridge, Massachusetts: Ballinger Publishing Company, 1981.

Offering yet another energy futures scenario, Hafele's edited book represents the culmination of the efforts of the energy group of the world-renowned IIASA to model future world energy needs and possibilities. Though somewhat weighed down with systems jargon and statistics, it is impressive in its attempt to blend scientific and humanistic perspectives into an analysis animated by trans-national values and global concerns.

Hayes, Denis. Rays of Hope: The Transition to a Post-Petroleum World. New York: W. W. Norton & Company, 1977.

In twelve lucidly-written, easily comprehensible chapters, Hayes examines the broad range of energy alternatives available for shaping national and global energy policy in the dawning post-petroleum era. Overall, this former member of the Worldwatch Institute and former director of the Solar Energy Research

Institute offers a compelling case for the feasibility of a safe, sustainable energy future based upon conservation and renewable energy sources.

Henderson, Hazel. The Politics of the Solar Age: Alternatives to Economics. New York: Doubleday & Company, 1981.

The basic argument of this superb collection of essays is that a new era of post-economic policy-making is dawning as a consequence of the exhaustion of the vision of unlimited economic development and limitless energy and natural resources underpinning the ideological orthodoxies of the nineteenth and twentieth centuries. A human and intellectual imperative of this new era is the necessity of abandoning the futile strategy of redoubling old efforts and instead embracing the alternative of reconceptualizing our "problems" in new ways. Essential to achieving this paradigm shift in the realm of energy policy is negotiating the transition to societies based upon renewable resources in the emergent Solar Age.

Kalter, Robert J., and Vogely, William, A., eds. Energy Supply and Government Policy. Ithaca, New York: Cornell University Press, 1976.

The scope and clarity of presentation in this study have made it a valuable textbook for college courses on energy policy and an indispensable guide to governmental policy and energy supply for budding energy analysts. Basic policy issues – such as energy and taxing policies, energy and the environment, energy R&D and public involvement, and government regulation and various energy industries – are discussed with insight and illumination.

Kelley, Donald R., ed. The Energy Crisis and the Environment: An International Perspective. New York: Praeger Publishers, 1977.

Through a series of individual case studies, this incisive overview of the energy/environment interface examines how the relative scarcity or abundance of energy supplies has shaped environmental policies in industrial nations (such as the United States, Japan, the Soviet Union, and various West European nations) and developing nations (such as Brazil, Iran, and certain East European states). Special attention is given to conditions which support or undermine the political foundations of environmental protection measures.

Kendall, Henry, et al. Energy Strategies: Toward a Solar Future. Cambridge, Massachusetts: Ballinger Publishing Company, 1980.

This convincing blueprint for an alternative energy future is the product of the labors of the Union of Concerned Scientists. Abandoning the conventional "business as usual" approach to estimating future U.S. energy needs, Kendall and his fellow scientists use impressive technical sources and data to support their claim that systematic efforts to improve energy efficiency, combined with diverse renewable energy resources and technologies, could lay the foundations for a smooth and unpainful transition to a benign, sustainable energy future for the United States.

Landsberg, Hans, ed. Energy: The Next Twenty Years. Cambridge, Massachusetts: Ballinger Publishing Company, 1979.

This wide-ranging and often technical study, sponsored by the Ford

338

Foundation and administered by Resources for the Future, seeks to offer a balanced assessment of the difficulties confronting energy policy-makers. Offering a mixture of economic analysis and interpretation and policy recommendations for coping with the long-term energy crisis, it concludes that energy will become more costly, not more scarce, and that no easy technological fixes are available to liberate the United States from global ecological and technological interdependence or from continued dependency upon Middle East oil.

Lapp, Ralph E. The Radiation Controversy. Greenwich, Connecticut: Reddy Communications, Inc., 1979.
> This noted scientist and publicist of science and technology issues, after a long journalistic career of attacking the military-industrial complex and the madness of the strategic arms race, has apparently become a favorite consultant for and leading defender of the civilian nuclear industry. The result of this metamorphosis is a short, but pointed, critique of studies critical of the effects of low-level ionizing radiation upon public health and an encomium for nuclear power.

Lawrence, Robert M., and Heisler, Morris O., eds. International Energy Policy. Lexington, Massachusetts: Lexington Books, 1980.
> The growing interdependence of energy production and consumption globally and on the North American continent and the developing importance of comparative analysis of international energy policies to the social sciences provide the twin organizational foci of the ten disparate essays in this edited volume. The essays by Kenneth Erickson and E. Spencer Welhofer stand out among the book's chapters for their subtle blending of theory and comparison.

Lilienthal, David. Atomic Energy: A New Start. New York: Harper & Row, Publishers, 1980.
> After reminiscing about his stewardship of the TVA and then the Atomic Energy Commission, the author proposes a new beginning for atomic power in the United States. He advocates the chartering of a New Atom Corporation, accountable to the President, which would conduct nuclear energy R&D, simultaneously chastened by awareness of the present hazards of the peaceful atom and somehow immunized from the promotional efforts of vested corporate and bureaucratic interests.

Lindberg, Leon N., ed. The Energy Syndrome: Comparing National Responses to the Energy Crisis. Lexington, Massachusetts: D. C. Heath and Company, 1977.
> Working from a political economy framework outlined by its editor, this work collects a series of case studies of energy policy-making in seven diverse political systems (the United States, Great Britain, France, Canada, Hungary, India, and Sweden). The authors highlight wide variations in domestic situations and political responses influenced by certain common international circumstances (e.g., the era of inexpensive and abundant foreign oil supplies followed by shifts in the international political economy of oil after the Arab embargo). In the concluding chapters, Lindberg presents a masterful comparative political analysis which places into theoretical perspective the varying political responses of these nations to the end of cheap and easy petroleum and which identifies the shared

and idiosyncratic obstacles faced by these nations in designing energy policies appropriate to new international economic and ecological realities.

Lipschutz, Ronnie. Radioactive Waste: Politics, Technology, and Risk. Cambridge, Massachusetts: Ballinger Publishing Company, 1980.

This timely study of the complexities of the radioactive waste problem explores the technical, social, and political issues bearing on disposal and management of radioactive waste from civilian and military nuclear programs. Acutely sensitive to the bureaucratic and political problems associated with formulating a safe and enduring solution to the public policy issue, Lipschutz offers a series of recommendations for developing a successful program based on maximizing the participation of the public in the policy formulation process and on buffering any administrative authority charged with overseeing such a program from political pressures or organizational constraints.

Lovins, Amory B. Soft Energy Paths: Toward a Durable Peace. Cambridge, Massachusetts: Ballinger Publishing Company, 1977.

The importance of this work lies in its success in redefining the terms of the debate over energy policy and in reconceptualizing the energy problem as preeminently a social and political one. Along with introducing the hard energy path and soft energy path distinction into the parlance of energy policy debates, Lovins' book presents a theoretical and a social vision in which an alternative energy future based upon renewable energy sources and technical fixes seems reasonable and, indeed, compelling. In addition, his familiarity with a prodigious volume of energy literature and his adeptness in handling the technical facets of energy questions have made this classic statement a tough target for criticism by partisans of the hard path.

Lovins, Amory, B., and Lovins, L. Hunter. Energy/War: Breaking the Nuclear Link. New York: Harper & Row, Publishers, 1980.

With his spouse and collaborator, Amory Lovins places his arguments for soft energy paths in the context of a critical assessment of the tightly-knit connection between atomic power developments and nuclear weapons proliferation. Lovins and Lovins garner many powerful arguments from myriad studies to advocate a strategy of breaking the nuclear link and overcoming the increasingly frightening possibilities of a world of nuclear powers being shaped by the continuing spread of atomic weapons capability.

Mangone, G. Energy Policies of the World. 2 volumes. New York: Elsevier-North Holland Publishing Company, 1977.

This two-volume work probes the historical background, indigenous conditions, and broader influences which bear upon the energy policies of Canada, China, the Arab States of the Persian Gulf, Venezuela, and Iran (Volume 1), and Indonesia, the North Sea Countries, and the Soviet Union (Volume 2). While no common framework or pattern is imposed upon these two sets of nations or discerned in their individual responses to the energy crisis, certain shared interests guide the authors in their case studies. These include the impact of international trade considerations upon energy policy-making, the extent of interest in diversification of energy sources, and the implications of different energy paths for domestic harmony and international peace.

Martin, Daniel W. Three Mile Island: Prologue or Epilogue? Cambridge, Massachusetts: Ballinger Publishing Company, 1980.

This minute-by-minute account of the nuclear accident at Three Mile Island goes beyond detailed narrative to expose the many organizational and informational failures of the Nuclear Regulatory Commission and other public actors who, in the author's words, "helped turn a crisis into a near calamity." His vivid portrait of this episode, combined with his reflections on the larger organizational setting, confirm the inherent weakness of state administration over atomic energy development in the United States and suggest that the deeper lessons of this near-catastrophe have yet to be learned.

Miller, Saunders. The Economics of Nuclear and Coal Power. New York: Praeger Publishers, 1976.

In his now-classic comparative economic analysis of the relative costs of coal and nuclear power, Saunders Miller reaches three conclusions: that the global shortage of uranium supplies will become serious; that the performance of coal-fired plants will continue to improve and remain superior to that of nuclear plants; and that heavy reliance on nuclear power as a major electric-generating source would lead to economic disaster. Since the publication of this book, studies by careful energy analysts like Charles Komanoff and Irvin Bupp have tended to confirm the validity of Miller's findings and predictions.

Mitchell, Edward J., ed. Perspectives on U.S. Energy Policy: A Critique of Regulation. New York: Praeger Publishers, 1976.

Published in cooperation with the American Enterprise Institute, the four studies comprising this volume advance the new conventional wisdom of the Reagan Administration that the inconsistent policy of government intervention in the market, rather than energy over-consumption, diminishing supplies, or corporate dominance over our political economy, is the foremost cause of America's energy woes. Among the topics considered are U.S. energy policy, natural gas, electric power, and the performance of the Federal Energy Office.

Murray, Francis X. Where We Agree: Report of the National Coal Policy Project. 2 volumes. Boulder, Colorado: Westview Press, 1978.

This two-volume work constitutes the final report of the National Coal Project, a one-year project sponsored by the Georgetown University Center for Strategic and International Studies. Intended to foster consensus on important national policy issues bearing on the use of coal in an environmentally- and economically-acceptable fashion, this project brought together for discussion, debate, and dialogue leading representatives from environmental and industrial groups and organizations. In addition to task force reports on transportation, air pollution, fuel utilization and conservation, energy pricing, and emission standards, the work delineates over 150 recommendations on which group consensus was reached.

Nader, Ralph, and Abbotts, John. The Menace of Atomic Energy. New York: W.W.
   Norton and Company, 1979.
   Assisted by the nuclear engineering expertise of his co-author, citizen's
   advocate Ralph Nader attacks nuclear power and its place in our energy future
   by probing its operation and shortcomings as well as the larger institutional
   setting in which it has developed and been promoted. The authors argue that
   energy policy need not be a special province of technical experts, bureaucrats,
   and political and economic elites, but that in fact it is already being influenced
   by citizen action groups around the world dedicated to advancing renewable and
   more benign alternatives to the nuclear option.

National Research Council. Energy in Transition, 1985-2010: Final Report of the
   Committee on Nuclear and Alternative Energy Systems. San Francisco, California: W. H. Freeman Company, 1980.
   This long-awaited final report of the Committee on Nuclear and Alternative Energy Systems addresses the management of the medium-term future
   comprising the transition from major dependence upon fossil fuels to an energy
   future based upon alternative energy sources. Beginning with a survey of major
   energy supply sources (including conservation measures), it explores in considerable depth their promise and risks in the energy transition and examines
   U.S. energy policy in the international economic context. In a final section, it
   sketches an assortment of different energy scenarios without prescribing any one
   as the most preferable.

Newman, Dorothy K., and Day, Dawn. The American Energy Consumer. Cambridge,
   Massachusetts: Ballinger Publishing Company, 1975.
   One of a series of studies commissioned by the Ford Foundation as part
   of its Energy Policy Project, this volume explores the many relationships existing
   between consumers and types and levels of energy usage and seeks to communi-
   cate these findings in a manner understandable to the general public. Drawing
   upon data generated from two surveys, this study presents a portrait – really a
   snapshot -- of energy use by Americans. Valuable both as a standard reference
   for students and a mine for empirically-minded sociologists and public opinion
   analysts, this work contributes to the interpretation of energy usage attitudes
   and behavioral patterns prior to the dawning of the Age of Resource Scarcity.

Novick, Sheldon. The Electric War: The Fight Over Nuclear Power. San Francisco,
   California: Sierra Club Books, 1979.
   Alternating between chapters offering historical insight into the background to the nuclear power debate and chapters interviewing noted partisans in
   the controversy, Novick graphically shows how the present fight over nuclear
   power has evolved out of a series of incidents and decisions in the past relating
   particularly to the outcome and aftermath of America's wartime struggle to
   build an atomic bomb as well as to the success of the electric utility industry in
   achieving monopoly status. This work is perhaps the best single study of the
   nuclear power controversy and is required reading for anyone interested in
   understanding this dispute in its historical depth and political complexity.

Parker, Sybil P., ed. <u>McGraw-Hill Encyclopedia of Energy</u>. 2nd ed. New York: Mc-
Graw-Hill, 1981.

Among the reference books on energy published in recent years, this
thick tome is distinguished by its superior quality, breadth and readability. Leading energy specialists from government, industry, and academia address the
diverse aspects of energy alternatives for the eighties and beyond. With more
than 850 illustrations, a detailed index, numerous appendices, and an extensive
cross-reference, this single-volume encyclopedia will serve an an informative and
invaluable guide for students, scholars, architects, engineers, and the general
public for years to come.

Reader, Mark, comp. and ed. <u>Atom's Eve: Ending the Nuclear Age</u>. New York:
McGraw-Hill, 1980.

This passionate anti-nuclear tract collects the writings of over forty
nuclear critics and addresses the full panoply of issues relating to the hazards of
nuclear energy. Yet, its underlying concern is less with indicting the safety of
atomic reactors than with criticizing the elitist, garrison-state structure of society
which Americans would be compelled to inhabit in a nuclear-dominated energy
future. A checklist of energy alternatives and groups, a calendar of major events
of the nuclear era, and a bibliography of energy books and audio-visual materials
supplement this useful undergraduate text.

Reece, Ray. <u>The Sun Betrayed: A Study of the Corporate Seizure of Solar Energy
Development</u>. Boston, Massachusetts: South End Press, 1979.

The Department of Energy R&D budget has been skewed overwhelmingly
toward high-technology solar demonstration projects by megacorporations in
the energy and aerospace fields, which design only solar systems modelled after
centralized/capital-intensive nuclear technology. As a result, Reece argues, the
solar energy industry has been hobbled and the democratic promise of decentralized solar technology systems remains unrealized.

Ridgeway, James. <u>Energy-Efficient Community Planning</u>. Erasmus, Pennsylvania:
J. G. Press, 1979.

This illustrated and referenced book focuses on communities which have
taken the initiative in responding to the energy crisis and have devised concrete
and often novel local energy plans promoting conservation and solar options.
Case studies of four leaders in community energy programs are followed by five
brief histories of other community innovations in energy saving. While the political dimensions of these programs are entirely neglected, the success of these
local initiatives points to the vitality and feasibility of energy actions at the local
level.

Rosenbaum, Walter A. <u>Coal and Crisis: The Political Dilemmas of Energy Management</u>.
New York: Praeger Publishers, 1978.

This lean, but perceptive, book by a long-time scholar of the politics of
energy and the environment focuses on the "ambiguous abundance" of coal as
a major energy resource in the United States. Examining the difficult choices
and inevitable trade-offs involved in any political commitment to vastly increasing the production and use of coal, the author suggests what changes in the

federal regulatory structure and environmental standards might be necessary -
and what political risks might be incurred -- in trying to realize such a national
objective.

Ruedisili, L. C., and Firebaugh, Morris W. Perspectives on Energy: Issues, Ideas, and
Environmental Dilemmas. New York: Oxford University Press, 1978.
   This comprehensive collection of thirty-five essays, edited by a physicist
and geologist, explores the multiple dimensions of the energy crisis, the plurality
of suggested solutions, and the environmental consequences and dilemmas
stemming from diverse alternative energy sources. Now in its second edition, this
popular book of readings for college courses has been expanded and updated with
the inclusion of twenty-six new articles focusing upon changing problems and
shifting emphases in the energy debate. Students new to the field will find this
work essential reading.

Ross, Marc H., and Williams, Robert H. Our Energy: Regaining Control. New York:
McGraw-Hill, 1981.
   Heavily influenced by soft energy paths and post-economic arguments
and assumptions, this book by two distinguished energy policy analysts presents
a blueprint for the United States to reassert control of its energy system and, by
implication, its economic future. In a manner reminiscent of Hazel Henderson's
recent musings, Ross and Williams argue that the era of apparently unlimited
material growth is coming to an end. They further claim that opportunities for
energy efficiency are so vast that the pursuit of negative energy growth over the
next few decades need not adversely affect the economy. In their view, the
alternative solar/conservation strategy which they proceed to outline is so compelling that it can become the basis for a new consensus among consumers,
producers, and environmentalists.

Sawhill, John C., ed. Energy Conservation and Public Policy. Englewood Cliffs,
New Jersey: Prentice-Hall, 1979.
   Under the editorship of former N.Y.U. President and Federal Energy
Administrator, John C. Sawhill, this collection of essays by twelve noted experts and energy policy-makers probes the methods of enhancing energy efficiency in the industrial, agricultural, transportation, and residential sectors;
examines the changes in efficiency which will likely occur in response to higher
energy prices; and sketches a variety of public policy options to facilitate the adjustment of market forces to increases in energy prices.

Schurr, Sam H., et al. Energy in America's Future: The Choices Before Us. Baltimore
Maryland: Published for Resources for the Future by the Johns Hopkins University Press, 1979.
   This weighty, but excellent, compendium of engineering and economic
analyses of energy trends, environmental, health and safety impacts, and energy
policy-making processes, is a useful reference for students of energy policy and
resource economics. The many charts and voluminous data should not divert the
reader from the authors' recognition and explicit acknowledgment that the key
problems of our energy situation are predominantly political and institutional,
not technical and economic.

Shrader-Frechette, K. S. Nuclear Power and Public Policy: The Social and Ethical Problems of Fission Technology. Hingham, Massachusetts: Kluwer-Boston, 1980.

This exercise in applied logical analysis and rigorous moral philosophy cuts through many of the faulty arguments and fallacious assumptions of current technology assessments supporting nuclear fission in order to expose the dangers to civil liberties and democratic practices inherent in nuclear technology. The author brings the full range of her powers as a trained philosopher to bear upon an often emotion-laden debate; the results are devastating in their impact.

Soneblum, Sidney. The Energy Connection: Between Energy and Economic Growth. Cambridge, Massachusetts: Ballinger Publishing Company, 1978.

The author of this thought-provoking study grapples with a central issue in the energy debate: the relationship between energy consumption and economic growth. Using U.S. energy trends over the past 50 years and other available data of a cross-national character, Soneblum interrelates the role of energy in alternative models of economic growth and then proposes a series of recommendations for future policy based upon his findings. Energy policy analysts and resource economists alike will find much to reflect upon in this controversial, but rewarding, book.

Steinhart, John, et al. Pathway to Sustainable Energy: The 2050 Study. San Francisco, California: Friends of the Earth, 1979.

This slim, readable study outlines an energy future scenario convergent with the socio-political vision of Amory Lovins and his Friends of the Earth compatriots. It argues that with only modest changes in lifestyle, Americans could get by with significantly less energy (64%) than 1975 levels were energy conservation measures taken seriously. The implications of such a low energy plan for individual, business, and public sectors are discussed in separate chapters.

Sternglass, Ernest J. Secret Fallout: Low-Level Radiation from Hiroshima to Three Mile Island. New York: McGraw-Hill, 1981.

This revised and updated version of the book, Low-Level Radiation, is both a personal statement and a public account of this scientific gadfly's effort to expose past and continuing attempts by the nuclear industry and governmental regulatory agencies to conceal the dangers of low-level radiation to the public health. Long the center of controversy in the nuclear debate, Sternglass provides in a new chapter to the original edition evidence of significant infant and fetal deaths caused by the effects of the Three Mile Island nuclear accident.

Stever, Donald W., Jr. Seabrook and the Nuclear Regulatory Commission. Hanover, New Hampshire: University Press of New England, 1980.

From his vantage point as New Hampshire's Assistant Attorney General for the Environmental Protection Division, the author draws a living and perceptive picture of the Seabrook licensing proceedings over a six-year period. Highly critical of the complicated licensing process and the often vague and overlapping jurisdictional divisions of authority among federal and state agencies, Stever argues that the nuclear regulatory process is skewed in favor of industrial interests over health and safety concerns, making future TMI's a real possibility.

345

His prescription combines reforms, involving procedural streamlining and greater public involvement, in order to promote better balance between industry and citizen interests and to restore public trust in the regulatory process.

Stobaugh, Robert, and Yergin, Daniel, eds. Energy Future: Report of the Energy Project at the Harvard Business School. New York: Random House, 1979.
      Emerging from the corporate elite's educational training ground, this bestseller demonstrates the superiority of the soft energy paths approach even as it tries to co-opt its argument and defuse its decentralized sociopolitical vision. Along the way, Stobaugh and Yergin and their associates persuasively demonstrate the need for the United States to plan for the transition to an energy future beyond oil and to abandon the illusions of the Nuclear Dream and King Coal.

Teller, Edward. Energy from Heaven to Earth. San Francisco, California: W. H. Freeman Company, 1979.
      Energy policy, according to the author, is in a stage comparable to adolescence. In order to avoid extremist appeals or policy despair, Teller offers a retrospective examination of energy from its cosmic origins to the near future. His proposed solution to the energy crisis is a balanced energy approach, incorporating elements from both the soft and hard energy paths. In the end, however, his praise of nuclear power and fusion betrays his life-long fascination with unfettered technological innovation as the solution to complex political problems.

Unseld, Charles T., et al., eds. Sociopolitical Effects of Energy Use and Policy. Washington, D.C.: National Academy of Sciences, 1979.
      Fifteen reports from the Sociopolitical Effects Resource Group to the Committee on Nuclear and Alternative Energy Systems comprise this noteworthy volume. Organized around five key aspects of the social and political impacts of energy use and policy, these studies collectively offer a major contribution to a woefully neglected area of energy policy-making. Of special value is a selected and annotated bibliography of social science research on energy-related issues conducted since the Arab oil embargo.

Vietor, Richard H. K. Environmental Policies and the Coal Coalition. College Station, Texas: Texas A&M University Press, 1980.
      This close student of government policy and the coal industry examines the development of federal and state regulation over coal production and use attendant upon the rise of an environmental coalition in the sixties and seventies. His analysis of changing social values and government-business relationships in this area of energy policy takes on special relevance in an era when political and economic forces threaten the regulations and reforms affecting public health and worker health and safety instituted during the previous decades.

Warkov, Seymour, ed. Energy Policy in the United States: Social and Behavioral Dimensions. New York: Praeger Publishers, 1978.
      Twenty-one social scientists train their analytic sights upon fifteen assorted energy/society issues and, in the process, disclose the critical nature of the social and behavioral dimensions of the energy crisis. As a sign of the growing maturity

of the social science literature on energy, this collection of conference papers is impressive; as a pointer to where the social sciences must go in order to contribute significantly to the energy debate and its resolution, this volume is suggestive. Three essays – those on energy regionalism in the United States, equity impacts of some energy alternatives, and social impact assessment of energy policy – are particularly noteworthy.

Wasserman, Harvey. Energy War: Reports from the Front. Westport, Connecticut: Lawrence Hill & Company, 1979.

      Quite literally a series of essays and eye witness news accounts, this hybrid of contemporary history and partisan journalism chronicles the rise of the anti-nuclear movement in America from Seabrook to Three Mile Island. It is absorbing reading for both general reader and energy expert, not so much for the sociological insights it conveys as for the richly-textured and lively portrait of the spirit, passion, and animating vision of the widening network of no-nuke alliances permeating the country. Articles outlining the movement's halting efforts to forge more broadly-based coalitions and essays projecting a positive view of a Solar Age foreshadow the possible transformation of the no-nuke movement from a movement of resistance to one of social reconstruction.

Willrich, Mason. Energy and World Politics. New York: The Free Press, 1975.

      This cogent and concise analysis of the global energy situation serves as a basic introduction to the implications of energy for national security, the world economy, the global environment, and international relations. Its easy style and non-preaching tone only add to the pleasure of reading a book which raises but does not force answers to complex and difficult questions springing from the global politics and ecology of the energy crisis.

Wilson, Carroll. Coal – Bridge to the Future. [World Coal Study] Cambridge, Massachusetts: Ballinger Publishing Company, 1980.

      Representing the collective efforts and findings of its members, this volume is the final report and summary of the World Coal Study, an international project involving participants from 16 major coal using and producing countries aimed at estimating the potential of coal. Composed of two parts, this work presents optimistic findings regarding coal's promise as the basic swing fuel to the future, and then presents technical, economic, and logistical issues and arguments underpinning this major conclusion. Challenging environmentalist views and worries, it urges swift governmental action to foster this central role for coal in the energy transition.